AF774931

Waste Management and Environmental Concerns

Waste Management and Environmental Concerns

Edited by Dave Whittaker

SYRAWOOD
PUBLISHING HOUSE
New York

Published by Syrawood Publishing House,
750 Third Avenue, 9th Floor,
New York, NY 10017, USA
www.syrawoodpublishinghouse.com

Waste Management and Environmental Concerns
Edited by Dave Whittaker

© 2018 Syrawood Publishing House

International Standard Book Number: 978-1-68286-561-3 (Hardback)

This book contains information obtained from authentic and highly regarded sources. Copyright for all individual chapters remain with the respective authors as indicated. All chapters are published with permission under the Creative Commons Attribution License or equivalent. A wide variety of references are listed. Permission and sources are indicated; for detailed attributions, please refer to the permissions page and list of contributors. Reasonable efforts have been made to publish reliable data and information, but the authors, editors and publisher cannot assume any responsibility for the validity of all materials or the consequences of their use.

The publisher's policy is to use permanent paper from mills that operate a sustainable forestry policy. Furthermore, the publisher ensures that the text paper and cover boards used have met acceptable environmental accreditation standards.

Trademark Notice: Registered trademark of products or corporate names are used only for explanation and identification without intent to infringe.

Cataloging-in-Publication Data

Waste management and environmental concerns / edited by Dave Whittaker.
p. cm.
Includes bibliographical references and index.
ISBN 978-1-68286-561-3
1. Refuse and refuse disposal--Environmental aspects. 2. Salvage (Waste, etc.)--
Environmental aspects. 3. Hazardous wastes--Environmental aspects. I. Whittaker, Dave.
TD793 .W37 2018
628.44--dc23

Printed in China.

TABLE OF CONTENTS

PREFACE

This book has been a concerted effort by a group of academicians, researchers and scientists, who have contributed their research works for the realization of the book. This book has materialized in the wake of emerging advancements and innovations in this field. Therefore, the need of the hour was to compile all the required researches and disseminate the knowledge to a broad spectrum of people comprising of students, researchers and specialists of the field.

Waste is any material or byproduct that has either been discarded after use or is of no use. The management, collection, transportation and treatment of waste is known as waste management. The growing environmental concern in the field of waste management is the recycling of waste. The benefits of recycling waste are numerous, as it is socially, economically and environmentally friendly. This book includes some of the vital pieces of work being conducted across the world, on various topics related to waste management and environmental concerns. The topics included in this book on waste management are of utmost significance and are bound to provide incredible insights to readers.

At the end of the preface, I would like to thank the authors for their brilliant chapters and the publisher for guiding us all-through the making of the book till its final stage. Also, I would like to thank my family for providing the support and encouragement throughout my academic career and research projects.

Editor

1

Practical remediation of the PCB-contaminated soils

Akiko Ido[1], Miki Niikawa[1,2], Shinji Ishihara[2], Yoshinari Sawama[2], Tsuyoshi Nakanishi[1], Yasunari Monguchi[2], Hironao Sajiki[2] and Hisamitsu Nagase[1*]

Abstract

A practical method for the elimination of PCBs from PCB-contaminated soil has been developed by the combination of Soxhlet extraction using a newly-developed modified Soxhlet extractor possessing an outlet valve on the extraction chamber with the chemical degradation. Various types of PCBs contaminated in soils could be completely extracted in refluxing hexane, and the subsequent hydrodechlorination could also be completed within 1 h in a hexane–MeOH (1 : 5) solution in the presence of Pd/C and Et_3N under ordinary hydrogen pressure and temperature without the transfer of the extracted PCBs to other reaction container (a complete one-pot procedure). The present system is quite useful as a simple, safe, mild and reliable remediation method of PCB-contaminated soil.

Keywords: Remediation, PCB-contaminated soil, Soxhlet extraction, Palladium on activated carbon (Pd/C), Hydrodechlorination

Introduction

Polychlorinated biphenyls (PCBs) were commercially produced as mixtures of 209 possible congeners beginning in 1929 through the mid-1970s and are non-polar chlorinated hydrocarbons with a biphenyl nucleus, on which one to ten hydrogen atoms have been replaced with chlorine atoms. PCBs were mainly used as electric insulating oils, lubricants, and coolants in electrical devices, such as transformers and capacitors, all over the world due to their chemical and thermal stabilities. However, their production, import and novel use were banned during the mid-1970s, as such properties led to serious environmental pollution including bioaccumulations and biomagnifications through the food chain [1-4]. While PCBs in general could be decomposed by combustion using high-temperature incinerators (above 1100°C), any drop in temperature below 1100°C creates the possibility to generate more toxic dioxins [5]; hence, it is quite difficult to reach an agreement by local communities for the construction of such incinerators. Therefore, tons of PCBs have been stored by each facility under the strict conditions in accordance with each country's laws. However, a huge expense and fear of improper disposal and accidental leaks of the PCBs, have increased with the longer periods of storage [6,7].

On the other hand, the fact is that most of the environmental PCB mass has been found in soil including marine, rivered, and lacustrine muck due to their high specific gravity and highly hydrophobic properties [8]. Although many methods for the remediation of PCB-polluted soils have been reported in the literature, such as ultrasonication [9,10], photochemical degradation [11-13], reductive dechlorination using metals [14-16], base-catalyzed decomposition [17-19], hydrogen-transfer hydrodechlorination [20,21], and fungous and bacterial treatments [22-25], most of these remediation methods require severe reaction conditions, such as high heat, high pressure and/or strongly basic conditions, due to the superior chemical stability of PCBs and the difficulty of complete extraction from the contaminated soil due to the strong affinity between the PCBs and soil based on a hydrophobic interaction. While direct methods for the PCB degradation without the PCB-extraction process from soil have been reported [13,18,19,22-25], a PCB extraction process was involved in most remediation parts for the reliable and reproducible PCB clean up using chemical degradation methods [9-11,14-16]. PCBs are basically extracted from their polluted soils

* Correspondence: nagase@gifu-pu.ac.jp
[1]Laboratory of Hygienic Chemistry and Molecular Toxicology, Gifu Pharmaceutical University, 1-25-4 Daigaku-nishi, Gifu 501-1196, Japan
Full list of author information is available at the end of the article

by the separatory funnel method using dichloromethane [16] or hexane–acetone [15], the ultrasonic extraction method in a hexane–acetone mixed solvent [14], and the Soxhlet extraction method [11,26-28]. The United States Environmental Protection Agency (U.S. EPA) recommends Soxhlet extraction using hexane and acetone (1 : 1, v/v) as a standard method to isolate PCBs from soil samples [26,27]. The Soxhlet extraction using toluene as a solvent was adopted as an official method to survey the dioxin concentration including co-planar PCBs in soil samples by the Ministry of the Environment of the Japan government [28].

We recently established an efficient method for the palladium on carbon (Pd/C)-catalyzed hydrodechlorination of aromatic chlorides using triethylamine (Et_3N) as a single electron donor at ambient temperatures and pressures [29,30], which has been successfully used for the degradation of dichlorodiphenyltrichloroethanes (DDTs) [31] and PCBs [32,33] based on the removal of the chlorine atoms from the aromatic nuclei, and a pilot study using a 50 L vessel for the practical PCB degradation was achieved [34]. Furthermore, magnesium metal was also used as a single electron donor instead of Et_3N for the Pd/C-catalyzed PCB degradation methods [35,36].

In this paper, we describe an easy and reliable remediation method of PCB-polluted soils by PCB extraction and continuous Pd/C–Et_3N-mediated complete hydrodechlorination of the extracted PCBs.

Methods

Materials

Quartz sand, diatomaceous earth, and bentonite were purchased from Nacalai Tesque, Inc. (Kyoto, Japan), and the silica sand (30–50 mesh) was purchased from Koso Chemical Co., Ltd. (Tokyo, Japan). The soil samples, Akadama-tsuchi and Kanuma-tsuchi, are clean, nonpollutant, mildly acidic, and produced in Tochigi prefecture, Japan. They were purchased as 2 L packs from Tachikawa Heiwa Nouen Co., Ltd. (Tochigi, Japan). The soil was ground by a mortar and pestle before use. Also no contamination by PCBs of all purchased soil samples (quartz sand, diatomaceous earth, bentonite, silica sand, Akadama-tsuchi and Kanuma-tsuchi) was confirmed by gas chromatography/mass spectrometry (GC/MS) analysis after Soxhlet extraction using hexane as an extraction solvent.

Aroclor 1242 was previously manufactured by the Monsanto Chemical Co. (St. Louis, MO). 10% Pd/C (K-type, wet), which contained approximately 50% H_2O, and Et_3N were obtained from N.E. Chemcat Co., Ltd. (Tokyo, Japan) and Wako Pure Chemical Industries, Ltd. (Osaka, Japan), respectively. Both hexane and MeOH were purchased from Nakalai Tesque, Inc. [Product number: 04338–93 (hexane) and 04338–41 (MeOH), Kyoto, Japan]. All other reagents were purchased from commercial sources and used without further purification.

A Soxhlet extractor, which has an outlet valve on the extraction chamber, was designed for the easy removal of the collected (distilled) solvents without taking apart the Soxhlet extraction setup (Figure 1).

Preparation of PCB-contaminated soils

A diethyl ether solution of Aroclor 1242 (1 mg/mL) was spread evenly to the purchased typical soil (quartz sand, silica sand, diatomaceous earth, bentonite, Akadama-tsuchi or Kanuma-tsuchi) to prepare the PCB-contaminated soils (250 μg PCBs/g). Each PCB-contaminated soil was stored at 35°C (oil bath) for 0.5–1 h to allow evaporation of the diethyl ether.

PCB extraction from PCB-contaminated soils using Soxhlet extractor

Figure 2, Entry 1: The Soxhlet extraction of PCBs from the PCB-contaminated quartz sand (30 g), which was placed in the extraction thimble, was conducted using

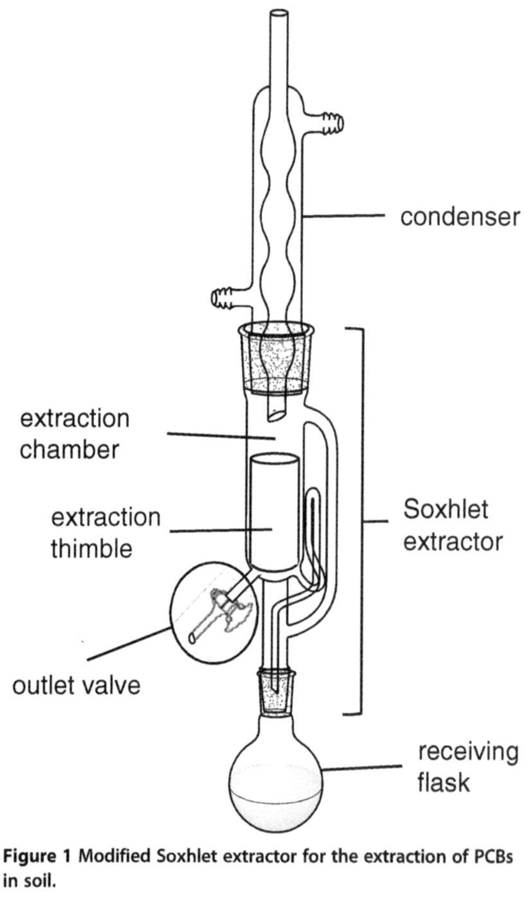

Figure 1 Modified Soxhlet extractor for the extraction of PCBs in soil.

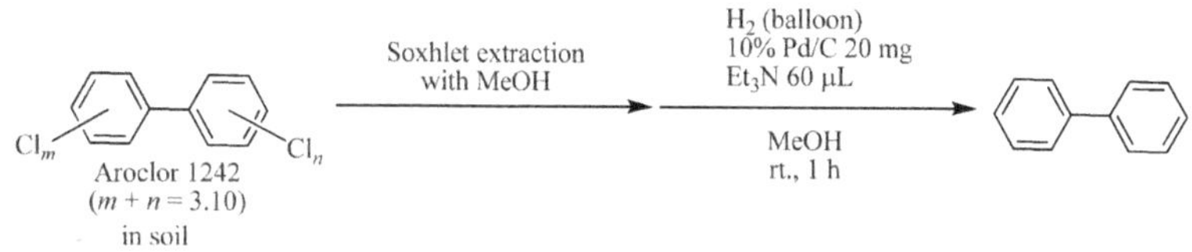

Entry	Soil	Extraction time [a]	Residual PCBs in the extraction chamber (%) [b]	Conversion by hydrodechlorination (%) [b]
1	Quartz sand	2 h	0	98
2	Silica sand	2 h	2	100
3	Silica sand	4 h	0	100
4	Diatomaceous Earth	2 h	7	100
5	Diatomaceous Earth	4 h	0	100
6	Bentonite	2 h	12	30

Figure 2 PCB extraction from PCB-contaminated soils with MeOH and hydrodechlorination under 10% Pd/C–Et_3N–H_2 conditions. **a** Using a modified Soxhlet extractor (Figure 1) **b** Determined by GC/MS.

MeOH (150 mL) as an extraction solvent for 2 h using a modified Soxhlet extractor that has an outlet valve connected on the extraction chamber for the easy removal of the collected (distilled) solvents after completion of the extraction (Figure 1). The PCB extract in the receiving flask was then heated at 90°C, and the distilled MeOH (ca. 35 mL) was collected in the extraction chamber and removed through the outlet valve. The removed MeOH solution (1.5 mL) was concentrated to dryness, and the residue was dissolved in hexane (1.5 mL) and analyzed by GC/MS. The residual PCBs in the extraction chamber were estimated based on the comparison of the total peak area of the PCBs between the PCB contents in the original PCB-contaminated soils and the PCB contents in the recovered soils after extraction. A part (1.5 mL) of the Soxhlet extract in the receiving flask [MeOH solution (100 mL)] was also concentrated to dryness and dissolved in hexane (1.5 mL). The hexane solution (800 μL) was diluted with additional hexane (400 μL), and the GC/MS analysis was conducted to confirm the recovery of the Aroclor 1242 in the receiving flask. The recovery ratio of Aroclor 1242 was estimated to be 93% by comparison of the peak area of the prepared hexane sample using the recovered PCBs in MeOH in the receiving flask with the PCBs in the original PCB-contaminated quartz sand (7.5 mg).

Figure 2, Entry 2: The same procedure used in Entry 1 was followed except for using silica sand (30 g) in place of the quartz sand (30 g).

Figure 2, Entry 3: The same procedure used in Entry 2 was followed, and the extraction time was extended to 4 h. The silica sand in the extraction thimble was recovered and extracted again with fresh hexane (150 mL) using the Soxhlet extractor for 2 h. No residual PCBs in the extraction chamber and receiving flask were detected by GC/MS.

Figure 2, Entry 4: The procedure was the same as the method in Entry 1 except for using diatomaceous earth (10 g) in place of the quartz sand (30 g). The amount of soil was determined according to its bulkiness due to the limited volume of the extraction thimble.

Figure 2, Entry 5: The same procedure used in Entry 4 was followed, and the extraction time was extended to 4 h. The diatomaceous earth in the extraction thimble was recovered and extracted again with fresh hexane (150 mL) using the Soxhlet extractor for 2 h. No residual PCBs in the extraction chamber and receiving flask were detected by GC/MS.

Figure 2, Entry 6: The procedure was the same as the method in Entry 1 except for using bentonite (20 g) in place of the quartz sand (30 g).

Figure 3, Entry 1: The extraction procedure was the same as the method in Figure 2, Entry 1 except for the use of hexane (150 mL) as an extraction solvent and a 1-h extraction time. The PCB extract in the receiving flask was heated at 90°C, and the distilled hexane (ca. 35 mL) was collected in the extraction chamber and removed through the outlet valve. The removed hexane solution

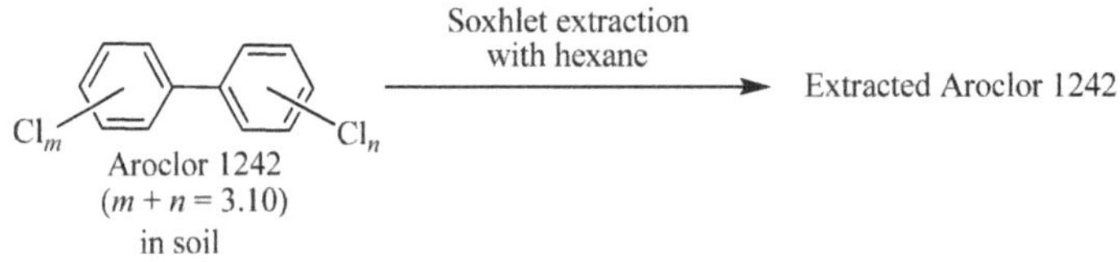

Entry	Soil	Extraction time	Residual PCBs in the extraction chamber (%)[a]
1	Quartz sand	1 h	0.2
2[b,c]	Quartz sand	2 h	0
3[c]	Silica sand	2 h	0
4[c]	Diatomaceous Earth	2 h	0
5	Bentonite	2 h	3
6[c]	Bentonite	4 h	0
7[c]	Akadama-tsuchi	2 h	0
8[c]	Kanuma-tsuchi	2 h	0

Figure 3 PCB extraction from PCB-contaminated soils with hexane using a modified Soxhlet extractor. a Determined by GC/MS. **b** The extraction and hydrodechlorination of PCBs were investigated three times. **c** The extracted PCBs in the receiving flask were used for the hydrodechlorination, see Figure 7.

(1 μL) from the extraction chamber was analyzed by GC/MS without dilution.

Figure 3, Entry 2: The extraction procedure was the same as the method for Figure 3, Entry 1 except for the extraction time (2 h).

Figure 3, Entry 3: The extraction procedure was the same as the method for Figure 3, Entry 2 except for the use of silica sand (30 g) in place of the quartz sand (30 g).

Figure 3, Entry 4: The extraction procedure was the same as the method for Figure 3, Entry 2 except for the use of diatomaceous earth (10 g) in place of the quartz sand (30 g).

Figure 3, Entry 5: The extraction procedure was the same as the method for Figure 3, Entry 2 except for the use of bentonite (20 g) in place of the quartz sand (30 g).

Figure 3, Entry 6: The extraction procedure was the same as the method for Figure 3, Entry 1 except for the extraction time (4 h) and use of bentonite (20 g) in place of the quartz sand (30 g).

Figure 3, Entries 7 and 8: The extraction procedure was the same as the method for Figure 3, Entry 2 except for the use of Akadama-tsuchi (20 g) (Entry 7) or Kanuma-tsuchi (20 g) (Entry 8) in place of the quartz sand (30 g).

In Figure 3, Entries 2–4 and 6–8, the Soxhlet extracts in the receiving flask was used for the PCB degradation under the 10% Pd/C–Et_3N–H_2 conditions using the procedure described in the method section "Hydrodechlorination of PCB extracts from PCB-polluted soils under 10% Pd/C–Et_3N–H_2 conditions".

The solvent effect on the hydrodechlorination of the PCBs under 10% Pd/C-Et_3N-H_2 conditions

Figure 4, Entry 1: A mixture of Aroclor 1242 (7.5 mg), 10% Pd/C (5 mg) and Et_3N (30 μL) in hexane (60 mL) in a round-bottom flask was vigorously stirred using a stirring bar under a hydrogen atmosphere (balloon) at ambient temperature (ca. 25°C). After 1, 2, and 3 h, the reaction mixture (1 mL) was sampled, filtered through a 0.45 μm membrane filter (Millipore, USA), and washed with H_2O (1 mL). The conversion yield (%) was determined by GC/MS analysis of the resulting hexane solution.

Figure 4, Entry 2: The procedure was the same as the method for Figure 4, Entry 1 except for the solvent [hexane–MeOH (1 : 1, 60 mL)] and the amount of the

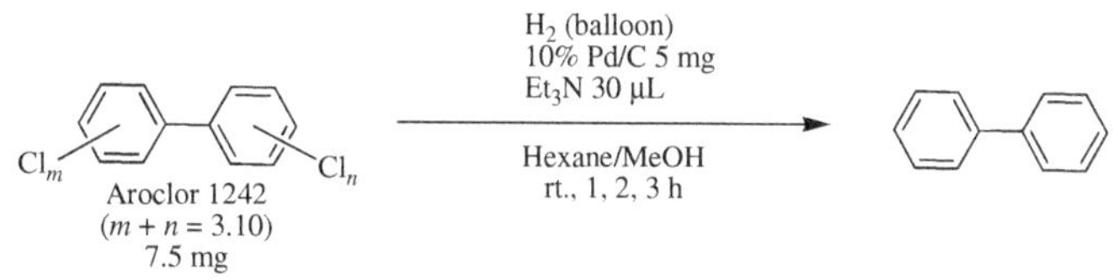

Entry	Reaction solvent	Conversion (%)[a]		
		1 h	2 h	3 h
1	Hexane	4	5	6
2	Hexane–MeOH, 1 : 1 (v/v)	46	56	57
3[b]	Hexane–MeOH, 1 : 5 (v/v)	98	100	100
4	MeOH	100	100	100

Figure 4 Solvent effect on the hydrodechlorination of PCBs under 10% Pd/C–Et_3N–H_2 conditions. a Determined by GC/MS. **b** The GC charts after 1, 2, and 3-h reactions are shown in Figure 5.

sampled reaction mixture (1.5 mL), which was successively washed with H_2O (1.5 mL). The conversion yield (%) was determined by GC/MS analysis using the doubly diluted hexane layer.

Figure 4, Entry 3: The procedure was the same as the method for Figure 4, Entry 1 except for the solvent [hexane–MeOH (1 : 5, 60 mL)] and the amount of the sampled reaction mixture (1.5 mL). Hexane (1.5 mL) and H_2O (1.5 mL) were added to the sampled reaction mixture, the mixture was shaken, then the hexane layer was used for the GC/MS analysis. The GC charts after 0, 1, and 2 h are shown in Figure 5.

Figure 4, Entry 4: The procedure was the same as the method for Figure 4, Entry 1 except for the solvent [MeOH (60 mL)]. The reaction mixture (1.5 mL) was sampled, filtered, and concentrated to dryness, then the residue was dissolved in hexane (1 mL). The solution was washed with H_2O (1 mL) and the hexane layer was used for the GC/MS analysis after a double dilution.

Evaluation of residual PCBs in distilled hexane in the extraction chamber after the Soxhlet extraction

The Soxhlet extraction of PCBs from the PCB-contaminated quartz sand (30 g) using hexane (150 mL) was carried out for 2 h. The extract in the receiving flask was concentrated at 90°C. The distilled and collected hexane solution (ca. 30 mL, first fraction) was removed from the extraction chamber through the outlet valve (Figure 1), and the first fraction (1 μL) was analyzed by GC/MS (Figure 6a). The remaining extract in the receiving flask was gently and further concentrated at 90°C, the distilled and collected hexane in the extraction chamber (ca. 30 mL) was removed through the outlet valve as the second fraction, and its GC/MS analysis (1 μL) was also performed (Figure 6b). Two more hexane fractions (third and fourth fractions, each ca. 30 mL) were collected in the extraction chamber from the extracts in the receiving flask by heating at 90°C, removed through the outlet valve, and analyzed using GC/MS in the same manner as described for the first and second fractions (Figure 6c and d).

Hydrodechlorination of PCB extracts from PCB-polluted soils under 10% Pd/C–Et_3N–H_2 conditions

Figure 2: 10% Pd/C (20 mg) and Et_3N (60 μL) were added to the PCB extract in MeOH (ca. 100 mL) in the receiving flask (see the method section for Figure 2, "PCB extraction from PCB-contaminated soils using Soxhlet extractor"). The mixture was vigorously stirred using a stirring bar under a hydrogen atmosphere (balloon) at ambient temperature (ca. 25°C) for 1 h. The reaction mixture (ca. 5 mL) was filtered through cotton, and the filtrate (3 mL) was concentrated to dryness. The residue was diluted with hexane (3 mL) and washed with H_2O (3 mL). The hexane layer (800 μL) was diluted with additional hexane (400 μL), and the resulting hexane solution was analyzed by GC/MS to analyze the degradation.

Figure 7: Each extracted PCB in hexane in the receiving flask (ca. 100 mL, see the method section for Figure 3, Entries 2–4 and 6–8, "PCB extraction from PCB-contaminated soils using Soxhlet extractor") was very gently concentrated to ca. 10 mL using a rotary evaporator (40°C, ca. 20 mmHg), and no PCBs were detected in the hexane collected in the rotary evaporator trap [b) in Figure 8] and receiving flask [a) in Figure 8] by GC/MS. MeOH (50 mL), 10% Pd/C (20 mg), and Et_3N (60 μL) were added to the receiving flask. The

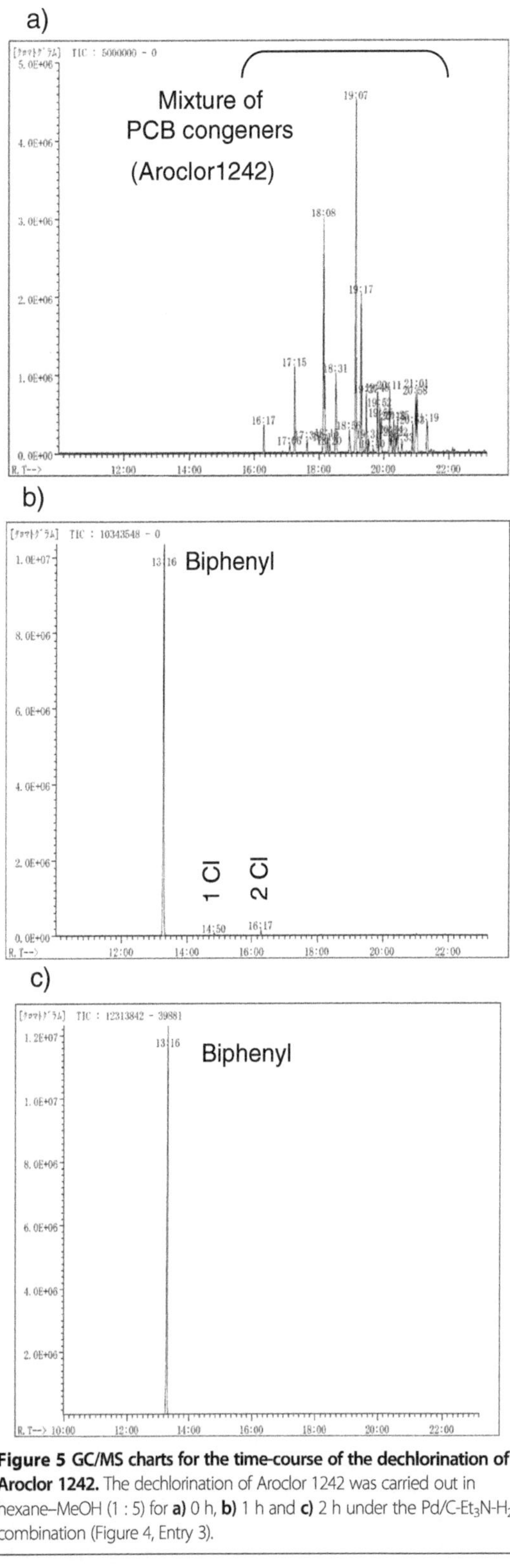

Figure 5 GC/MS charts for the time-course of the dechlorination of Aroclor 1242. The dechlorination of Aroclor 1242 was carried out in hexane–MeOH (1 : 5) for **a)** 0 h, **b)** 1 h and **c)** 2 h under the Pd/C-Et_3N-H_2 combination (Figure 4, Entry 3).

mixture was vigorously stirred using a stirring bar under a hydrogen atmosphere (balloon) at ambient temperature (ca. 25°C) for 1 h. The reaction mixture (ca. 6 mL) was filtered through a 0.45 μm membrane filter (Millipore, USA) and concentrated to dryness. The residue was diluted with hexane (5 mL) and washed with H_2O (5 mL). The hexane layer (400 μL) was diluted with additional hexane (600 μL), and the resulting hexane solution was analyzed by GC/MS to analyze the degradation.

Instrumental analysis

A JMS Q1000 GC [7890A gas chromatography (Agilent Technologies, USA) equipped with a JEOL MK II mass spectrometer (JEOL Co., Ltd., Japan)] and an Inert Cap 5MS/sil + GD capillary column (30 m × 0.25 mm i.d., 0.25 μm film thickness: GL Science, Japan) were used for the PCB analysis. Helium was employed as a carrier gas at the flow rate of 1.3 mL/min. The injector and detector temperatures were 280°C. The column temperature was programmed to ramp from 70°C (5 min hold) to 280°C (4 min hold) to the rate of 10°C/min. One μL of the sample solution was injected. The retention time of biphenyl was 13.01 min and those of the PCBs (Aroclor 1242) were 16.26–21.29 min. The products were identified by their GC/MS retention times in comparison to those of authentic commercial samples.

Results and discussion

Soxhlet extraction of PCBs from PCB-contaminated soil

PCBs were extracted from a variety of artificially PCB-contaminated soils (quartz sand, silica sand, diatomaceous earth and bentonite) for 2 h using a modified Soxhlet extractor that has an outlet valve attached to the extraction chamber for the easy removal of the extraction solvent (Figure 1). MeOH was first examined as an extraction solvent for the present Soxhlet extraction (Figure 2), since MeOH was found to be an effective and preferable solvent for the Pd/C-catalyzed PCB degradation based on the dechlorination reaction in the presence of Et_3N under an H_2 atomosphere [29,30,32,33]. Thus, the use of MeOH for the extraction could avoid the solvent exchange process for the PCB degradation after the extraction. When the 2-h Soxhlet extraction of the PCBs from the PCB-contaminated quartz sand was carried out, no PCBs were detected from the MeOH collected in the extraction chamber (Entry 1), and the nearly quantitative extraction of the PCBs was confirmed by the GC/MS analysis of the PCBs in the receiving flask (93%). We determined that the PCBs were nearly quantitatively extracted in the receiving flask when no PCBs were detected in the solvents obtained from the extraction chamber after the Soxhlet extraction. Although the MeOH extract in the receiving flask

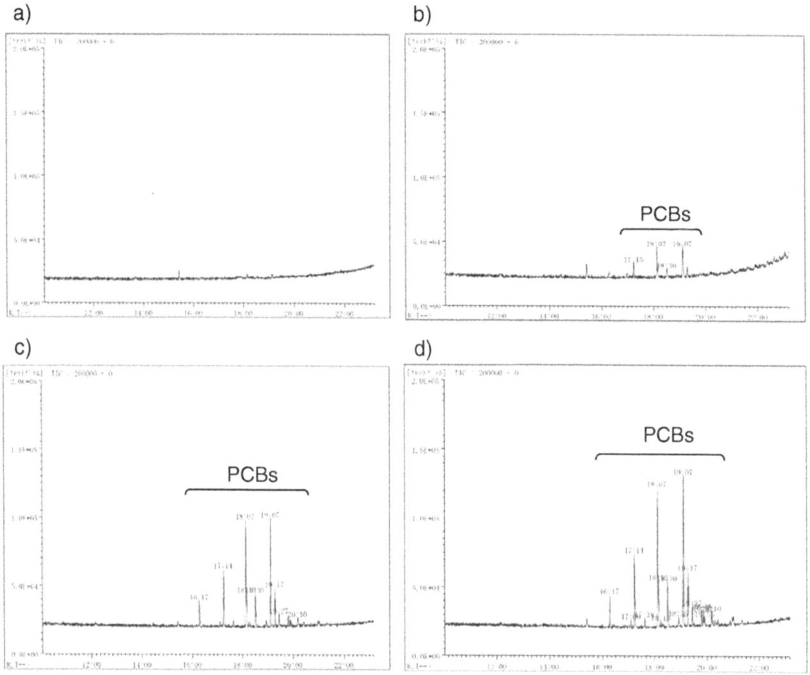

Figure 6 GC spectra of the collected hexane in the extraction chamber after the 2 h-Soxhlet extraction. a) first fraction: distilled hexane collected in the extraction chamber; **b)** second fraction: distilled hexane collected in the extraction chamber from the remaining hexane solution in the receiving flask after the removal of the first fraction; **c)** third fraction: distilled hexane from the remaining hexane solution in the receiving flask after the removal of the second fraction; **d)** fourth fraction: distilled hexane from the remaining hexane solution in the receiving flask after the removal of the third fraction.

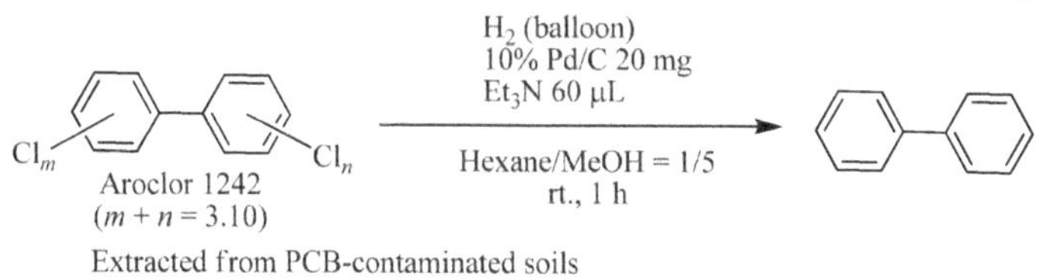

Entry	Model soil	Conversion by hydrodechlorination (%) [a]
1	Quartz sand	100
2	Silica sand	100
3	Diatomaceous Earth	100
5	Bentonite	100
6	Akadama-tsuchi	100
7	Kanuma-tsuchi	100

Figure 7 Hydrodechlorination of the extracted PCBs under 10% Pd/C–Et_3N–H_2 conditions. a Determined by GC/MS.

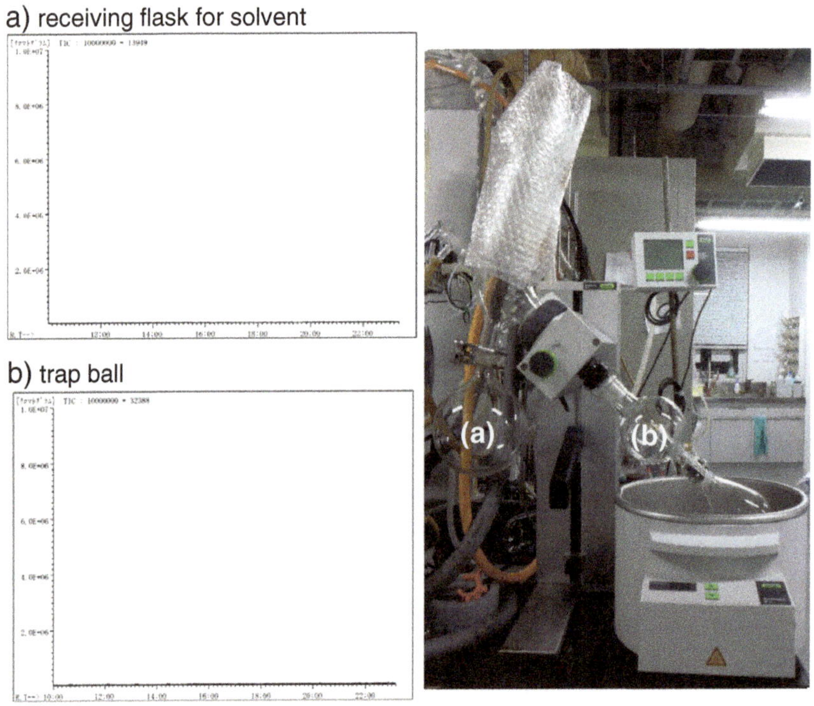

Figure 8 The GC/MS charts of the recovered hexane in the evaporator after the concentration. The GC/MS charts of the recovered hexane **a)** in the receiving flask (shown by (a) in the panel) and **b)** in the trap ball (shown by (b) in the panel) of the rotary evaporator. The panel shows rotary evaporator used in this study.

was vigorously stirred with 10% Pd/C and Et_3N under an H_2 atmosphere (balloon) at room temperature (ca. 25°C) for 1 h, the hydrodechlorination of the PCBs in the extract was not completed (98% conversion) contrary to our expectation. The incomplete reaction would be attributed to the sulfur contaminants in the original quartz sand, since significant amounts of sulfur contaminants [1800 μg/g by the oxygen bomb combustion ion chromatography (Sumika Chemical Analytical Service, Ltd., Japan)] were extracted by MeOH from the quartz sand. Such sulfur components could function as catalyst poisons toward the Pd/C-catalyzed hydrodechlorination based on the significant coordination activity of sulfur atoms to palladium atoms. Therefore, MeOH was not an appropriate solvent for the Soxhlet extraction from the PCB-polluted quartz sand.

Although a small amount of PCBs was still extracted in MeOH in the extraction chamber after the 2-h extraction of the PCB-contaminated silica sand (2%, Figure 2, Entry 2) and diatomaceous earth (7%, Entry 4), the lengthening of the extraction time to 4 h certainly enhanced the extraction efficiency, and no PCBs were detected in MeOH in the extraction chamber (Entries 3 and 5). The soils in the extraction thimble after 4-h extraction (Entries 3 and 5) were recovered, and further extraction was conducted with fresh hexane using the Soxhlet extractor for 2 h, since hexane was generally employed as an effective solvent for the PCB extraction [11,14]. Consequently, no PCBs were detected from both the extraction chamber and the receiving flask, indicating that the PCBs were completely extracted from soils by the first 4-h Soxhlet extraction. Therefore, we estimated that the complete PCB extraction was achieved when no PCBs were detected in the solvent in the extraction chamber. Furthermore, PCBs in the MeOH extracts of the PCB-contaminated silica sand and diatomaceous earth underwent a complete degradation under Pd/C–Et_3N-mediated hydrodechlorination conditions (Entries 3 and 5). On the other hand, the Soxhlet extraction of the PCB-contaminated bentonite using MeOH was unsuccessful probably due to the strong adsorptive property, affinity for MeOH and/or viscosity of the bentonite, and a significant amount of PCBs (12%) still remained in MeOH in the extraction chamber after 2-h extraction. Furthermore, the subsequent 10% Pd-catalyzed hydrodechlorination proceeded in only 30% (Entry 6). Bentonite is originally derived from volcanic ash and contains a substantial amount of sulfur concomitants, such as sulfates and sulfides. The sulfur materials would be extracted from the bentonite together with

PCBs by the Soxhlet extraction with MeOH and strongly suppress the PCB degradation as catalyst poisons in analogy (more significantly) with the hydrodechlorination of the MeOH extracts of quartz sand (Entry 1). Therefore, we concluded that MeOH was not effective as a solvent for the Soxhlet extraction of PCBs.

Hexane has been mainly used as an efficient solvent for the extraction of PCBs from the PCB-contaminated solid materials and water in the literature [11,14,37,38]. Furthermore, the low solubility of sulfur contaminants in hexane is also easily predicted compared to MeOH. Therefore, the Soxhlet extraction of PCBs with hexane from the PCB-contaminated quartz sand was carried out to establish the effective PCB extraction method (Figure 3, Entries 1 and 2). Although the 1-h Soxhlet extraction was incomplete for the perfect removal of PCBs from the PCB-contaminated quartz sand and 0.2% of the PCBs still remained in hexane in the extraction chamber (Entry 1), the PCB-residue completely disappeared from the extraction chamber by extending the extraction time to 2 h (Entry 2). The Soxhlet extraction procedure for PCBs with hexane was reliable since it could be reproduced three times (Entry 2), and the amounts of the extracted sulfur contaminants from the PCB-contaminated quartz sand were notably decreased compared to the extraction with MeOH [880 μg/g by oxygen bomb combustion ion chromatography (Sumika Chemical Analytical Service, Ltd., Japan)]. These results indicated that the PCBs were completely and selectively extracted from the polluted quartz sand by hexane. The Soxhlet extraction method using hexane could also be applied to extraction from the PCB-contaminated silica sand and diatomaceous earth (Entries 3 and 4), although the extraction of PCB-contaminated bentonite was incomplete within 2 h (Entry 5) probably because bentonite could strongly adsorb PCBs due to its strong viscosity property. A quantitative extraction of PCBs from PCB-contaminated bentonite was successfully achieved by extending the extraction time to 4 h (Entry 6).

The method for the remediation of PCB-polluted soils was also used for the cleanup of horticultural soils, Akadama-tsuchi and Kanuma-tsuchi (Entries 7 and 8). The PCBs in both the PCB-contaminated Akadama-tsuchi (Entry 7) and Kanuma-tsuchi (Entry 8) were completely extracted with hexane within 2 h using the modified Soxhlet extractor.

The extracted Aroclor 1242 in the receiving flask from polluted-quartz sand (Entry 2), -silica sand (Entry 3), -diatomaceous earth (Entry 4), -bentonite (Entry 6), -Akadama-tsuchi (Entry 7), and -Kanuma-tsuchi (Entry 8) were successively employed for the Pd/C-catalyzed hydrodechlorination, which will be discussed later (Figure 7).

Solvent effect on the PCB dechlorination

Since hexane was found to be appropriate as a solvent for the Soxhlet extraction of PCBs, the use of hexane as a solvent for the degradation of PCBs based on the hydrodechlorination of aromatic chloride under Pd/C–Et_3N–H_2 conditions would be desired due to the safe and easy handling on the basis of no need to exchange solvents after the extraction. However, the 10% Pd/C-catalyzed hydrodechlorination of Aroclor 1242 in the presence of Et_3N hardly proceeded in hexane (Figure 4 Entry 1). The detailed study of the Pd/C-catalyzed hydrodechlorination in the presence of Et_3N under an H_2 atmosphere has already revealed that MeOH was an effective and preferable solvent [29,30,32,33]. Aroclor 1242 was completely hydrodechlorinated at room temperature within 1 h in MeOH instead of hexane (Entry 4). The mixed solvent of hexane and MeOH was then investigated, since the complete removal of hexane from the receiving flask for the exchange of solvents after the Soxhlet extraction includes the risk of the possible evaporation or leak of PCBs together with hexane (see the discussion in the section "Procedure for the concentration of the PCB extract in the receiving flask"). The higher MeOH ratio of the mixed solvents more effectively promoted the hydrodechlorination (Figure 4 Entries 2 and 3), and it could be completed in hexane–MeOH (1 : 5) within 2 h under ambient pressures and temperatures (Entry 3) as shown in Figure 5.

Concentration of the PCB extract in the receiving flask

We next investigated the establishment of a procedure for the removal of hexane from the Soxhlet extract in the receiving flask for the subsequent Pd/C-catalyzed hydrodechlorination of PCBs in hexane–MeOH (1 : 5). After the 2-h Soxhlet extraction of PCBs from the PCB-contaminated quartz sand using hexane, the hexane was transferred to the extraction chamber from the receiving flask by the four-step distillation, and collected as different fractions (Figure 6). Although no PCBs were detected in the first fraction (Figure 6a), a small quantity of PCBs was found in the 2nd fraction (Figure 6b), and the detected PCB quantity increased in a fraction-number-dependent manner (Figure 6b–d). These results suggest that a small amount of PCBs could be vaporized together with the distilled hexane with decreasing amount of hexane in the receiving flask even under gentle refluxing conditions. On the other hand, when the further gentle concentration of the hexane extract in the receiving flask using a rotary evaporator under reduced pressure (ca. 20 mmHg) at 40°C was done, no PCBs were detected in the collected hexane in both the rotary evaporator trap [b) in Figure 8] and the solvent receiving flask [a) in Figure 8] by GC/MS, indicating that no PCBs had been vaporized during the gentle evaporation under a sufficiently reduced pressure.

Continuous operation of the PCB extraction using the Soxhlet extractor and dechlorination using the 10% Pd/C–Et_3N–H_2 conditions

The PCB extraction from a variety of PCB-contaminated soils (quartz sand, silica sand, diatomaceous earth, bentonite, Akadama-tsuchi and Kanuma-tsuchi) with hexane (Figure 3) and the subsequent dechlorination using Pd/C–Et_3N–H_2 conditions in hexane–MeOH were carried out under ambient H_2 pressures (balloon) and temperatures (Figure 7). While the PCBs in the MeOH extract from the PCB-contaminated quartz sand were not completely hydrodechlorinated under 10% Pd/C–H_2–Et_3N conditions (Figure 2, Entry 1), the degradation of the hexane extract (Figure 3, Entry 2) was completed (Figure 7, Entry 1) because the amount of sulfur-contaminants in the hexane extract was sufficiently low (880 μg/g) compared to that of MeOH (1800 μg/g). Furthermore, all other PCB extracts derived from the silica sand, diatomaceous earth, bentonite, Akadama-tsuchi and Kanuma-tsuchi were also smoothly and completely degraded within 1 h (Figure 7, Entries 2–7).

The present soil remediation method using a modified (newly-developed) Soxhlet extractor that bears an outlet valve on the extraction chamber is able to extract PCBs from PCB-contaminated soils completely and the subsequent chemical degradation is successfully accomplished by a one-pot procedure. In reported remediation methods in the literature involved extraction process, the PCB extraction from soil and its chemical degradation could be achieved totally independent manner with the exchange of the reaction containers [9-11,14-16], to our knowledge. The present method is quite safe by reason that the operators could prevent the extracted PCBs from the direct contacting. Furthermore, PCBs extracted from PCB-contaminated soils quantitatively degrade based on a Pd/C-catalyzed dechlorination in the presence of Et_3N under ambient hydrogen pressure and temperature in a short time. Moreover, the Pd/C might be recovered and reused by the simple wash in water and hexane sequentially [32]. The present remediation method achieved extensive improvement in simplicity, safety and time-efficiencies over the previously reported remediation processes.

Conclusions

We developed a practical method for the remediation of PCB-contaminated soils. One of the notable features of the PCB remediation of PCB-contaminated soils is the use of a receiving flask for the Soxhlet extractor as a reaction flask for the hydrodechlorination of PCBs. Therefore, the present remediation method never requires the transfer of the extracted PCBs to another reaction container and the degradation is successfully accomplished by a one-pot procedure. Thus, the possible secondary pollution by the extracted (concentrated) PCBs does not occur. PCBs are efficiently extracted with hexane from various types of polluted soils, and their dechlorination could be easily achieved in a hexane–MeOH (1 : 5) solution using the 10% Pd/C–Et_3N–H_2 conditions under ambient pressures (balloon) and temperatures (ca. 25°C). The present method is quite simple and would be useful for the mild and practical remediation of PCB-polluted soils.

Abbreviations

PCBs: Polychlorinated biphenyls; Pd/C: Palladium on activated carbon; Et_3N: Trimethylamine; U.S.EPA: The United States environmental protection agency; DDTs: Dichlorodiphenyltrichloroethanes; GC/MS: Gas chromatography/mass spectrometry.

Competing interests

The authors declare that they have no competing interests.

Authors' contributions

YM, HS and HN conceived the work; AI, MN and SI performed the experiments; AI, MN, SI, YS, TN, YM, HS and HN designed the experiments and analyzed the data; and AI, YM, HS and HN wrote the manuscript. All authors read and approved the final manuscript.

Acknowledgements

We sincerely appreciate the support by the New Energy and Industrial Technology Development Organization (NEDO), Japan. We also thank the N.E. Chemcat Corporation for the gift of Pd/C.

Author details

[1]Laboratory of Hygienic Chemistry and Molecular Toxicology, Gifu Pharmaceutical University, 1-25-4 Daigaku-nishi, Gifu 501-1196, Japan. [2]Laboratory of Organic Chemistry, Gifu Pharmaceutical University, 1-25-4 Daigaku-nishi, Gifu 501-1196, Japan.

References

1. U.S. Department of Health and Human Services, Public Health Service, Agency for toxic substances and disease registry: Toxicological profile for polychlorinated biphenyls. 2000. [http://www.atsdr.cdc.gov/toxprofiles/tp17.pdf]
2. Blais JM, Schindler DW, Muir DCG, Kimpe LE, Donald DB, Rosenberg B. Accumulation of persistent organochlorine compounds in mountains of western Canada. Nature. 1998;395:585–8.
3. Epling GA, Florio EM, Bourque AJ, Qian XH, Stuart JD. Borohydride, micellar, and exciplex-enhanced dechlorination of chlorobiphenyls. Environ Sci Technol. 1988;22:952–6.
4. Erickson MD. Analytical Chemistry of PCBs. 2nd ed. Boca Raton: CRC press; 1997.
5. Tundo P, Facchetti S, Tumiatti W, Fortunati UG. Chemical degradation of 2,3,7,8-TCDD by means of polyethylene glycols in the presence of weak bases and an oxidant. Chemosphere. 1985;14:403–10.
6. The Environment Programme of the United Nations: Inventory of world-wide PCB destruction capacity; second issue. 2004. [http://www.chem.unep.ch/pops/pcb_activities/pcb_dest/PCB_Dest_Cap_SHORT.pdf]
7. Meharg AA, Osborn D. Dioxins released from chemical accidents. Nature. 1995;375:353–4.
8. Travis CC, Hester ST. Global chemical pollution. Environ Sci Technol. 1991;25:814–9.
9. Okitsu K, Yoshioka Y, Tanabe S. Effect of transition metal powder and low-energy ultrasound frequency on degradation rate of chlorinated hydrocarbons in water. Bunseki Kagaku. 2006;55:567–71.
10. Okuno H, Yim B, Mizukoshi Y, Nagata Y, Maeda Y. Sonolytic degradation of hazardous organic compounds in aqueous solution. Ultrason Sonochem. 2000;7:261–4.

11. Achari G, Jakher A, Gupta C, Dhol A, Langford CH. Practical method to extract and dechlorinate PCBs in soils. Pract Period Hazard, Toxic, Radioact Waste Manage. 2010;14:98–103.
12. Koshioka M, Kanazawa J, Iizuka H, Murai T. Photodegradation of decachlorobiphenyl. Bull Environ Contam Toxicol. 1987;38:409–15.
13. Kouta S, Mizuki S, Wataru A, Yukio T. Photodecomposition of dioxin in ash and paddy field soil. Organohalogen Compounds. 2007;69:435/431–4.
14. He N, Li P, Zhou Y, Ren W, Fan S, Verkhozina VA. Catalytic dechlorination of polychlorinated biphenyls in soil by palladium-iron bimetallic catalyst. J Hazard Mater. 2009;164:126–32.
15. Huang Q, Hong C-S. TiO_2 photocatalytic degradation of PCBs in soil-water systems containing fluoro surfactant. Chemosphere. 2000;41:871–9.
16. Mitoma Y, Egashira N, Simion C. Highly effective degradation of polychlorinated biphenyls in soil mediated by a Ca/Rh bicatalytic system. Chemosphere. 2009;74:968–73.
17. Noma Y, Mitsuhara Y, Matsuyama K, Sakai S-I. Pathways and products of the degradation of PCBs by the sodium dispersion method. Chemosphere. 2007;68:871–9.
18. Pittman CU, He J. Dechlorination of PCBs, CAHs, herbicides and pesticides neat and in soils at 25°C using Na/NH_3. J Hazard Mater. 2002;92:51–62.
19. Taniguchi S, Murakami A, Hosomi M, Miyamura A, Uchida R. Chemical decontamination of PCB-contaminated soil. Chemosphere. 1997;34:1631–7.
20. Ukisu Y, Iimura S, Uchida R. Catalytic dechlorination of polychlorinated biphenyls with carbon-supported noble metal catalysts under mild conditions. Chemosphere. 1996;33:1523–30.
21. Ukisu Y, Miyadera T. Hydrogen-transfer hydrodehalogenation of aromatic halides with alcohols in the presence of noble metal catalysts. J Mol Catal A Chem. 1997;125:135–42.
22. Wu Q, Sowers KR, May HD. Microbial reductive dechlorination of Aroclor 1260 in anaerobic slurries of estuarine sediments. Appl Environ Microbiol. 1998;64:1052–8.
23. Murinova S, Dercova K, Dudasova H. Degradation of polychlorinated biphenyls (PCBs) by four bacterial isolates obtained from the PCB-contaminated soil and PCB-contaminated sediment. International Biodeterioration & Biodegradation. 2014;91:52–9.
24. Petric I, Hrsak D, Fingler S, Udikovic-Kolic N, Bru D, Martin-Laurent F. Insight in the PCB-degrading functional community in long-term contaminated soil under bioremediation. Journal of Soils and Sediments. 2011;11:290–300.
25. Song M, Luo C, Li F, Jiang L, Wang Y, Zhang D, et al. Anaerobic degradation of Polychlorinated Biphenyls (PCBs) and Polychlorinated Biphenyls Ethers (PBDEs), and microbial community dynamics of electronic waste-contaminated soil. Sci Total Environ. 2015;502:426–33.
26. U.S. Environmental Protection Agency: USEPA Method 3540C: Soxhlet extraction. 1996. [http://www.epa.gov/osw/hazard/testmethods/sw846/pdfs/3540c.pdf]
27. U.S. Environmental Protection Agency: USEPA Method 8028A: Polychlorinated biphenyls by gas chromatography. 2007. [http://www.epa.gov/wastes/hazard/testmethods/sw846/pdfs/8082a.pdf]
28. Ministry of the Environment of the Japan Government: Manual of the survey measurement of dioxins in soils. 2009. [http://www.env.go.jp/chemi/dioxin/manual/dojo-manual/main.pdf] (Japanese only)
29. Monguchi Y, Kume A, Hattori K, Maegawa T, Sajiki H. Pd/C-Et_3N-mediated catalytic hydrodechlorination of aromatic chlorides under mild conditions. Tetrahedron. 2006;62:7926–33.
30. Sajiki H, Kume A, Hattori K, Hirota K. Mild and general procedure for Pd/C-catalyzed hydrodechlorination of aromatic chlorides. Tetrahedron Lett. 2002;43:7247–50.
31. Monguchi Y, Kume A, Sajiki H. Facile and catalytic degradation method of DDT using Pd/C-Et_3N system under ambient pressure and temperature. Tetrahedron. 2006;62:8384–92.
32. Kume A, Monguchi Y, Hattori K, Nagase H, Sajiki H. Pd/C-catalyzed practical degradation of PCBs at room temperature. Appl Catal, B. 2008;81:274–82.
33. Sajiki H, Kume A, Hattori K, Nagase H, Hirota K. Complete and truly catalytic degradation method of PCBs using Pd/C-Et_3N system under ambient pressure and temperature. Tetrahedron Lett. 2002;43:7251–4.
34. Monguchi Y, Ishihara S, Ido A, Niikawa M, Kamiya K, Sawama Y, et al. Pilot-plant study of the PCB degradation at ambient temperature and pressure. Org Process Res Dev. 2010;14:1140–6.
35. Ido A, Ishihara S, Kume A, Nakanishi T, Monguchi Y, Sajiki H, et al. Practical method for PCB degradation using Pd/C-H_2-Mg system. Chemosphere. 2013;90:57–64.
36. Ishihara S, Ido A, Monguchi Y, Nagase H, Sajiki H. Pd/C-catalyzed dechlorination of polychlorinated biphenyls under hydrogen gas-free conditions. J Hazard Mater. 2012;229–230:15–9.
37. Adenugba AA, McMartin DW, Beck A. Environmental contamination of ready meals by polychlorinated biphenyls (PCBs). J Environ Sci and Health, A Tox Hazard Subst & Environ Eng. 2012;47:2230–40.
38. Marku E, Nuro A, Murtaj B. Polychlorinated biphenyls and organochlorine pesticide residues in drinking water of Tirana city (Albania). J Environ Prot Ecol. 2011;12:7–15.

2

Application of geographical information system in disposal site selection for hazardous wastes

Mehdi Rezaeimahmoudi[1], Abdolreza Esmaeli[2*], Alireza Gharegozlu[1], Hassan Shabanian[3] and Ladan Rokni[4]

Abstract

Background: The aim of this study was to provide a scientific method based on Geographical Information System (GIS) regarding all sustainable development measures to locate a proper landfill for disposal of hazardous wastes, especially industrial (radioactive) wastes.

Methods: Seven effective factors for determining hazardous waste landfill were applied in Qom Province, central Iran. These criteria included water, slope, population centers, roads, fault, protected areas and geology. The Analysis Hierarchical Process (AHP) model based on pair comparison was used. First, the weight of each factor was determined by experts; afterwards each layer of maps entered to ARC GIS and with special weight multiplied together, finally the best suitable site was introduced.

Results: The most suitable sites for burial were in northwest and west of Qom Province and eventually five zones were introduced as the sample sites.

Conclusion: GIs and AHP model is introduced as the technical, useful and accelerator tool for disposal site selection. Furthermore it is determined that geological factor is the most effective layer for site selection. It is suggested that geological conditions should be considered primarily then other factors are taken into consideration.

Keywords: Geographical information system (GIS), Disposal site selection, Hazardous waste, AHP, Locating criteria, Iran

Introduction

Protecting human health and the environment in a world rapidly becoming more and more complicated and considered as a major challenge for the international community. Developing in technology and industry leaves some materials, which are hazardous for human life. These materials are potentially dangerous or later cause some risks in environment.

There are some materials used by refineries, power factories, chemical and petrochemical industries that create hazardous wastes. In general, experts recognize that hazardous materials and waste are resulted from the activities of various sectors such as industry, agriculture, trade, and services [1]. Rosenfeld believes that" If a material has been discarded and can cause substantial harm to humans or the environment, it is considered a hazardous waste" [2]. Additionally, hazardous material is, by another definition, "a matter which is itself hazardous for human, animal, and environment health" [3]. According to the suggested definition by Iran Department of Environment (IDOE), hazardous wastes are wastes, which have one of the properties, including toxic, flammable, corrosive and pathogenic – for human, animal, or environment [4].

Radioactive materials are used in medicine, research centers and operation of nuclear power plants, and fuel recycling lead to the production of radioactive wastes, which has negative effect on the surrounding environment. Now, more than 50 countries have consumed their stored fuel and are waiting for reprocessing or disposal them [5]. Various approaches have been investigated in different countries, such as application of iron nanaoparticles in landfill leachate treatment [6], or Converting them into a value added product [7] for usual waste; and for dangerous waste: burying radioactive

* Correspondence: aesmaeli@aeoi.org.ir
[2]Plasma and Nuclear Fusion Research School, Nuclear Science and Technology Research Institute, AEOI, Tehran, Iran
Full list of author information is available at the end of the article

wastes in shallow pits, waste disposal deep underground, burial in halite formations, burial and disposal at sea and in space. Remaining these wastes has been a challenge and global growing issue, hence finding a suitable site for burial them is being deterioration because there are numerous factors and parameters involved. Besides, it is a complex, costing and time-consuming issue.

It is necessary for burial sites to have some geological characteristics. Based on the IDOE Standards these sites should contain the following items: [8-10]

- Suitable substrate of low permeability and high strength with minimal groundwater trend
- Homogeneous rock mass slope with low and stable hydraulic steep,
- Having a low vibration with low frequency
- Stay away from earthquake fault line
- Stay away from populated areas and protected areas and national reserves
- Stay away from lakes and rivers and groundwater discharge areas

Although no research has been done on nuclear technology due to its novelty in Iran, but there are some studies on hazardous wastes, which the common point among all of them is using GIS techniques to locating a proper site for wastes disposal [11-13]. Similarly, GIS techniques and Analysis Hierarchical Process (AHP) has combined to select the best disposal location [14] and an environmental approach was applied to provide the disposal measures [15]. In other countries, some researches in this field have been conducted using GIS; for instance it is said that GIS has a good accuracy to reduce the errors in spatial databases [16]. Furthermore, GIS have been used for the suitability analysis of nuclear waste [17].

The aim of this study was to provide a scientific method, and prevent of risks, occurred due to incorrect disposal site selection. This selection should be done based on the specific circumstance of each area. To clarify the issue, new devices have been established, among which, geographical information system (GIS) is of great prominence. This tool can introduce a suitable site for burning waste with spending less time and money using spatial data. This study focused on industrial (radioactive) wastes.

Materials and methods

Study area

The Qom Province, central Iran is located at the west of Salt Lack (Figure 1). Area of this province is 0.68% of Iran [18].

Data collection

A compendium of methods were implemented to proceed the study as follows: determining the measures for identifying disposal location, reviewing relevant previous

Figure 1 Location of Qom Province, central Iran.

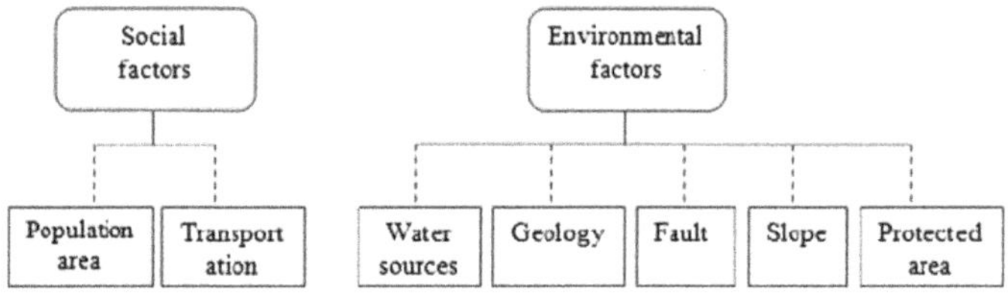

Figure 2 Studied criterias and sub criterias.

studies, study of review articles by the relevant international organizations, and interviews with experts in environmental organizations and Nuclear Energy Organization of Iran.

With reconciliation of the obtained data, the most effective measures based on the studying area were determined, including water resources, slope, population centers, communication lines, protected zones, fault and geology. The measures for mapping were obtained from data of various agencies, such as topographic maps scaled 1/50000 of National Cartographic Center (NCC), geological maps scaled 1/250000, the map of political divisions and rural and urban areas of the country, the map of protected areas of IDOE, provincial road maps provided by Ministry of Road and Urban Development, and ETM images of LandSat. Much of the collected information was raw data, which converted into useful date by using information processing methods of various software packages, including ARCGIS and ENVI. Since the AHP model [19] has been used in the study. Delphi questionnaire was prepared to determine the weights of measures, which was completed by experts related to each measure and incorporated into the final weights.

Data analysis

Base maps were prepared by using GIS & RS analysts, then replaced with weights determined by the experts using AHP model, and finally the appropriate sites were determined based on the identified priorities.

Positioning

The environment is a complex system, which needs a model to investigate all its aspect, and influencing factors and sub-factors [20]. To advance this goal, by reviewing the organizational resources, information, and procedures, and despite all limitations, some measures have been identified.

The measures were divided into environmental measures and social measures, which have their own sub-measures; moreover, the sub-measures were also interrelated while there were associations between two classes (Figure 2).

Ethical approval

The study was approved by the Ethics Committee of Science and Research Branch, Islamic Azad University, Tehran, number 92290.

Results

To choose the landfill, we adopted a holistic, in which quantitative methods were used for concluding. Selected measures were divided into two major areas based on research problem: Environmental and socioeconomic as shown in Table 1. Since the environmental measures for locating the landfill are more important, they were assigned higher weights and more important than others.

Table 1 Determining the criteria for disposal site selection of hazardous wastes using previous studies [16,17,21,22]

Reference No.				
[16]	**[17]**	**[21]**	**[22]**	**Present study**
Geology	Slope	Heidrology	Geology	Geology
Transportation	Geology	Geology	Accessibility	Slope
Environment	Protected areas	Environmental	Population Distribution	Protected areas
Population	Water source	Technical	Hydrology	Water source
	Population centers	Climate		Population centers
	Communication lines	Social		Communication lines
				Faults

Table 2 The expert's opinion about the criteria's weight based on AHP method

	Expert 1	Expert 2	Expert 3	Expert 4	Weight
Geology	0.287305	0.242083	0.260185	0.26766	0.264308
Slope	0.152447	0.159808	0.178881	0.153448	0.161146
Population centers	0.103835	0.117281	0.108006	0.118455	0.111894
Protected areas	0.107581	0.100433	0.096106	0.09548	0.0999
Water source	0.118128	0.115308	0.106863	0.128787	0.117272
Distance from road	0.106083	0.097759	0.09382	0.09359	0.097813
Faults	0.124621	0.167328	0.156139	0.14258	0.147667
Total	1	1	1	1	1

The weights were set according to the experts' idea. Table 1 indicates the method of determining measure in the study using comparative deductions of previous studies [16,17,21,22].

After comparing pair-wise environmental and socioeconomic measures based on AHP method [23], each expert evaluated independently measures of research problem, and calculated the final weight by using approximate arithmetic mean methods (Table 2).

After determining weights and preparing standard maps of selected measures, all layers incorporated into the final map were created in the ArcGIS software based on the following formula.

$$R = \left(\sum_{c=1}^{nc} W_C C_C \right)$$

Were, C is: Criteria, NC: Number of Criteria, WC: The weight of each criterion, $C_{C:}$ Scores for each criterion.

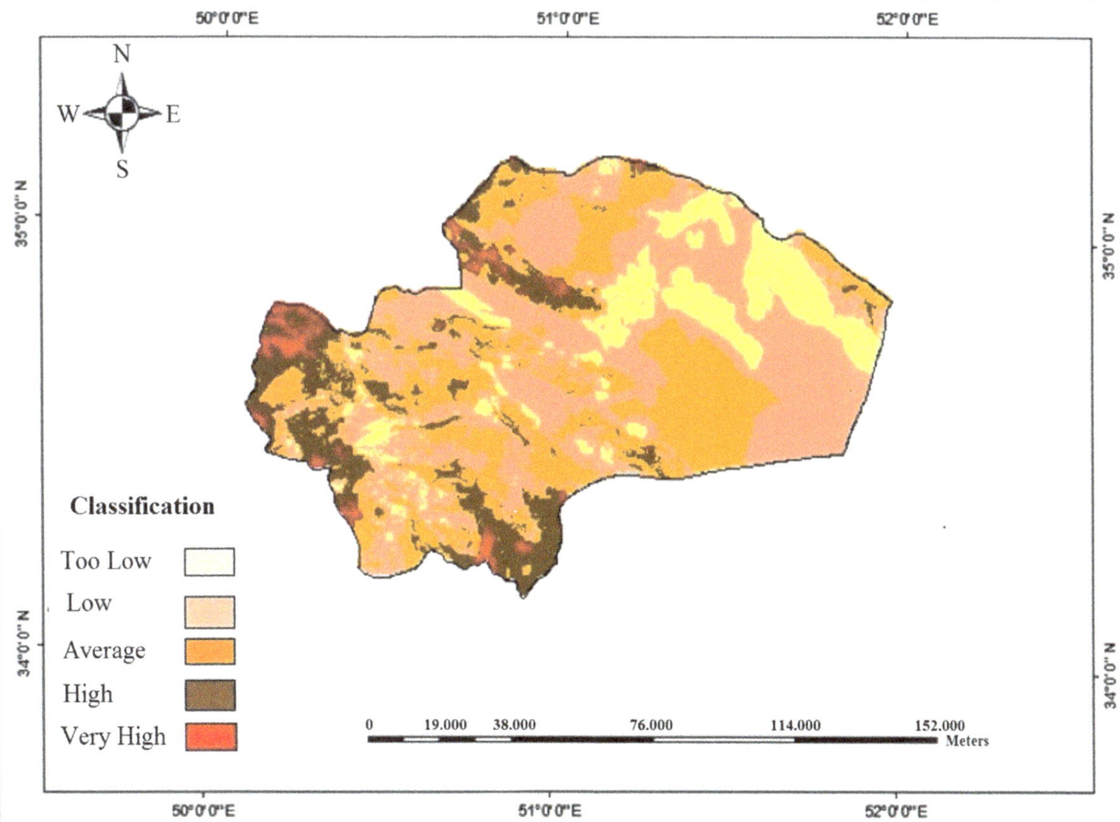

Figure 3 Zoning of appropriate locations for disposal wastes in the region.

Results indicated that the studied province had relatively proper situation for disposal of nuclear wastes in terms of six criteria of water resources, slope, population centers, roads, protected areas and fault. However, given the constraints of the studied area in relation to geological formations, the map indicated that very limited areas in the north, northwest, southwest and south of Qom Province are appropriate for disposal of nuclear wastes; means that geological measure is considered as the most important limiting factor in the identification of nuclear waste disposal in the city of Qom. Figure 3 shows the most appropriate locations for disposal wastes in the region.

Discussion

In this study, we could introduce a technical and useful tool for Disposal Site Selection.

It is believed that site selection is the vital part of burying hazardous waste. This stage, known as the final stage of hazardous waste management, is very complex and technologically demanding [2]. Chen et al. considers the impact of environmental, social and economic problems, caused by burying disposals, as the considerable parts [24].

Our results showed that GIS could be used as an analyst in site selecting for nuclear waste disposal. Using GIS, similar studies have been conducted in other countries, for example NIREX (Nuclear Industry Radioactive Waste Executive, UK), which is a British company with responsibility for identifying radioactive waste disposal site, did a useful investigation and concluded that GIS was an accelerator tool for reaching the goal in this arena [25].

Furthermore, considering seven main criteria in this paper, and using them in incorporation of AHP model and GIS, lead us to understand that selected area in Qom Province has limited potential based on geology; therefore, very limited areas of Qom surface are suitable for nuclear waste disposal. In a relevant study done by Openshaw, using GIS, at first, some main layers were determined, then he tried to choose the most effective layer for site selection, which finally found the geology layer as the most important one [16]. Furthermore, it is confirmed that all criteria related to geology have the most effective role [26]. On the other hand AHP model is introduced as an auxiliary and functional tool in combination, rating and disposal site selection [17]. Combine GIS, AHP in Serbia as a strategic approach to waste management has replaced with traditional methods [27].

Conclusion

Geological factor is introduced as the most important element in disposal site selection for hazardous waste and since it acts like a limiting factor, we suggest that firstly geological conditions should be examined and in next steps other factors be considered. In addition, Landfill sites should be located in homogeneous bedrock, Lands that are less permeable and where groundwater levels are low.

The main limitation in such investigation is obtaining necessary base information. Consequently need for comprehensive research is undeniable in this arena.

Competing interests

The authors declare that they have no competing interest.

Authors' contributions

MR contributed in conceptualization, design, implement, and preparing and results interpretation of the article. AE contributed in conceptualization and design of the study as well as interpretation of the results. AG contributed in design, implement, and interpretation of the results. LR contributed in implement, preparing and results interpretation of the article. All authors undertake equally the liability of the contents of the article. All authors read and approved the final manuscript.

Acknowledgements

This study was part of a M.S thesis; done in Science and Research Branch, Islamic Azad University, Tehran, Iran. The authors declare that there is no conflict of interest.

Author details

[1]Department of GIS & RS, School of Environment and Energy, Science and Research Branch, Islamic Azad University, Tehran, Iran. [2]Plasma and Nuclear Fusion Research School, Nuclear Science and Technology Research Institute, AEOI, Tehran, Iran. [3]Department of GIS & RS, School of Geography, University of Tehran, Tehran, Iran. [4]Department of Tourism Planning, School of Human Geography, University of Tehran, Tehran, Iran.

References

1. Samadi S, Ghaffari H, Khodadadi M, Karimian K: **Environmental risks of accident of hazardous materials transportation.** In *Proceedings of the First hazardous materials transportation National Conference on Tehran.* Tehran: Ministry of Roads; 2008:9–15.
2. Rosenfeld PE, Feng L: *Risks of Hazardous Wastes.* Boston: William Andrew Publishing; 2011.
3. Zekrallahi M, Zekrallahi M: **Classification of hazardous materials.** In *Proceedings of the First hazardous materials transportation National Conference on Tehran.* Tehran: Ministry of Roads; 2008:65–72.
4. IDE: *Roles and Terms of Human Environment.* [http://www.doe.ir/portal/File/ShowFile.aspx?ID=e9b54981-abd1-4f3b-9f9f-083a1438f625], [In Persian].
5. UIC: *Uranium Information Centre, International Nuclear Waste Disposal Concepts.* http://www.uic.org/spip.php?rubrique1592.
6. Esfahani Kashitarash Z, Samadi MT, Naddafi K, Afkhami A, Rahmani A: **Application of iron nanaoparticles in landfill leachate treatment - case study: Hamadan landfill leachate.** *J Environ Health Sci Eng* 2012, doi:10.1186/1735-2746-9-36.
7. Kumari Muniyandi S, Johan S, Azman H, Siti SM: **Converting non-metallic printed circuit boards waste into a value added product.** *J Environ Health Sci Eng* 2013, **11**:2. doi:10.1186/2052-336X-11-2.
8. NWMO: *Nuclear Waste Management Organization, Choosing a Way Forward: The Future Managementof Canada's Used Nuclear Fuel.* Toronto: Nuclear Waste Management Organization; 2005:451.
9. Davison CC, Brown A, Everitt RA, Gascoyne M, Kozak ET, Lodha GS, Martin CD, Soonawala CD, Stevenson DR, Thorne GA, Whitaker SH: *The disposal of Canada's nuclear fuel waste: site screening and site evaluation technology,* Atomic Energy Canada Limited Research Report, AECL-10713, COG-93-3; 1994:255.
10. Sheng G, Almasi I: *Application of GIS in siting for high-level nuclear waste disposal – regional watershed approach,* Proceedings of the 2001 International High-Level Radioactive Waste Management Conference, Las Vegas; 2001.

11. Karimi A: **Site selection with overlaying.** In *Proceedings of the Second Environmental Engineering National Conference on Tehran: 2008; Tehran;* 2008.
12. Sartaj M: **Using GIS in disposal site selection.** *Disposal Manag* 2007, **8**:152–157.
13. Sheikh Narani T, Hafezi N: **Site selection of disposal waste management.** In *Proceedings of the Fifth Engineering geology and environmental issues National Conference on Tehran.* Tehran: Geological Society of Iranian Engineers; 2008:80–87.
14. Saeidi M, Abesi A, Sarpak M: **Locating a suitable place to bury hazardous waste using GIS rating and AHP: case report of Shahid Rajaee Powerhouse.** *Environ Sci Tech* 2009, **11**(1):231–241.
15. Fathi T: **Environmental criteria to locate hazardous waste landfill sites.** In *Proceedings of the Third Disposal waste management National Conference on Tehran: 2007.* Tehran: Municipalities Organization; 2008:415–419.
16. Openshaw S: **Learning to live with errors in spatial databases.** In *the accuracy of spatial databases.* Edited by Goodchild MF, Gopal S. London: Talor and Francis; 1989:263–276.
17. Xinglai HL, Sheng G: **GIS-based Hierarchy Process for the Suitability Analysis of Nuclear Waste. Disposal Site.** *Environ Informat Arch* 2006, **13**:286–296.
18. Location of Qom Province, central Iran 2011; Available at: [http://amar.sci.org.ir/index_e.aspx]
19. Saaty TL: *The Analytic Hierarchy Process: Planning, Priority Setting.* New York: Resource Allocation, McGraw-Hill; 1980.
20. Ellen KC, Sara M: *GIS and Public Health.* New York: Guilford; 2002.
21. Bromand M, Khamechian M, Nikoodel N: **Site selection for hazardous material by GIS in Zanjan.** In *Proceedings of the Forth Civil Engineering National Conference on Tehran.* Tehran: University of Tehran; 2008:9–15.
22. Tajvidi G: *An introduction to geographical information system*, Mapping organization of Iran. 2002;
23. Sener S, Karagüzel R: **Solid waste disposal site selection with GIS and AHP methodology: a case study in Senirkent–Uluborlu (Isparta) Basin, Turkey.** *Environ Assess* 2010, **173**:533–554.
24. Chen YW, Wang CH, Lin SJ: **A multi-objective geographic information system for route selection of nuclear waste transport.** *Omega* 2008, **36**(3):363–372.
25. Nuclear Decommissioning Authority Annual Report & Accounts. 2010/2011 (/[http://www.nda.gov.uk/news/arac-2010-2011.cfm.
26. Alavi N, Godarzi N, Babaey AA, Jafarzadeh N, Hossein Zadeh M: **Municipal solid waste ladfill site selection with GIS and analytical hierarchy process: a case study in Mahshahr, Iran.** *Waste Manag Res* 2013, **31**(1):98–150.
27. Tang X, Yang J: **Long-term stability and risk assessment of lead in mill waste treated by soluble phosphate.** *Sci Total Environ* 2012, **1**:299–303.

3

Heavy metals and hydrocarbon concentrations in water, sediments and tissue of *Cyclope neritea* from two sites in Suez Canal, Egypt and histopathological effects

Hesham M Sharaf[1] and Abdalla M Shehata[2*]

Abstract

Heavy metals and hydrocarbons are of the most common marine pollutants around the world. The present study aimed to assess the concentration of petroleum hydrocarbons and heavy metals in tissues of the snail *cyclope neritea,* water and sediments from two sites of the study area (Temsah lake and Suez canal) represent polluted and unpolluted sites respectively. The results showed that, the levels of the heavy metals (Pb, Cd, Co, Mg and Zn) in the polluted area have reached harmful limits recorded globally. Lead in water, sediment and tissue of the snail reached to 0.95 ppm, 4.54 ppm and 7.93 ppm respectively. Cadmium reached 0.31 ppm, 1.15 ppm and 3.08 ppm in the corresponding samples. Cobalt was not detected in water, but it reached 1.42 ppm and 10.36 ppm in the sediment and snails tissue respectively. Magnesium in water, sediment and tissue of the snail reached 3.73 ppm, 9.44 ppm and12.6 ppm respectively. Zinc reached 0.11 ppm, 3.89 ppm and 12.60ppm in the corresponding samples. Meanwhile, hydrocarbons in the polluted area (site1) reached 110.10 μg/L, 980.15 μg/g and 228.00 μg/g in water sediment and digestive gland tissues of the snails respectively. Whereas, hydrocarbons in the unpolluted area (site2) were estimated as 14.20 μg/L, 55.60 μg/g and 22.66 μg/g in water, sediment and tissue of the snails respectively. The combination of histopathological image with monitoring of the metal level in the digestive gland of the present snail provides an important tool for early detection of impending environmental problems and potential public health issues. Petroleum hydrocarbons are toxic to the marine fauna when present above certain limit in the marine water. The major detoxification organ in molluscs is the digestive gland, which has been used as a bioindicator organ for toxicity assessment. The effect of high crude oil on the digestive gland tubules of exposed snails when examined microscopically reveals a series of histological changes which indicates that the cellular compensatory mechanism is activated by hydrocarbons. These changes include vacuolation and presence of pyknotic nuclei.

Keywords: Heavy metals, Hydrocarbons, Timsah Lake, Suez canal, *Cyclope neritea*

Introduction

Adverse anthropogenic effects on the coastal environment include eutrophication, heavy metals, organic, oil spills and microbial pollution. Consequently, levels of contaminants in the marine environment are increasing continuously. In order to establish adequate coastal management programs, it is important to characterize the environment of concern chemically. The extent of contamination can be assessed by measuring pollutant concentration in water, sediments and exposed animal tissues samples. Trace metals can be divided to essential and non essential elements. Essential elements occur naturally in all organisms, essential elements in high doses can be poisonous causing hazardous effects on organisms. Non essential elements do not have any positive effects on organisms and they are harmful already in low doses. They can inhibit an essential element to bind

* Correspondence: ashehataeg@yahoo.com
[2]Chemistry Department, Faculty of Science, Suez Canal University, Al-arish, Egypt
Full list of author information is available at the end of the article

to enzyme and disturb the normal enzymatic function in the body [1-9].

As compared with the open sea, lagoons are more subject to pollution, particularly by heavy metals from industrial, agricultural and urban origin. Near shore sediments are found in a wide variety of environments (bays, lagoons). The water sediments interface is more important to biological fauna as compared with surface sediments, since meiofauna live about the reduced zone in sediment. Therefore, the composition of sediment has a significant influence on the living conditions of marine organisms. The trace metal results were obtained by sediment analysis, unlike sea water analysis, where the detection limit and contamination risks are significantly reduced [9].

It is well known that molluscs accumulate organic and metal pollutants at concentrations several order of magnitude above those observed in the field environment. Fewer studies have been done on gastropod molluscs, some of which are considered as useful biomonitors of certain metals [10]. Most metals are generally concentrated many times within an organism's soft tissue.

Advancement in technology as well as increase in population have led to environmental concerns relating from indiscriminate dumping of refuse and discharge of industrial effluents, petroleum waste water and crude oil spills replete with most common heavy metals in our environment [7,11].

Histopathological patterns represent a rapid, sensitive, reliable and comparatively inexpensive tool for assessment of stress response to xenobiotics [12-14]. These cascades of stress related responses including histopathology are now increasingly being used as biomarkers of environmental stress since they provide a definite biological endpoint of historical exposure [15-17].

Heavy metal pollution of terrestrial and aquatic ecosystems has long been recognized as a serious environmental concern. This is largely due to their non biodegradability and tendency to accumulate in plants and animals tissues. As a result, metal bioaccumulation is a major route through which increased levels of the pollutants are transferred across food chains, creating public health problems [18-20]. Therefore, it is important to determine the bioaccumulation capacity for heavy metals by certain organisms in order to assess potential risk to human health.

Several authors [9,13,21] have reported the importance of molluscs as good indicators for monitoring heavy metal pollution even through abnormally high environmental concentrations, since heavy metals affect numerous biological processes involved in the development and maintenance of molluscan populations such as feeding, growth, reproduction and general physiological activities [22].

Contamination of the sea with petroleum hydrocarbons, especially in shipping channels and ports, where crude and refined petroleum products are transported in significant quantities of oil entering the water column and sediment. The hepatopancreas of molluscs is the major site of petroleum hydrocarbons (PHC) detoxification. The main route for elimination of hydrocarbon metabolism is through faecal matter [23].

This study aimed to investigate the concentration of Pb, Cd, Co, Mg, Zn and hydrocarbons in water, sediment and soft tissues of *Cyclope neritea* snail from two different sites at Suez Canal, Egypt, map(1). The first site is

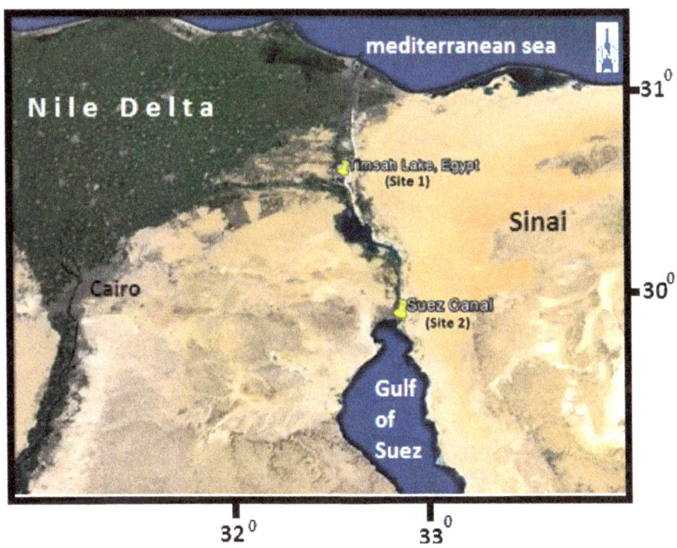

Figure 1 A map of Suez Canal showing the two investigated sites. The first site; Timsah Lack and the second site; Suez Canal.

Table 1 Concentration of heavy metals (ppm) in water, sediment and digestive gland tissue of *Cyclope neritea* snails (Mean ± SD) at the two investigated sites

Metal	Timsah Lake (site1)			Suez Canal (site2)		
	Mean ± SD			Mean ± SD		
	Water	Sediment	Digestive gland tissue	Water	Sediment	Digestive gland tissue
Pb	0.95 ± 0.04	4.54 ± 0.05	7.93 ± 0.03	0.92 ± 1.04	2.11 ± 0.82	3.93 ± 0.50
Cd	0.31 ± 0.04	1.15 ± 0.17	3.08 ± 0.20	0.50 ± 0.02	1.00 ± 0.13	1.72 ± 0.07
Co	0.00 ± 0.00	1.42 ± 0.17	10.63 ± 0.21	0.00 ± 0.00	0.67 ± 0.96	2.66 ± 0.25
Mg	3.73 ± 0.20	9.44 ± 1.18	12.6 ± 0.96	2.63 ± 0.01	3.92 ± 0.37	5.86 ± 0.15
Zn	0.11 ± 0.04	3.89 ± 0.20	12.6 ± 0.96	0.13 ± 0.01	1.33 ± 1.00	4.44 ± 0.80

located in Timsah Lake which receives effluents from many industrial sources site (1) where as the second site is Suez Canal where there is limited industrial or anthropogenic activity site (2).

Moreover, an attempt has been made to elucidate the response of the digestive gland of *Cyclope neritea snails* to heavy metals and PHCs and to identify histopathological biomarkers.

Materials and methods

Heavy metal content (Pb, Cd, Co, Mg and Zn) and hydrocarbons were analyzed in the water, sediment and soft tissues of *Cyclope neritea* snails from two different sites at Suez canal, the first was polluted, site1 (Timsah lake) while the second unpolluted site2 (Suez Canal) from the main canal (Figure 1).

Water and sediment

From both sites water samples were collected directly in a precleaned polyethylene bottles then sealed until analysis, sediment samples were collected using a precleaned PVC corer and immediately kept in polyethylene bags which were sealed and kept in an ice box for further analysis in the laboratory. Sediment samples were washed with bidistilled water and dried at 60°C then ground in a glass mortar and reduced into fine particles.

Samples collections

Snails were collected from both sites by hand picking. The soft tissues were removed from snail's shells with a sharp knife and dried at 60°C. The dried tissue was ground into fine powder and stored in a desiccator for further analysis.

Chemicals and method of analysis

Estimation of trace metals in water and sediment was carried out by suitable volume of water or digestion of one gram of sediment with conc. Nitric acid (HNO3) and conc. Perchloric acid (HClO4) (4:l) and analyzed in Atomic absorption specterophotometers perkin-Elmer, Analyzed TM 300 (USA) at suitable current and wave length for all studied heavy metals [24,25]. The values were expressed in ppm and the standard deviation was calculated. The hydrocarbons in all samples were

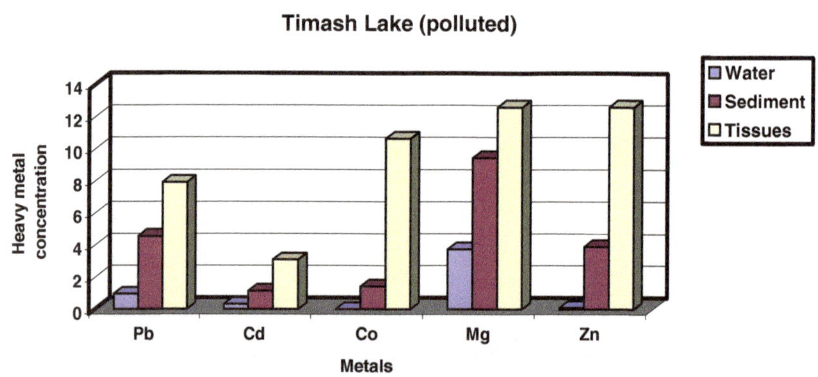

Figure 2 Concentration of heavy metals (ppm) in water, sediment and digestive gland tissue of *Cyclope neritea* snails in Timsah Lake (polluted area).

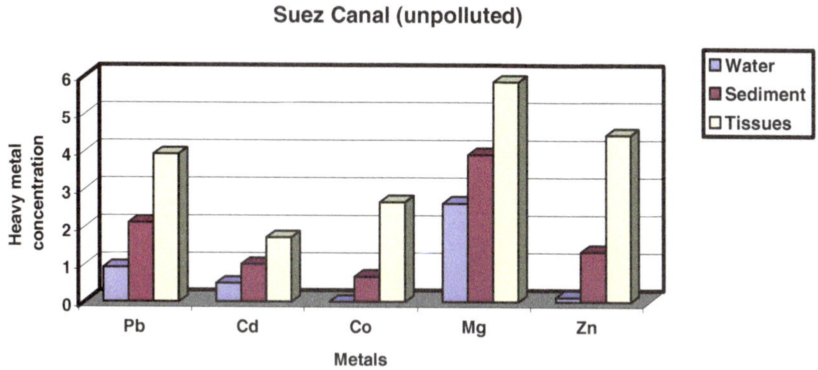

Figure 3 Concentration of heavy metals (ppm) in water, sediment and digestive gland tissue of *Cyclope neritea* snails in Suez Canal (unpolluted area).

extracted by Soxhlet instrument and measured by a Spectrofluorophotometer.

Statistical analysis

SPSS13 for Windows software was used for the statistical analysis of the heavy metal contents in water, sediment and digestive gland tissue of snails at the two investigated sites by using Mann–Whitney test.

For histolopathological examination the digestive gland of *Cyclope neritea* was taken out of their shells and dropped immediately into Alcoholic Bowin's, fixative then processed in the usual manner for histological investigation and stained with Haematoxylin and Eosin.

Results

Mean ± SD of the heavy metals at the two investigated sites were tabulated in Table 1. Statistical Mann–Whitney test was applied for all parameters at the two sites.

Heavy metals in Site 1 (Timsah Lake, polluted area) are shown in Table 1, Figure 2. The results revealed that the values of heavy metals in water are 0.95 ± 0.04, 0.31 ± 0.04, 0.00 ± 0.00, 3.73 ± 0.20 and 0.11 ± 0.04 ppm for Pb, Cd, Co, Mg and Zn respectively. Meanwhile, heavy metals values in sediment are 4.54 ± 0.05, 1.15 ± 0.17, 1.42 ± 0.17, 9.44 ± 1.18 and 3.89 ± 0.20 ppm for Pb, Cd, Co, Mg and Zn respectively. Besides, the values of heavy metals in digestive gland tissues are 7.93 ± 0.03, 3.08 ± 0.20, 10.63 ± 0.21, 12.6 ± 0.96 and 12.6 ± 0.96 ppm for Pb, Cd, Co, Mg and Zn respectively.

On other hand, heavy metals in Site2 (Suez Canal, unpolluted area) are shown in Table 1, Figure 3. The results revealed that, the values of heavy metals in water are 0.92 ± 1.04, 0.50 ± 0.02, 0.00 ± 0.00, 2.63 ± 0.01 and 0.13 ± 0.01 ppm for Pb, Cd, Co, Mg and Zn respectively. Meanwhile, heavy metals values in sediment are 2.11 ± 0.82, 1.00 ± 0.13, 0.67 ± 0.96, 3.92 ± 0.37 and 1.33 ± 1.00 ppm for Pb, Cd, Co, Mg and Zn respectively. And the values of heavy metals in digestive gland tissues are 3.93 ± 0.50, 1.72 ± 0.07, 2.66 ± 0.25, 5.86 ± 0.15 and 4.44 ± 0.80 ppm for Pb, Cd, Co, Mg and Zn respectively.

The Mann–Whitney test was made between the heavy metals concentrations of the digestive gland tissue of the snails and both water and sediment in the two sites, asymptotic significance (2-tailed) = 1.00 and 0.18 respectively was not significant at $P < 0.05$ level. Also, the Mann–Whitney test was made between the heavy metals concentrations of the digestive gland tissue of the snails in the two sites, asymptotic significance (2-tailed) = 0.047 was significant at $P < 0.05$ level.

On the other hand, the concentrations of petroleum hydrocarbons were tabulated in Table 2, Figure 4. The results revealed that concentration of petroleum hydrocarbons are 110.1, 980.15 and 228.00 ug/g for water, sediment and digestive gland tissues respectively in site (1). Meanwhile, they are 14.2, 55.6 and 22.66 ug/g for water, sediment and digestive gland tissues respectively in site (2). The Mann–Whitney test was made between the petroleum hydrocarbons concentrations of the digestive gland tissues and both of water and sediment from the two sits, asymptotic significance (2-tailed) = 0.0499 was significant at $P < 0.05$ level.

Histpathological changes

The digestive gland of normal *C. neritea* is a large, tubulo-acinar gland which occupies the greater part of the cavity of the shell spire. The gland is covered by

Table 2 Concentration of hydrocarbons in water, sediment and digestive gland tissue of *Cyclope neritea* snails in the study area

Site	Water (µg/L)	Sediment (µg/g)	Digestive gland tissue (µg/g)
Timsah Lake	110.10	980.15	228.00
Suez Canal	14.20	55.60	22.66

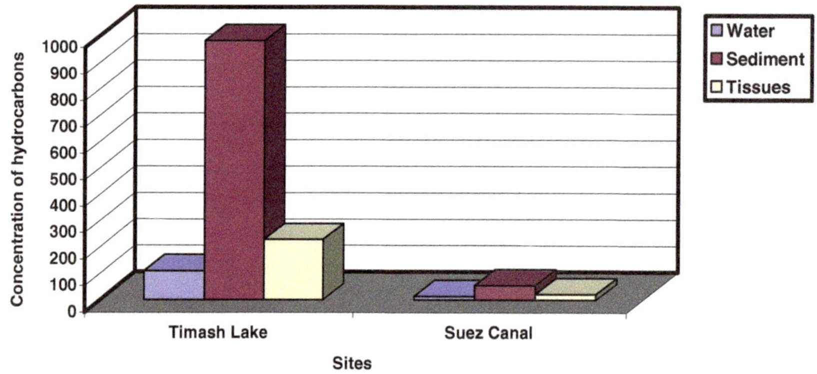

Figure 4 **Concentration of hydrocarbons in water, sediment and digestive gland tissue of *Cyclope neritea* snails in the study area.**

squamous epithelium resting on a thin layer of fibrous connective tissue Figure 5. The digestive gland tubules are lined with simple epithelium. This epithelium consists of two main cell types, digestive cells and secretory cells.

The digestive cells Figure 5 are by far the most numerous elements in the wall of the digestive gland tubules. They are long columnar or cube shaped with domed distal apices and flat bases by which they rest on a very thin basement membrane. They vary greatly in length within the same tubule. The nucleus is basal, usually oval but may be spheroidal or even irregular. Inside the major part of the cell body the cytoplasm is lightly stained and shows various degrees of vaculation and different contents.

The secretory cells Figure 5 are present in much smaller number than the digestive cells. They are shorter pyramidal or cone-shaped, but may sometimes be columnar. They are markedly shorter than the digestive cells and therefore appear wedged in between groups of the digestive cells. Their cytoplasm is basophilic and the body of the excretory cells is usually crowded with numerous spherules (ex.s) of regular form but different sizes. Each excretory cell contains oval or spherical excretory bodies (dark granules) which are usually chromophobic, usually retain their natural colour and appear located inside colourless vacuoles.

These bodies may represent the final stage in elaboration of the excretory material within the excretory cells. The digestive gland obtained from the polluted area shows an increase in the secretion of the digestive cells, where the cytoplasm becomes more acidophilic and their nuclei start unusual division Figure 6. It can be seen that the dark granules increased greatly in number and most of the digestive cells become completely degenerated and lysing while most of the tubules are damaged Figures 7, 8 and 9. At much higher concentrations of pollutants the digestive gland tubules become

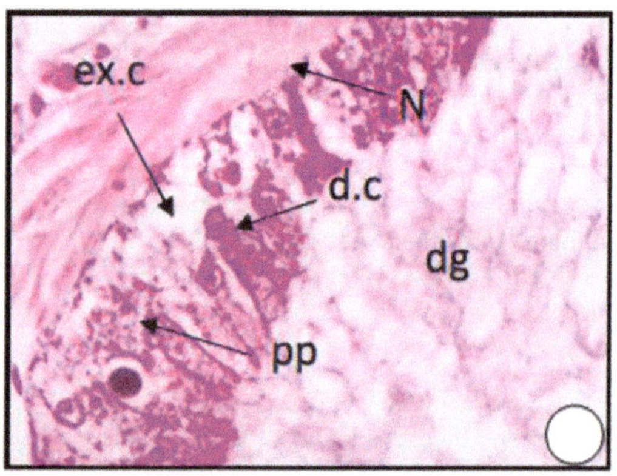

Figure 5 **A photomicrograph of T.S. of digestive gland of *Cyclope neritea* from the unpolluted site (Suez Canal) showing digestive tubules without hydrocarbon precipitations in the digestive gland cells.** N (Nucleus); dg (digestive gland); d.c (digestive cell); ex.c (excretory cell); PP (hydrocarbon precipitations). X400.

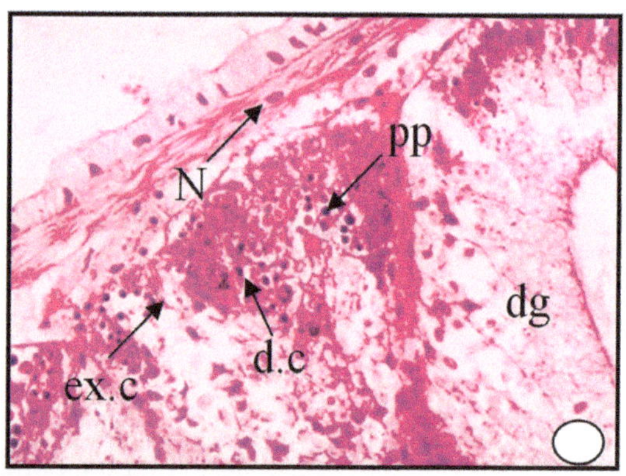

Figure 6 **Digestive tubules T.S. of digestive gland in snail *Cyclope neritea* from polluted area (Timsah Lake) showing hydrocarbon precipitations in the digestive gland cells.** X200.

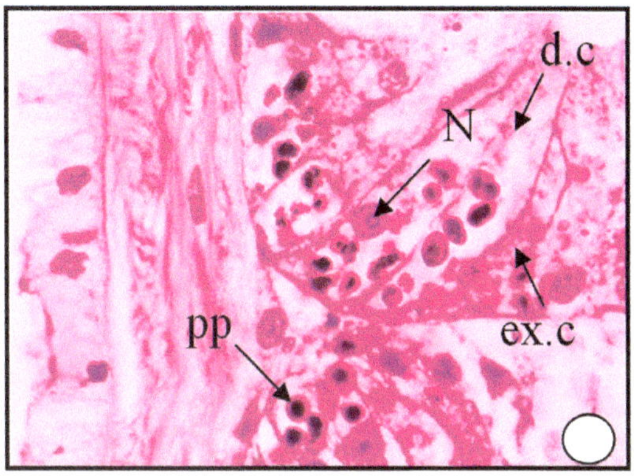

Figure 7 Higher magnification of Digestive tubules in T.S. of digestive gland of the snail *Cyclope neritea* from polluted area (Timsah Lake) showing hydrocarbon precipitations in the digestive gland cells. N (Nucleus); dg (digestive gland); d.c (digestive cell); ex.c (excretory cell); PP (hydrocarbon precipitations); V (vacuolation). X400.

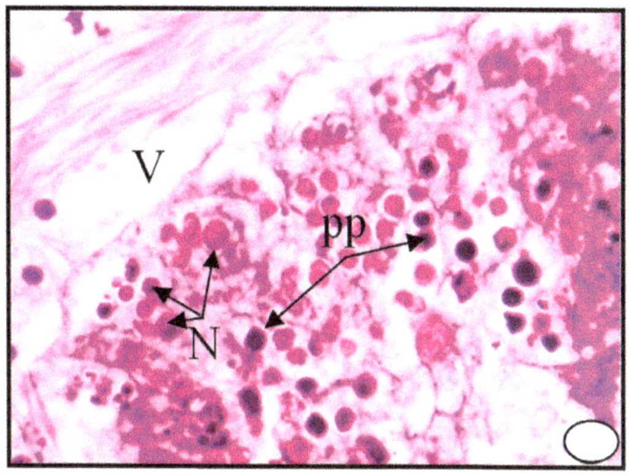

Figure 9 A photomicrograph of digestive gland of the snail *Cyclope neritea* from polluted area (Timsah Lake) showing that most of the digestive cells became polygonal and bi and multinucleated and finally degenerated and lysed due to hydrocarbon precipitations. N (Nucleus); PP (hydrocarbon precipitations); V (vacuolation). X400.

mostly damaged, necrotic and the cells take cluster shape and become bi and multinucleated and finally lysed Figures 8 and 9.

Discussion

This study shed light on some of the environmental factors that might have an impact directly or indirectly on the concentration ability of heavy metals into the bodies of the investigated host. The concentration of water with a wide range of heavy metals has become a matter of great concern over the last few decades [26] and a lot of studies have been published on the heavy metals at all levels of aquatic ecosystem [27-30]. Many authors associated the heavy pollution in water with industrial and municipal discharges. These heavy metals may be taken up by living organisms, deposited in the sediments or remain for some period in the water itself [31,32].

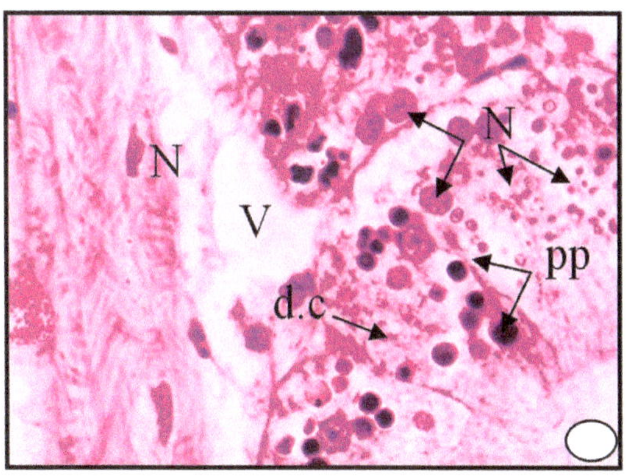

Figure 8 A photomicrograph of digestive gland of the snail *Cyclope neritea* from polluted area (Timsah Lake) showing the accumulations of hydrocarbons particles which lead to damaging and vacuolation of most digestive cells of the digestive gland cells. N (Nucleus); d.c (digestive cell); PP (hydrocarbon precipitations); V (vacuolation). X400.

In the present study concentrations of heavy metals in water, sediment and tissues of molluscs from two sites of Suez Canal were determined. The highest concentrations for all studied heavy metals were recorded at site (1) when compared with site (2). The differences in heavy metals concentration between site 1 and site 2 might be attributed to the highly discharge of mixture of industrial, municipal and agricultural drains into site (1), this opinion agrees with [33,34]. Several authors reported that the variation in heavy metals concentration in water might be attributed to the contaminated sediment; these sediments reflect the quality of water current and form the major repository of heavy metals in aquatic system. They added that the rate of accumulation depends mainly on the environmental parameters. Therefore, sediments can be used to detect the presence of contamination that does not remain soluble after the discharge into water [35,36].

The data of the heavy metals in the present work revealed that there is a highly significant ($P < 0.01$) difference between the two sites, the increase in values of site 1 may be attributed to a number of factors such as industrial effluents, agricultural drainage and waste municipal [31]. Water of the two sites showed higher concentrations of Lead (0.95, 0.92 ppm); Cadmium (0.31, 0.50 ppm), Cobalt (0.00, 0.00 ppm); Manganese (3.73, 2.63 ppm); and Zinc (0.11, 0.13 ppm). Meanwhile, sediment of the two sites showed higher concentrations of Lead (4.54, 2.11 ppm);

Cadmium (1.15, 1.00 ppm), Cobalt (1.42, 0.67 ppm); Manganese (9.44, 3.92 ppm); and Zinc (3.89, 1.33 ppm), Also, heavy metals concentration in the digestive gland tissues of *C. neritea* snail of the two sites showed higher concentrations of Lead (7.93, 3.93 ppm); Cadmium (3.08, 1.72 ppm), Cobalt (10.63, 2.66 ppm); Manganese (12.6, 5.86 ppm); and Zinc (12.6, 4.44 ppm), these finding are in agreement with [37] where they determined the levels of heavy metals Lead (Pb), Cobalt (Co), Manganese (Mg), Zinc (Zn) and Cadmium (Cd) in coastal water, sediment and soft tissues of the gastropod limpet *Patella caerulea* and the bivalvae *Barbatus barbatus* from seven different stations in the western coast of the Gulf of Suez. The highest accumulated metals were Fe, Zn and Mn in both *Patella caerulea* than *Barbatus barbatus*. In the present study, metals such as Pb, Mg and Zn, exhibited modreate concentrations. Cadmium and Cobalt metals were below the detectable limit at both sites.

In concordance with the present study, several relevant studies have been made earlier. The heavy metal accumulation in the gastropod *Cerithium scabridium* from Kuwait coast has been analyzed [38] where the concentration of the cadmium in the gills ranged between 7.06 ppm and 0.09 ppm, which is comparatively lower than that in the present study. The concentrations of the analyzed heavy metals exhibited variations in water, sediments and tissues of the studied animal from both sites.

The assessment and description of toxicants on organisms is attracting the interest of the researchers working with marine organisms [2,9,39]. Accumulated hydrocarbons were found to interact with cellular hydrocarbons leading to different histopathological lesions with the highest necrosis index of cells [18]. Cellular and subcellular histopathological changes have proved to be reliable biomarkers in toxicological assays [20]. In the present study, the accumulations of hydrocarbons lead to damaging and vacuolation of most digestive cells, where they became polygonal and bi and multinucleated and finally degenerated and lysed. These results agree with those of [2,17,40-42].

Further studies are still needed to investigate the effect of heavy metal contamination on the biochemical parameters and fine structure to evaluate their potentiality as sentinel organisms to heavy metal contamination in aquatic habitats.

Competing interests
The authors declare that they have no competing interests.

Authors' contributions
HS and AS are the main investigators, performed all experimental work and paper writing, carried out data analysis the authors read and approved the final manuscript.

Authors' information
Sharaf H.M, Zoology Department, Faculty of Science, Zagazig University, Egypt.

Shehata A.M.A, Chemistry Department, Faculty of Science, Suez Canal University, Al-arish, Egypt.

Acknowledgements
The authors acknowledge Prof. Dr. Abdallah Ibrahium, Faculty of Science, Ain Shams, Cairo, Egypt, for his guidance, help, read and approved the final manuscript.

Author details
[1]Department of Zoology, Faculty of Science, Zagazig University, Zagazig, Egypt. [2]Chemistry Department, Faculty of Science, Suez Canal University, Al-arish, Egypt.

References
1. Samiulallah Y. Biological effects of marine oil pollution. Oil Petrochem Pollut. 1985;2:235–64.
2. Alyami M Y. Ecological Study and Petroleum Hydrocarbons Estimation of Bivalve (Tridacna maxima) from the coastal area of Jeddah Red Sea, Msc. Faculty of Marine Science, King Abdulaziz University 2002.
3. Azza khaled. Heavy metals concentrations in certain tissues of five commercially important fishes from El-max bay, Alexandria, Egypt, Environment Division, National institute of Oceanography and fisheries, kayet bay Alexandria, Egypt 2003.
4. Hassn MA, Heba M, Al Edresi AM. Background level of heavy metal in Dissolved particulate passes of water and sediment of Al Hodeidah Rea Sea coast of Yemen, Faculty of marine science and Environment Alhodeidah university city republic of Yemen JKAU. Mar Sci. 2004;15:53–71.
5. Bu-olayan AH, Thomas BV. Validating species diversity of benthic organisms to trace metal pollution in Kuwait bay, of the Arabian Gulf. Appl Ecol Environ Res. 2005;3(2):93–100.
6. Al Sayad HA, Dairi M. Metal accumulation in the edible snail *Turbo coronatus* (Gmelin)) from different 10 cautions in Bahrain, Arab Gulf. J Sci Res. 2006;24:48–57.
7. Gbaruko BC, Friday OU. Bioaccumulation of heavy metals in some fauna and flora. Int J Environ Sci Tech. 2007;4(2):197–202.
8. Chukwujindu MA, Iwegbue Francis O, Arimoro Godwin E, Nwa jh Osa Eguavoen. Heavy metal content in the African giant snail *Archachatina marginata* (Swainson,1821) (Gastropoda: Pulmonata: Achatinidae) in Southern Nigeria. Folia Malacologica. 2008;16(1):31–4.
9. Kesavan K, Murugan V, Venkatesan V, Vijay Kumar BS. Heavy metal accumulation in mollusks and sediment from uppanar estuary, southeast coast of India. An Int J Mar Sci. 2013;29(2):15–21.
10. Bryan GW, Langston WJ, Hummerstone LG, Burt GR, Ho YB. An assessment of the gastropod, *Litorina lgtorea* ,as an indicator of heavy metal contamination in United Kingdom estuaries. J Mar Biol Assoc UK. 1983;63:327–45.
11. Wills J. A survey of offshore oil field drilling wastes and disposal techniques to reduce the ecological impact of sea dumping. Sakhalin Environ. Watch. 2000;13:23–9.
12. Balkas IT, Tugrel S, Salhogln I. Trace metals levels in fish and crustacean from Northern Mediterranean coastal waters. Mar Environ Res. 1982;6:281–9.
13. Bryan GW, Langston WJ. Bioavailability, accumulation and effects of heavy metals in sediments with special reference to United Kingdom estuaries: a review. Eviron Pollut. 1992;76(2):89–131.
14. Otitoloju AA, Don-Pedro KN. Integrated laboratory and filed assessments of heavy metals accumulation in edible periwinkle, *Tympanotonus fuscatus* var radula (L.). Ecotoxicol Environ Saf. 2004;57(3):354–62.
15. Matthiessen P, Thain JE, Law RJ, Fileman TW. Attempts to assess the environmental hazard posed by complex mixtures of organic chemicals in UK estuaries. Mar Pollut Bull. 1993;26:90–5.
16. Stentiford GD, Feist SW. A histopsthological survey of shore crab (*Carcinus maenas*) and brown shrimp (*Crangon crangon*) from six estuaries in the United Kingdom. J Inert Pathol. 2005;88:136–46.
17. Ravera O, Beone GM, Trincherine PR, Riccardi N. Seasonal variations in metal content of two Union *Pictorum mancus* (Mollusca, Unionidae) populations from two lakes of different tropic state. J Limnol. 2007;66(1):28–39.
18. Berthou FG, Balouet G, Bodennee, Marchand M. The occurrence of hydrocarbons and histopathological abnormalities in oysters for seven years

following the wreck of Amoco Cadiz in Brittany (France). Mar Environ Res. 1987;23:103–33.
19. Tuzen M. Determination of heavy metals in fish samples of the middle Black Sea (Turky) by graphite furnance atomic absorption spectromertry. Food Chem. 2003;80:19–23.
20. Otitoloju AA, Ajikobi DO, Egonmwan RI. Histopathology and bioaccumulation of heavy metals (Cu&Pb) in the giant land snail, *Archachatina marginata* (Swainson). Open Environ Pollut Toxicol J. 2009;1:79–88.
21. Kiffney PM, Clement WH. Bioaccumulation of heavy metals by benthic Invertebrates at the Arkansas River. Colorado. Environ Toxicol Chem. 1993;12(8):1507–17.
22. Coughtre PJ, Martin MH. The distribution of Pb, Zn and Cu within the pulmonate molluscs *Helix aspersa*. Oeclogia. 1976;23:315–22.
23. Lee RF, Sauerheber R, Benson A. Petroleum hydrocarbons uptake and discharge by the marine mussel *Mytilus edulis*. Science. 1972;177:344–6.
24. Chester R, Hughes MJ. A chemical technique for separation of ferromanganese minerals, carbonate minerals and absorbed trace elements from pelagic sediments. Chem Geol. 1967;2:249–62.
25. Clescerl L, Greenberg A, Eaton A. Standard methods for the examination of waters and wastewaters. 20th edition. Washington DC: American Public Health Association; 1999. p. 1325. ISBN 0875532357.
26. Yilmaz F, Ozdemir N, Demirak A, Tuna AL. Heavy metal levels in two fish species *Leuciscus cepalus* and *Lepomis gibbosus*. Food Chem. 2007;100 (2):830–5.
27. Ravera O. Monitoring of the aquatic environment by species accumulator of pollutants: a review. J Limnol. 2001;60(Suppl1):63–78.
28. Rainbow PS. Trace metal concentrations in aquatic invertebrates, why so what? Enivron Pollut. 2002;120:497–507.
29. Dugong G, Pera LL, Bruzzese A, Pellicano T, Turco V. Concentration of Cd (II), Cu (II), Pb (II), Se (IV) and Zn (II) in cultured sea bass (Dicentrarchus labrax) tissues from Tyrrhenian Sea and Sicilian Sea by derivative stripping potentiometry. Food Control. 2006;17:146–52.
30. Jayakumar P, Paul VI. Patterns of cadmium accumulation in selected tissues of the catfish *Claiias batrachus* (Linn.) exposed to sublethal concentration of cadmium chloride. Vet Arhiv. 2006;76:167–77.
31. Emara AM, Belal AH. Marine fouling in Suez Canal. Egypt J Aquat Res. 2004;30(A):189–206.
32. Santos IR, Silva-Filho MR, Campos LS. Heavy metals contamination in coastal sediments and soils near the Brazilian Antarctic Station, King George Island. Mar Poll Bull. 2005;50:85–194.
33. Haggag AM, Marie MAS, Zaghloul KH. Seasonal effects of the industrial effluents on the Nile catfish, Clarias gariepinus. J Egypt Ger Soc Zool. 1999;28(A):365–91.
34. Zaghloul KH. Effect of different water sources on some biological and biochemical aspects of the Nile tilapia; *Oreochromis niloticus* and the Nile catfish, *Clarias gariepinus*. Egypt. J Zool. 2000;34:353–77.
35. Chapman PM, Wang FY. Assecing sediment contamination in estuaries. Environ Toxicol Chem. 2001;20:3–22.
36. Shahat MA, Amer OSO, AbdAllah AT, Abdelsater N, Moustafa MA. The distribution of certain heavy metals between intestinal parasites and their fish hosts in the River Nile at Assuit Province, Egypt. Egypt J Hosp Med. 2011;43:241–57.
37. Mohamed AH, Ahmed ME. Marine molluscs as biomonitors for heavy metal levels in the Gulf of Suez, Red Sea. J Mar Syst. 2006;60:220–34.
38. Bu-olayan AH, Thomas BV. Heavy metal accumulation in the gastropod, *Cerithium scabridum* L., from the Kuwait coast. Environ Mon Assess. 2001;68:187–95.
39. Miriam Paul S, Menon NR. Histopathological changes in the hepatopancreas of the penaeid shrimp *Metapenaeus dobsoni* exposed to petroleum hydrocarbons. J Mar Biol Ass India. 2005;47(2):160–8.
40. Sorenen EMB, Ramirez-Mitchell R, Harlan CW, Bell JS. Cytological changes in the fish liver following chronic, environmental arsenic exposure. Bull Environ Contam Toxicol. 1980;25:93–9.
41. Auffret M. Histopathological changes related to chemical contamination in *Mytilus edulis* from field and experimental conditions. Mar Ecol Prog Ser. 1988;64:101–7.
42. Stentiford GD, Longshaw M, Lyons BP, Jones G, Green M, Feist SW. Histopathological biomarkers in estuarine fish species for the assessment of biological effects of contaminants. Mar Environ Res. 2003;55:137–59.

4

Performance evaluation of a scoria-compost biofilter treating xylene vapors

Mohammad Mehdi Amin[1,2], Amir Rahimi[3], Bijan Bina[1,2], Mohsen Heidari[1,2,4*] and Fazel Mohammadi Moghadam[5]

Abstract

The removal of xylene vapors was studied in a biofilter packed with a new hybrid (scoria/compost) packing material at various inlet loads (IL) and empty bed residence times (EBRT) of 90, 60, and 40s. The best performance was observed for EBRT of 90s, where a removal efficiency of 98% was obtained under steady state condition for inlet xylene concentration of 1.34 g m^{-3}, while a maximum elimination capacity of 97.5 g m^{-3} h^{-1} was observed for IL of 199.5 g m^{-3} h^{-1}. Carbon dioxide production rates and the microbial counts for xylene-degraders followed xylene elimination capacities. Overall look to the results of this study indicates that the scoria/compost mixture could be considered as a potential biofilter carrier, with low pressure drop (here <4 mm H_2O), to treat air streams containing VOCs.

Keywords: Xylene, Biofilter, Scoria, Compost, Elimination capacity

Introduction

Xylene, together with benzene, toluene, and ethyl benzene, constitute the volatile organic compounds (VOCs) group BTEX. Xylene isomers namely *p*-xylene, *m*-xylene, and *o*-xylene are listed as hazardous and toxic atmospheric contaminants under Clean Air Act Amendments (CAAA) [1]. Chronic exposure to xylene is associated with adverse effects on the liver, kidneys and the central nervous system [2]. Xylene is a major constituent of gasoline and is used as a solvent in many production industries including printing, rubber, leather, painting and varnishing industries. In general, over 60% of the total emissions of xylene into the atmosphere originate from industrial facilities, especially petrochemical plants [1,3].

Among the air pollution control technologies, the biological treatment of VOCs provides an environmentally-friendly and low-cost alternative to other physicochemical treatment technologies [4-6]. The most widely used biological air treatment process is biofiltration [3]. The performance of biofilter significantly depends on the packing material properties [7]. Some of the desirable media properties include balanced chemical composition, relatively high water holding capacity, high microbial population density, and structural integrity [8,9]. These properties, except the later one, could be found in organic materials such as compost and peat, which allow the biofilter to have a quick startup and high elimination capacity. However, organic materials have a relatively low durability and tend to settle and compact, which in turn result in increased pressure drop and channeling [10]. On the other hand, the use of an inert material with rigid structure and large pores as filter bed allows a better gas distribution inside the reactor than organic carriers and minimizes the pressure drop build ups [11,12]. Consequently, using a mixture of organic and inert material inoculated with acclimated microbial consortium not only might improve the performance of biofilters, but also provides low pressure drops.

Many studies have been conducted on the biofiltration of xylene from waste air streams mainly in mixture with other VOCs. Most of them demonstrated that the removal of xylene isomers is always less efficient among the BTEX compounds [13-16]. In other hand, the biofiltration of xylene as the sole pollutant has been less considered. Saravanan and Rajamohan [1] studied the biofiltration of xylene vapors on a laboratory scale biofilter packed with press mud as filter material inoculated with activated sludge and reported a maximum elimination capacity of 67 g m^{-3} h^{-1} for inlet load of 103 g m^{-3} h^{-1} and EBRT of 42 s.

To the best of our knowledge, no attempt has been made on the removal of VOCs in biofilters packed with

* Correspondence: Moheidari84@gmail.com
[1]Environment Research Center, Isfahan University of Medical Sciences (IUMS) and Department of Environmental Health Engineering, School of Health, IUMS, Isfahan, Iran
[2]Student Research Center, School of Health, Isfahan University of Medical Sciences, Isfahan, Iran
Full list of author information is available at the end of the article

a mixture of scoria and compost. Scoria is a vesicular pyroclastic rock with basaltic composition, which is light in weight, porous, and cheap. It is abundant in many places worldwide [17]. In this study a mixture of scoria (pre-soaked in an adapted inoculum) and compost was used for the biofiltration of xylene. The aim of this study was to evaluate the performance of the biofilter under various operating conditions in terms of xylene removal efficiency and elimination capacity. The carbon dioxide evolution and microbial population distribution along the bed length were also monitored.

Materials and methods

Inoculum and packing media preparation

Scoria and compost were used as the biofilter media. The media characteristics are given in Table 1. Scoria, as the inert porous media, was used to ensure more homogeneous gas distribution across the filter bed, to maintain the structural integrity of the biofilter and to minimize the pressure drop. Activated sludge from a wastewater treatment plant located at Isfahan Petroleum Co. (Isfahan, Iran) was used as the inoculum source for the biofilter. Microorganisms in the activated sludge were acclimated to xylene in order to reduce the required adaptation time for the microorganisms in the biofilter. For acclimation, one liter of the activated sludge was placed in an aerated batch reactor and diluted with 3 L of nutrient solution. The used nutrient solution was composed of (g l^{-1} of tap water): 2.36 $(NH_4)_2SO_4$, 2.09 KH_2PO_4, 1.91 NH_4Cl, 0.27 $Na_2HPO_4.7H_2O$, 0.5 NaCl, 0.03 $CaCl_2.2H_2O$, 0.2 $MgSO_4.7H_2O$, 0.03 $ZnSO_4.7H_2O$, 0.03 $FeCl_3.7H_2O$, 0.03 $MnSO_4$. Xylene as sole carbon source was injected to the reactor at rates of 5 10 ml day^{-1}. In order to avoid metabolic by-product accumulation and nutrient deficiency, one liter of the mixture was substituted with fresh nutrient solution on weekly basis. After two months, 5 liter of solution nutrient was added to the mixture and shaken vigorously, and then around 3 kg of dried scoria was soaked in the mixture. Before packing, the prepared scoria stones were mixed with compost in a 3:1 volume ratio.

Table 1 Biofilter media characteristics

Characteristic	Fresh scoria	Compost
Particle size (mm)	6-13	1-6
Bulk density (kg m^{-3})	444	325
Initial moisture (%)	<2	26.5
Water holding capacity (kg kg^{-1} dry weight)	43	125
pH (–)	7.9	8.5
Porosity (%)	70	58
C/N ratio (–)	-	18
Special surface area (m^2 g^{-1})	0.282	-

Experimental setup

Experiments were carried out in a four-stage laboratory-scale biofilter made of stainless steel with an effective bed height of 80 cm and an internal diameter of 11.5 cm (Figure 1). Perforated stainless steel plates were located at the bottom of each section to support the packing material and to enhance the radial distribution of gas between the sections. A 10-cm space between adjacent sections allowed for gas sampling and redistribution of the contaminant streams. Gas sampling ports were installed at the inlet, outlet and at additional spaces between adjacent sections. Furthermore, a port was located at the middle of each section to recover bed particles for additional analyses. The filter bed was irrigated daily with 1 liter of the previously mentioned nutrient solution. The leachate was manually collected on a daily basis at the bottom section of the biofilter.

Compressed air was continuously passed through a granular activated carbon column to remove moisture, oil and particulate matter. The purified air was divided into two streams; the major one was passed through a water column in order to increase the relative humidity of inlet air and the minor one was bubbled through a glass bubbler containing xylene solution (Merck, Germany). The bubbler was held in a thermostatically-controlled water bath. Afterward the streams were mixed and fed to the bottom of the biofilter column in a counter-current flow mode. Xylene concentration was maintained at the desired value by adjusting the flow rate of air passed through the bubbler, temperature of water bath, and the level of pollutant in bubbler.

Analytical methods

Gas phase xylene was measured by a gas chromatograph equipped with flame ionization detector (GC-FID) (Agilent GC, 7890A). The column used was an Agilent 19091S-433 capillary column, 30 m 250 μm 0.25 μm. High purity helium gas (99.995%) was the carrier gas and supplied at a flow rate of 1.11 ml min^{-1}. The temperatures of column oven, injector and detector were 150, 230 and 250C, respectively. The method detection limit (MDL) for xylene analysis was estimated as 11.1 μg l^{-1}. The calibration curve was prepared by injecting known amounts of the xylene into a glass bottle sealed with Teflon septum according to the standard procedure [18]. Air samples of 100 μl were drawn from the bottle and various sampling ports along the biofilter column with a 1 mL gas tight syringe (Hamilton, USA), and directly injected into GC. A Guardian Plus Model D100 IR CO_2 analyzer (Edinburgh Sensors Ltd., Munchen, Germany) (0 3000 ppm) was used to analyze carbon dioxide concentration along the biofilter column. Total organic compounds (TOC) in the leachate were measured using a TOC analyzer (Beckman 915A, USA). In order to break microbial flocs into small

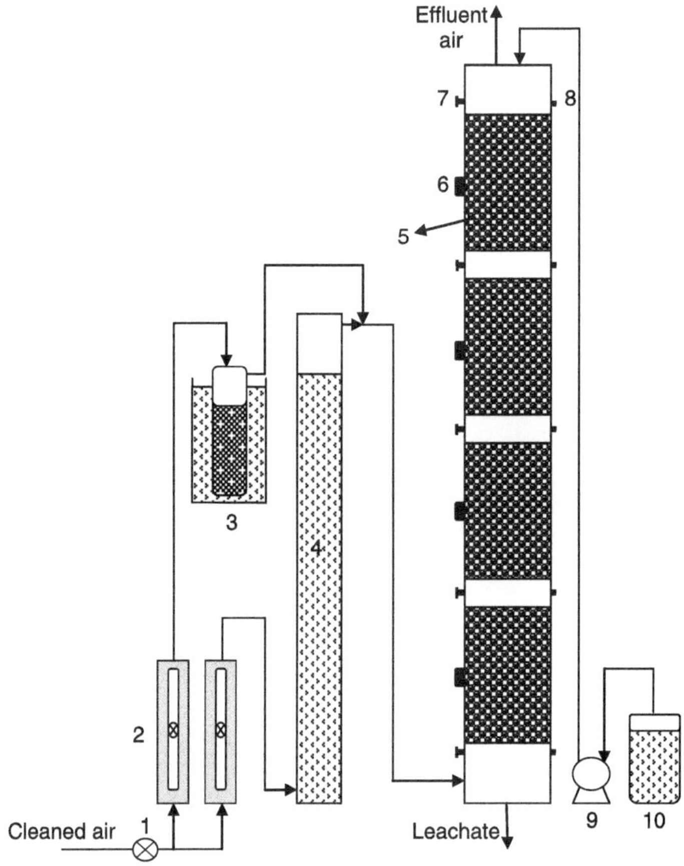

Figure 1 Schematic of the biofiltration set up. 1. Air pressure regulator, 2. Air rotameters, 3. Water bath and bubbler, 4. Humidifier, 5. Biofilter bed, 6. Bed sampling port, 7. CO_2, temperature and pressure drop measurement port, 8. Gas sampling port for xylene measurement, 9. peristaltic pump, and 10. Nutrient solution container.

pieces and achieve a homogeneous TOC concentration, the leachate sample was surged by ultrasonic wave. The moisture content of packing material was determined as weight loss after drying at 105C for 24 hr.

The temperatures at midlevel of each section and at gas sampling ports were measured by online K-type thermocouple and a digital thermometer, respectively. The pressure drop across the biofilter was measured using a water column manometer. Scanning electron micrographs of the scoria were carried out using a Vega/Tescan scanning electron microscope (SEM). For the sample taken at the end of experiment, the microorganisms were fixed with 3% glutaraldehyde aqueous solution overnight. The fixed sample was washed with phosphate buffer, dehydrated by placing in 30%, 50%, 70% and 100% ethyl alcohol, dried and then covered with a gold layer.

Microbial cell counts

Almost one gram of the packing material was taken from each section of the biofilter and mixed with 9 mL sterile saline solution (0.9% w/v NaCl). The samples were then vortexed for 3 min and serially diluted up to 10^{-10} in sterile saline solution. Aliquots of 0.1 ml were spread over the surfaces of agar plates. The Nutrient Agar amended with Nistatin (to inhibit fungal growth) and the Potato Dextrose Agar (PDA) amended with chloramphenicol (to inhibit bacterial growth) were used for culturing bacteria and fungi, respectively. Furthermore, a sterile mineral medium containing 1.5% (w/v) agar was used for the enumeration of xylene-degraders. Xylene vapor as sole carbon source was supplied through placing plates in a sealed container. The cultured plates were incubated at 28 30C for 5 7 days. The counts were reported as colony forming unit (CFU) g^{-1} of packing material (dry basis).

Performance evaluation

Biofilter performance was evaluated in terms of inlet xylene load, IL (g m^{-3} h^{-1}); elimination capacity, EC (g m^{-3} h^{-1}); removal efficiency, RE (%); and carbon dioxide production rate, PCO_2 (g m^{-3} h^{-1}). These parameters are defined as:

$$IL \quad \frac{QC_{in}}{V} \tag{1}$$

$$EC \quad \frac{Q\ C_{in}-C_{out}}{V} \tag{2}$$

$$RE \quad \frac{C_{in}-C_{out}}{C_{in}} \times 100 \tag{3}$$

$$PCO_2 \quad \frac{Q(CO_{2,\ out}-CO_{2,\ in})}{V} \tag{4}$$

where Q is the air flow rate (m^3 min^{-1}), V is the volume of packed bed (m^3), C_{in} and C_{out} are the inlet and outlet xylene concentrations (g m^{-3}), and $CO_{2,in}$ and $CO_{2,out}$ are the inlet and outlet carbon dioxide concentrations (g m^{-3}), respectively.

Results and discussion

Overall performance of biofilter

The biofiltration of xylene vapors was carried out during 101 days at various operating conditions, which are listed in Table 2. In each phase, the inlet concentration of xylene was increased gradually to the high levels in order to envisage the maximum performance of the biofilter. The variation of inlet and outlet xylene concentrations and removal efficiency of the biofilter as a function of time are presented in Figure 2.

The system was started up with a gas flow rate of 5.5 l min^{-1}, resulting in an EBRT of 90 s with an inlet xylene concentration of 0.17 g m^{-3}. The removal efficiency of the biofilter was 100% for first day. This means that no acclimation time was needed for the microorganisms to remove xylene effectively. The variation of xylene concentration through the bed length shows that for initial days of operation, all the incoming xylene was removed through the first and second sections of the bed (Figure 3). Therefore, based on the above mentioned observations and to ensure the biofilm growth on all surfaces of the media through the bed length, the inlet concentration of xylene was increased in a rapid manner. By increasing the concentration up to 1.42 g m^{-3}, the removal efficiency decreased to 92.6% at day 10. The complete removal of xylene in the initial days may be attributed to the presence of highly adapted microbial populations along with the adsorption of xylene on the packing medium of the biofilter. Inoculation of the biofilter media with adapted microbial aggregates minimizes the acclimation time of the biofilter [19]. As noted previously, scoria was inoculated with an adapted microbial culture before packing. This phenomenon increased the number of xylene-degraders up to 1 10^8 CFU g^{-1} dry scoria, which was almost 380-fold higher than the initial cell concentration reported for an inoculated compost-based biofilter treating p-xylene vapors (i.e. 2.65 10^5 CFU p-xylene degraders g^{-1} dry compost) [20].

During phase I, the inlet concentration was increased slowly in order to prevent shock loading of biofilter and to reach a maximum performance (EC_{max}). From day 10 to 25, the inlet concentration was varied from 1.32 to 1.58 g m^{-3} and the removal efficiency was maintained over 92.6%. The removal efficiency decreased significantly to 53.6% as the xylene concentration increased to 4.47 g m^{-3} by day 44. From days 46 to 56, the inlet concentration was raised and maintained at highest levels during the whole experiment, i.e. 4.7 to 5.3 g m^{-3} with corresponding loads of 187.8 to 212.8 g m^{-3} h^{-1}. In this period a pseudo steady state was achieved, since the xylene removal efficiencies varied from 45.3 to 47.9%. According to Rahul et al. [16] the pseudo steady state was presumed when the changes in the xylene removal efficiency were within 5% for three successive days.

The next two phases of operation were aimed at investigating the biofilter performance under lower EBRTs. In phase II, the EBRT was reduced to 60 s and the biofilter could provide over 80% removal of xylene for the inlet concentrations below 1.14 g m^{-3}. The removal efficiency sharply declined to an average of 33.4 4.1% during the days 68 76, when the inlet xylene concentration was increased to a range of 3.22 to 3.68 g m^{-3}, with loading rates of 193.4-220.9 g m^{-3} h^{-1}. The last phase of the experiment commenced on day 77; the EBRT was further decreased to 40 s and a removal efficiency of 81.0% was observed for the xylene concentration of 0.26 g m^{-3} (IL = 23.3 g m^{-3} h^{-1}). With the sudden increase in the loading rate (84.9 g m^{-3} h^{-1}), the removal efficiency of xylene decreased to 53.6%, but it recovered to 68.2% after 3 days. Similar to the previous phases, the biofilter experienced high xylene loading rates ranged from 192.7 to 204.6 g m^{-3} h^{-1} during days 90 95 and in accordance with this operating condition, an average xylene removal efficiency of 30.3 2.2% was obtained. At the end of this phase, the xylene concentration was lowered and the removal efficiency recovered to 89% for loading rate of 22.2 g m^{-3} h^{-1}.

In this study, a removal efficiency of 98% was obtained under steady state condition for the inlet concentration of 1.34 g m^{-3} at EBRT of 90s. Comparatively, Saravanan and Rajamohan [1] observed a maximum xylene removal

Table 2 Biofilter operating conditions

Phase	I	II	III
Days of operation	1-61	62-76	77-101
EBRT (s)	90	60	40
Inlet concentration range (g m^{-3})	0.17-5.32	0.54-3.68	0.14-2.27
Inlet loading rate range (g m^{-3} h^{-1})	6.9-212.8	32.3-220.9	12.6-204.6

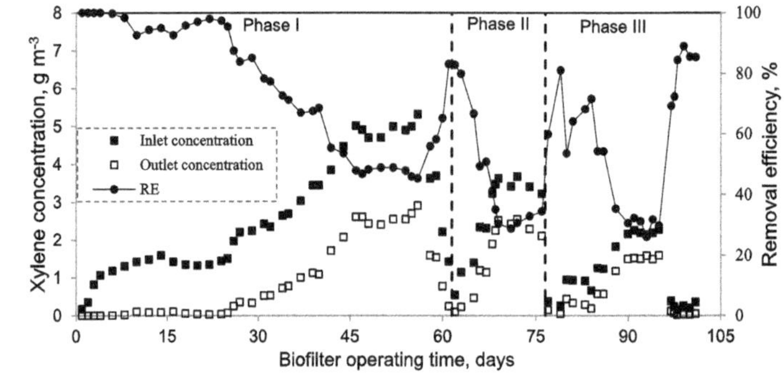

Figure 2 The variation of the inlet and outlet xylene concentrations and removal efficiency of the biofilter as a function of time.

efficiency of 50% for the inlet concentration of 1.2 g m^{-3} at an EBRT of 88.2 s in a biofilter packed with press mud.

The results obtained during this study indicate that the xylene removal efficiency decreased either by an increase in xylene concentration in the entering air or by a decrease in EBRT. At lower EBRTs, the microbial population on the surface of media gets less contact time for the degradation of xylene. Besides, the increase in xylene loads towards the highest levels during the last two phases, especially phase II, occurred in a more rapid manner than that of the phase I. Then, there was more time for microorganisms to adapt the increased xylene at phase I.

The elimination capacity (EC), which is the main parameter to describe the biofilter performance, is plotted as a function of the inlet load in Figure 4. Two different regimes are identified, as previously described in the literature [21,22]; the diffusion limitation regime (DLR) and the reaction limitation regime (RLR).

In case of EBRT 90 s; (i) under DLR, the elimination capacity increased to 94.3 g m^{-3} h^{-1} when the inlet loading rate was increased to 137.8 g m^{-3} h^{-1}. In DLR, an increase in the inlet concentration of xylene enhances the transfer rate of this pollutant from the gas-phase to the biofilm and more microorganisms actively participate in the biodegradation process. (ii) under RLR, the EC stabilized around 93.9 gm^{-3} h^{-1} for inlet loadings beyond 178.8 g m^{-3} h^{-1} and up to 212.8 gm^{-3} h^{-1}. Presumably, the RLR occurs when the amount of active xylene-specific microorganisms is insufficient to degrade all the gas-phase xylene that could possibly be transferred to the biofilm. Under this circumstance, the biofilm would be fully saturated with xylene and the metabolic activity of the cells in the biofilm would be the rate limiting step [3]. The maximum elimination capacity of the biofilter was 97.5 g m^{-3} h^{-1} for inlet xylene load of 199.5 g m^{-3} h^{-1} at EBRT of 90 s. The biofilter performance at EBRTs of 60 s and 40 s had similar trends to that of the EBRT of 90 s, with lower EC values. The maximum ECs of 79.8 and 65.5 g m^{-3} h^{-1} were achieved at EBRTs of 60 and 40 s, respectively. Jorio et al. [23] obtained maximum xylene elimination capacities of 67, 52, and 41 g m^{-3} h^{-1} at EBRTs of 158.9, 90.8, and 63.6 s, respectively, in a conventional biofilter packed with peat mixed with structuring and conditioning agents and initially inoculated with a microbial consortium. In other hand, Gallastegui et al. [24] reached a p-xylene elimination capacity as high as 130 g m^{-3} h^{-1} at

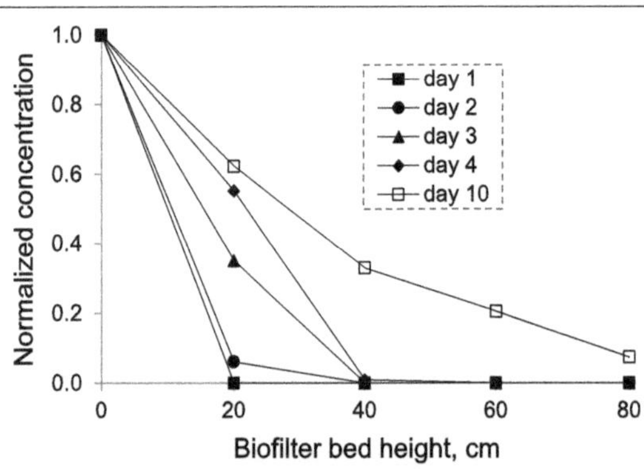

Figure 3 Normalized xylene concentration profiles along the bed height of biofilter at initial days of operation.

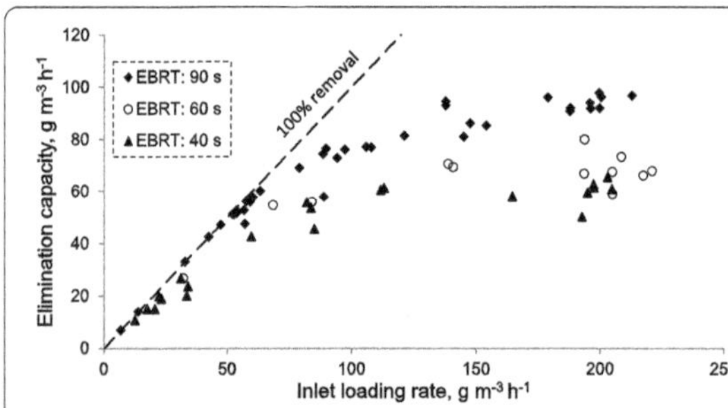

Figure 4 Effect of xylene inlet load on the elimination capacity of biofilter.

high EBRTs ranging from 180 to 270 s in a biofilter packed with pelletised sawdust and pig manure.

Higher elimination capacities at EBRT of 90 s compared to EBRTs of 60 and 40 s (for the same loading rate) may be attributed to the higher residence time of carrier gas, and availability of higher concentration of xylene at higher EBRTs in a given inlet loading rate.

Xylene profile along the biofilter length

The typical profiles of xylene along the length of the biofilter for different inlet concentrations at EBRTs of 90 and 40 s are illustrated in Figure 5. At day 18 (EBRT 90s), nearly 34% of the inlet xylene was removed in the first 0.2 m of the biofilter, 58% at 0.4 m, 81% at 0.6 m and 96% at 0.8 m of the biofilter height for an inlet xylene concentration of 1.34 g m^{-3} (loading rate of 53.8 g m^{-3} h^{-1}). A somewhat similar profile was observed for day 79 (EBRT 40s), where xylene concentration and loading rate were 0.26 g m^{-3} and 23.3 g m^{-3} h^{-1}, respectively. Therefore, in moderate inlet xylene concentrations or loading rates, compared to those reported in the literature using biofilter for handling xylene polluted air [1,3], removal is more efficient in the lower parts of the biofilter than in the upper parts. This could be explained by the presence of more carbon source (xylene) in sections near the inlet of the filter bed, which causes a higher metabolic reaction [25]. Since the biofilter was operated in the up-flow mode, then the highest local concentration of xylene was at the bottom of the bed and the lowest local concentration of xylene was at the top of the bed. With an increase in inlet xylene concentration to 5.0 g m^{-3} at day 52 (EBRT 90s), all sections had almost similar removal efficiencies ranging from 10.7% to 13.6%. Likewise, all sections showed somewhat similar performance with an average removal efficiency of 7.4% 1.7% at day 95 (C_{in} = 2.27 g m^{-3}, EBRT 40s). The xylene loads at days 52 and 95 were 199.5 and 204.6 g m^{-3} h^{-1}, respectively, which lay in reaction limitation regime (Figure 4). The xylene removal rate under this regime is nearly independent from inlet concentration variations.

At day 95, the inlet xylene concentration was about that of the day 31 (C_{in} = 2.43 g m^{-3}, IL = 97.1 g m^{-3} h^{-1}), while xylene load was about that of the day 52. As shown in Figure 5, the xylene removal profile at day 95 was very similar to that of the day 52. This indicates that the xylene removal profile along biofilter length is more dependent on the inlet load than on the inlet concentration.

Carbon dioxide production

In biofiltration process, the CO_2 production is an important parameter indicating the degree of VOCs degradation, because VOCs are aerobically biodegraded into water and carbon dioxide and utilized as carbon source for the microbial growth [16]. For complete chemical oxidation of xylene to water and CO_2, the mass-ratio of CO_2 produced to xylene degraded should be 3.32, according to the following stoichiometric reaction:

$$C_8H_{10} \quad 10.5\ O_2 \rightarrow 8CO_2 \quad 5H_2O \qquad 5$$

The carbon dioxide production rate (PCO_2) as a function of the xylene elimination capacity (EC) is shown in Figure 6. The alignment of the data series indicates that the quantity of carbon dioxide produced strongly correlated with the quantity of xylene eliminated and a linear regression provides the following equation:

$$PCO_2 \quad 1.93EC \quad 23 \qquad 6$$

Accordingly, the value of the slope in equation [6] shows that about 60% of the eliminated xylene was converted to CO_2. Wu et al. [26] obtained a mass ratio of PCO_2/EC of 1.65 when they treated p-xylene in a hybrid biofilter with an added nutrient solution containing ammonium salts. Li

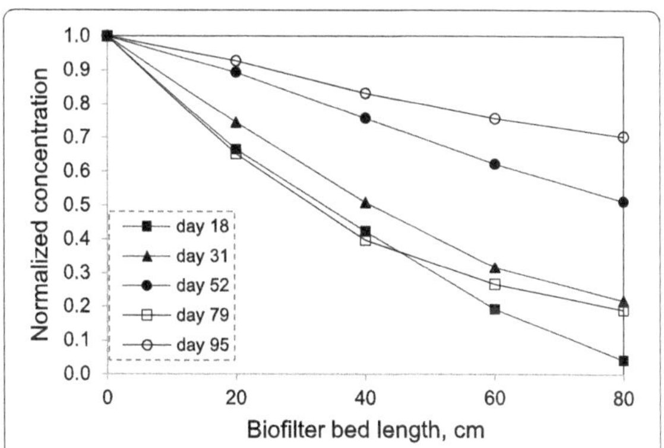

Figure 5 Normalized xylene concentration profiles along the bed length of biofilter at various days of operation.

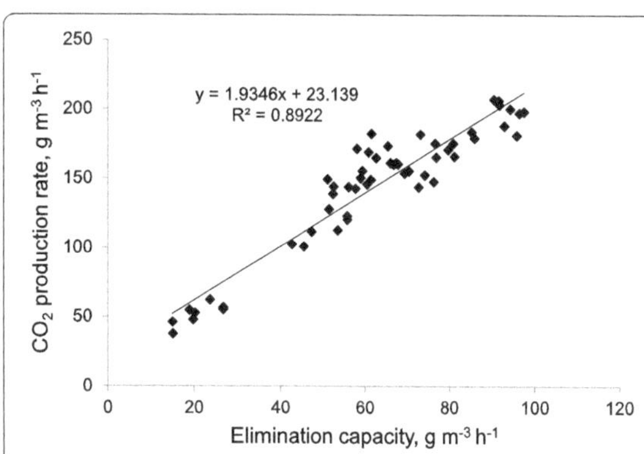

Figure 6 Relationship between the carbon dioxide production rate (PCO_2) and xylene elimination capacity (EC).

et al. [27] reported that 62% of removed xylene in a bacterial and fungal biofilter was converted to CO_2.

The discrepancy observed in CO_2 production in comparison with the case of complete chemical oxidation of xylene can be mainly explained by the biomass production. In addition, some of the CO_2 produced may partly accumulate in the liquid-phase in the form of HCO_3^-, H_2CO_3, and CO_2^{-3} [26].

By taking into account a general biomass composition formula as $C_5H_7NO_2$ and PCO_2/EC ratio of 1.93 (equation 6) (ignoring the CO_2 accumulated in the leachate), and considering ammonium as nitrogen source, equation [5] could be re-written as:

$$C_8H_{10} \quad 7.32O_2 \quad 0.67NH_4 \rightarrow 0.67C_5H_7O_2N \quad 4.65CO_2 \quad 3.99H_2O \qquad 7$$

Therefore, 0.71 g of dry biomass was produced per g of xylene consumed, which corresponds to a biomass yield coefficient value of 0.42 gC dry mass synthesized per gC xylene degraded.

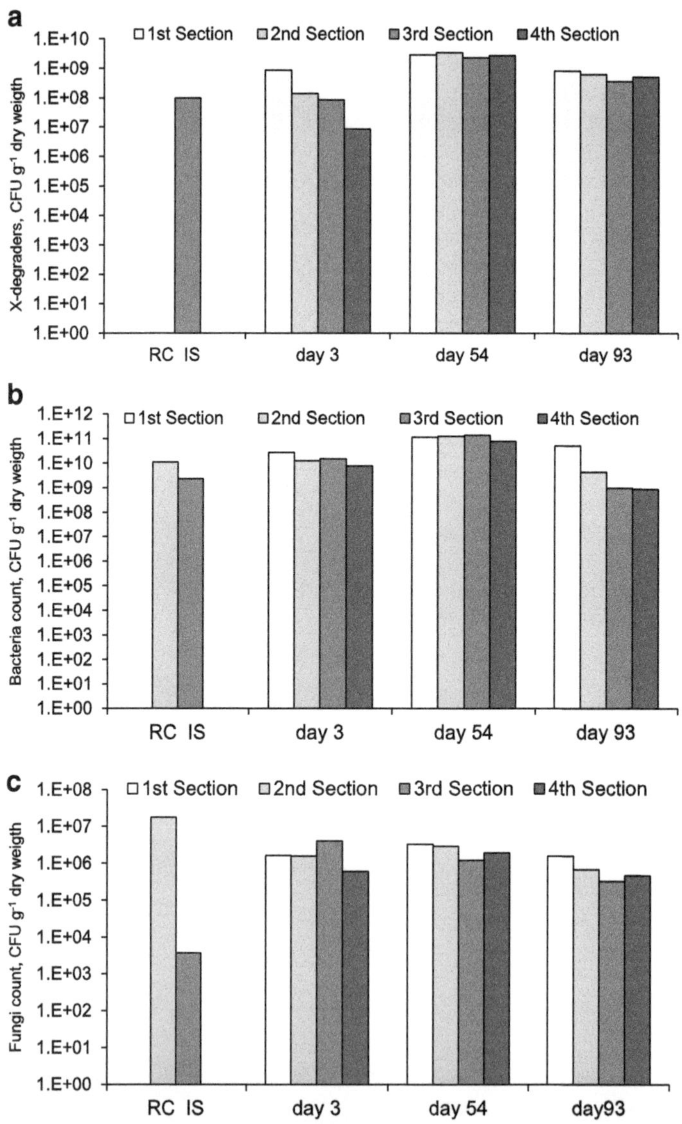

Figure 7 The microbial counts of (a) xylene degraders, (b) total bacteria, and (c) total fungi in the biofilter bed at different days of operation (RW: Raw Compost, IS: Inoculated Scoria).

Microbial aspects

Microbial counts in the filter material were regularly achieved for the follow-up of the microbial growth intensity inside the biofilter. The xylene-degraders, bacteria, and fungi counts in the packing material before biofilter assembly and in the samples withdrawn from each section of the filter bed at different days of biofilter operation are presented in Figure 7(a), (b), and (c). Inoculated scoria contained 10^8 CFUs of xylene degraders per gram dry weight, while no CFU was detected in one gram dry raw compost and scoria. This is taken as an evidence of the appropriate development of the inoculum, which probably resulted in good biofilter performance during the initial days of operation.

At day 3, the counts of xylene-degraders decreased around two orders of magnitude from bottom to the top of biofilter, which can be related to the complete removal of xylene at the first two sections. The xylene degraders increased by 10-fold to an average count of 2.87 10^{9} 4.8 10^{8} CFU g^{-1} at day 54 of operation and then decreased by 5-fold to 5.82 10^{8} CFU g^{-1} at day 93. These values are higher than those reported for p-xylene degraders in the biofilters packed with food waste compost (1.28 10^{8} g^{-1} of dry compost) and pig manure compost (2.58 10^{7} g^{-1} of dry compost) [20]. The microbial counts for xylene degraders follow xylene elimination capacities, which were 33.0, 93.7, and 50.2 g m^{-3} h^{-1} at days 3, 54, and 93, respectively. In addition, the number of xylene degraders in all four sections of the biofilter were within the range of the same order of magnitude at days 53 (9 log CFU g^{-1}) and 93 (8 log CFU g^{-1}), which are related to approximately similar performance of the sections at high inlet xylene loads.

The comparison of the number of microbial cells (average ratio of bacterial to fungal CFUs =2.4 10^{4}:1) implies that the bacteria were the dominant microorganisms responsible for the degradation of xylene in the biofilter. Irrespective of the day of sampling and biofilter section, the average total bacterial count was 4.8 10^{10} CFU (gram of dry mass)$^{-1}$ containing 10.5% xylene degraders.

A low magnification SEM image revealed that the scoria is very porous, and it showed a raw surface with big pores, which allowed the microbial attachment (Figure 8(a)). In addition, a sample was taken from the third section of biofilter at the end of operating period and analyzed using SEM. A dense biofilm covered the surface of the scoria and extended to the pores (Figure 8(b)). While the morphology is difficult to be observed, some coccoid and rod-shaped bacteria embedded in a matrix of extracellular polymeric substances (EPS) are present. Although such dense biofilm limits the diffusion of pollutants toward the inner particles, it may improve the resistance of microorganisms against high concentrations and shock loads of xylene.

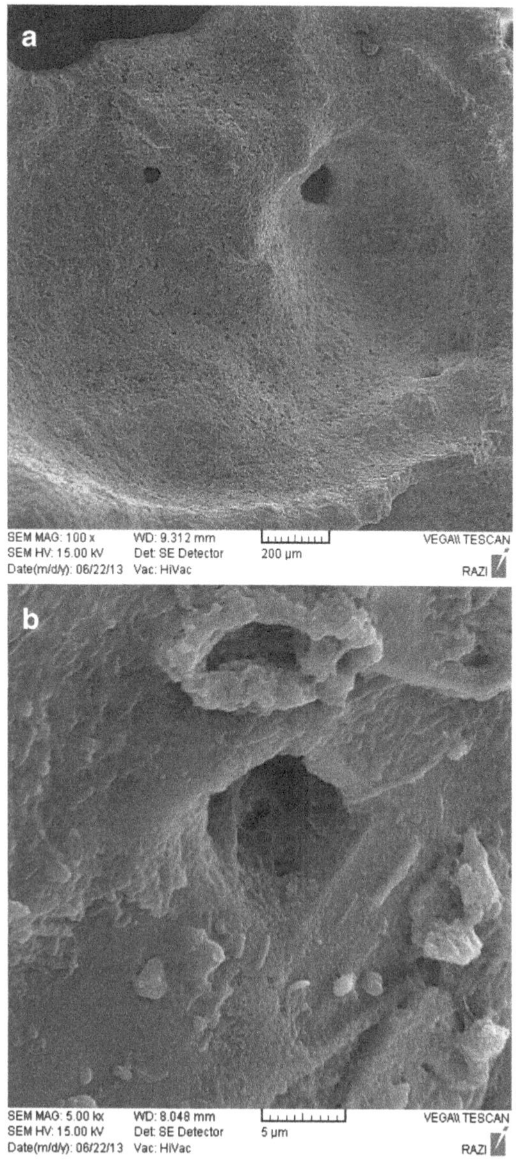

Figure 8 SEM micrographs of (a) raw scoria (100) and (b) filter bed taken from 3rd section at the end of biofilter operation (5000).

Temperature and pressure drop

The average bed temperature changed from 18.9 to 26.4C through the experiment, which is within the normal range reported for the biofiltration of gas phase VOCs [28]. The VOCs biodegradation by microorganisms in biofilters is an exothermic process, which increases the temperature inside of the biofilters [29]. Although the temperature of inlet gas sometimes increased up to 5.6C along biofilter, no significant correlation was found between EC and temperature ($p = 0.675$). This could be due to the heating

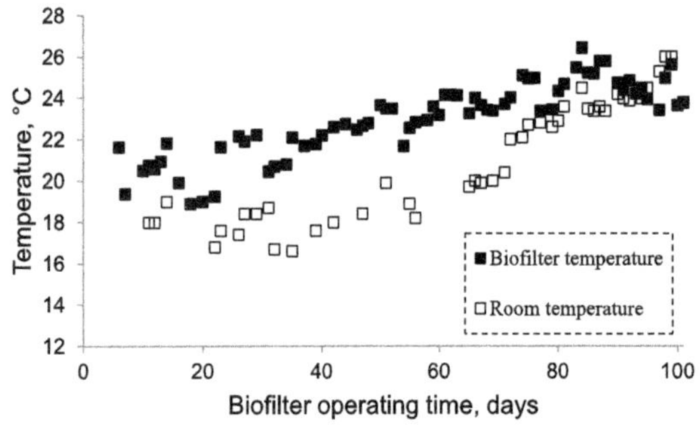

Figure 9 The biofilter bed and room temperature variations over biofilter operating time.

of the inlet gas stream in cool conditions and large changes in ambient air temperatures, ranging from 16.6 to 26.0C. The average filter bed temperature was mainly subject to room temperature (Figure 9) as well as a significant correlation was found between them ($p < 0.001$, $R^2 = 0.735$).

The pressure drop values were less than 4 mm H_2O during the entire experiments. In addition, negligible compaction and deterioration of the bed was observed, indicating a good mechanical strength of the media. The low pressure drop, which is the prime prerequisite for biofilter packing material, could be related to low velocity of carrier gas, the filter media characteristics, irrigation strategy and the nature of microbial community within biofilter. By volume, about 75% of the filter material was scoria (6 13 mm particle size), which is an inert media. One major advantage of inert material on organic ones is the low pressure drop due to minimum compaction over time and good air distribution. Excess biomass accumulation is a major factor causing an increase in the pressure drop [30]. Nevertheless, adequate irrigation of the filter bed made the removal of the excess of biomass possible [23]. Here the filter bed was irrigated with 1 liter nutrient solution, where, 728 and 1620 mg TOC were discharged from the system at days 59 and 94, respectively. The filter bed was dominated by bacteria. Usually pressure drop in bacterial biofilters is lower than that of observed in fungal biofilters. This is because of the occupation of the free space by the mycelia. Estrada et al. [31] reported that the bacterial biofiltration treating a VOC mixture exhibited a final pressure drop of 60% lower than that of the fungal biofilter due to mycelial growth.

Conclusions

A biofilter packed with scoria/compost was employed for the removal of xylene vapors. The best performance of biofilter ($EC_{max} = 97.5$ g m^{-3} h^{-1}) was observed at the highest EBRT, that may be due to higher residence time of carrier gas. The CO_2 production rate and the distribution of microbial populations in the biofilter were well correlated with the xylene removal rates, indicating the biodegradation of xylene in the biofilter. The packing media provided low pressure drop and long term stability, demonstrating its potential as media in the biofiltration of VOCs.

Competing interests

The authors declare that they have no competing interests.

Authors contributions

MH participated in the design of the study, implemented the experiments, conducted data analysis and drafted the manuscript, MMA, BB and AR were the supervisors, participated in the design of the study and in the interpretation of data, and helped to draft the manuscript, FMM participated in the design of the study. All authors read and approved the final manuscript.

Acknowledgements

This article is the result of Ph.D thesis approved in the Isfahan University of Medical Sciences (IUMS). The authors wish to acknowledge to Vice Chancellery of Research of IUMS for the financial support, Research Project, # 391249.

Author details

[1]Environment Research Center, Isfahan University of Medical Sciences (IUMS) and Department of Environmental Health Engineering, School of Health, IUMS, Isfahan, Iran. [2]Student Research Center, School of Health, Isfahan University of Medical Sciences, Isfahan, Iran. [3]Chemical Engineering Department, College of Engineering, University of Isfahan, Isfahan, Iran. [4]Department of Environmental Health Engineering, Faculty of Health, Hormozgan University of Medical Sciences, Bandar Abbas, Iran. [5]Department of Environmental Health Engineering, School of Health, Shahrekord University of Medical Sciences, Shahrekord, Iran.

References

1. Saravanan V, Rajamohan N: **Treatment of xylene polluted air using press mud-based biofilter.** *J Hazard Mater* 2009, **162:**981 988.
2. ATSDR: *Toxicological Profile Information Sheet.* USA: Agency for Toxic Substances and Disease Registry, Department of Health and Human Services, Public Health Service; 2003.
3. Rene ER, Murthy DVS, Swaminathan T: **Effect of flow rate, concentration and transient-state operations on the performance of a biofilter treating xylene vapors.** *Water Air Soil Pollut* 2010, **211:**79 93.
4. Cox HHJ, Deshusses MA: **Effect of starvation on the performance and re-acclimation of biotrickling filters for air pollution control.** *Environ Sci Technol* 2002, **36:**3069 3073.
5. Dehghanzadeh R, Aslani H, Torkian A, Asadi M: **Interaction of acrylonitrile vapors on a bench scale biofilter treating styrene-polluted waste gas streams.** *Iran J Environ Health Sci Eng* 2011, **8:**159 168.
6. Mehdinia SM, Abdul Latif P, Taghipour H: **A comparative evaluation of dried activated sludge and mixed dried activated sludge with rice husk silica to remove hydrogen sulfide.** *Iran J Environ Health Sci Eng* 2013, **10:**22.
7. Hernndez J, Prado J, Almarcha M, Lafuente J, Gabriel D: **Development and application of a hybrid inert/organic packing material for the biofiltration of composting off-gases mimics.** *J Hazard Mater* 2010, **178:**665 672.
8. Roshani B, Torkian A, Aslani H, Dehghanzadeh R: **Bed mixing and leachate recycling strategies to overcome pressure drop buildup in the biofiltration of hydrogen sulfide.** *Bioresour Technol* 2012, **109:**26 30.
9. Lee S, Li C, Heber AJ, Ni J, Huang H: **Biofiltration of a mixture of ethylene, ammonia, n-butanol, and acetone gases.** *Bioresour Technol* 2013, **127:**366 377.
10. Sun D, Li J, An T, Xu M, Sun G, Guo J: **Evaluation of the performance of structured mixed packing and inert packing materials in toluene biotrickle-filtration.** *Biotechnol Bioproc Eng* 2011, **16:**1009 1018.
11. Prado J, Veiga MC, Kennes C: **Treatment of gas-phase methanol in conventional biofilters packed with lava rock.** *Water Res* 2005, **39:**2385 2393.

12. Detchanamurthy S, Gostomski PA: **Biofiltration for treating VOCs: an overview.** *Rev Environ Sci Biotechnol* 2012, **11**:231 241.
13. Strauss JM, Riedel KJ, Plessis CA: **Mesophilic and thermophilic BTEX substrate interactions for a toluene-acclimatized biofilter.** *Appl Microbiol Biotechnol* 2004, **64**:855 861.
14. Gabaldn C, Martnez-Soria V, Martn M, Marzal P, Penya-roja JP, Alvarez-Hornos FJ: **Removal of TEX vapours from air in a peat biofilter: influence of inlet concentration and inlet load.** *J Chem Technol Biotechnol* 2006, **81**:322 328.
15. Gallastegui G, Ramirez A, Elas A, Jones JP, Heitz M: **Performance and macrokinetic analysis of biofiltration of toluene and p-xylene mixtures in a conventional biofilter packed with inert material.** *Bioresour Technol* 2011, **102**:7657 7665.
16. Rahul, Mathur AK, Balomajumder C: **Biological treatment and modeling aspect of BTEX abatement process in a biofilter.** *Bioresour Technol* 2013, **142**:9 17.
17. Kwon JS, Yun ST, Lee JH, Kim SO, Jo HY: **Removal of divalent heavy metals (Cd, Cu, Pb, and Zn) and arsenic (III) from aqueous solutions using scoria: kinetics and equilibria of sorption.** *J Hazard Mater* 2010, **174**:307 313.
18. Lodge JP: *Methods of Air Sampling and Analysis.* New York: Lewis Publishing Inc; 1989.
19. Jorio H, Payre G, Heitz M: **Mathematic modeling of gas-phase biofilter performance.** *J Chem Technol Biotechnol* 2003, **78**:834 846.
20. Hwang JW, Jang SJ, Lee EY, Choi CY, Park S: **Evaluation of composts as biofilter packing material for treatment of gaseous p-xylene.** *Biochem Eng J* 2007, **35**:142 149.
21. Jorio H, Jin Y, Elmrini H, Nikiema J, Brzezinski R, Heitz M: **Treatment of VOCs in biofilters inoculated with fungi and microbial consortium.** *Environ Technol* 2009, **30**:477 485.
22. Rene ER, Mohammad BT, Veiga MC, Kennes C: **Biodegradation of BTEX in a fungal biofilter: influence of operational parameters, effect of shock-loads and substrate stratification.** *Bioresour Technol* 2012, **116**:204 213.
23. Jorio H, Bibeau L, Viel G, Heitz M: **Effects of gas flow rate and inlet concentration on xylene vapors biofiltration performance.** *Chem Eng J* 2000, **76**:209 221.
24. Gallastegui G, Muoz R, Barona A, Ibarra-Berastegi G, Rojo N, Elas A: **Evaluating the impact of water supply strategies on p-xylene biodegradation performance in an organic media-based biofilter.** *J Hazard Mater* 2011, **185**:1019 1026.
25. Znad HT, Katoh H, Kawase Y: **High loading toluene treatment in a compost based biofilter using up-flow and down-flow swing operation.** *J Hazard Mater* 2007, **141**:745 752.
26. Wu D, Quan X, Zhao Y, Chen S: **Removal of p-xylene from an air stream in a hybrid biofilter.** *J Hazard Mater* 2006, **B136**:288 295.
27. Li L, Liu JX: **Removal of xylene from off-gas using a bioreactor containing bacteria and fungi.** *Int Biodeterior Biodegrad* 2006, **58**:60 64.
28. Kong X, Wang C, Ji M: **Analysis of microbial metabolic characteristics in mesophilic and thermophilic biofilters using Biolog plate technique.** *Chem Eng J* 2013, **230**:415 421.
29. Rene ER, Veiga M, Kennes C: **Performance of a biofilter for the removal of high concentrations of styrene under steady and non-steady state conditions.** *J Hazard Mater* 2009, **168**:282 290.
30. Dorado AD, Baeza JA, Lafuente J, Gabriel D, Gamisans X: **Biomass accumulation in a biofilter treating toluene at high loads-Part 1: experimental performance from inoculation to clogging.** *Chem Eng J* 2012, **209**:661 669.
31. Estrada JM, Hernndez S, Muoz R, Revah S: **A comparative study of fungal and bacterial biofiltration treating a VOC mixture.** *J Hazard Mater* 2013, **250 251**:190 197.

5

Study of physio-psychological effects on traffic wardens due to traffic noise pollution; exposure-effect relation

Shamas Tabraiz[1*], Saeed Ahmad[1], Iffat Shehzadi[2] and Muhammad Bilal Asif[1]

Abstract

Background: Noise pollution has increased to alarming extent in most of the urban areas in Pakistan. It is assumed even more perilous than air and water pollution due to its direct acute and chronic physio-psychological effects. The objective of this study is to analyze and evaluate the psychological and physiological effects caused by traffic noise on traffic wardens and to find relation type between exposure time and effect.

Methods: Three wardens check posts near roads were selected for survey in Taxila and Islamabad cities of Pakistan. Survey conducted included noise measurements at aforementioned check posts for one month and Performa based interviews of traffic wardens.

Results and conclusions: Analysis of results showed that noise levels varied between 85-106 dB hence violating OSHA regulations. Major psychological effects found in wardens were aggravated depression 58%, stress 65%, public conflict 71%, irritation and annoyance 54%, behavioral affects 59% and speech interference 56%. Physiological effects found were hypertension 87%, muscle tension 64%, exhaustion 48%, low performance levels 55%, concentration loss 93%, hearing impairment 69%, headache 74% and cardiovascular issue 71%. Relation between exposure time and effects were evaluated by using simple regression test in excel. Percentage of psychological and physiological effects in wardens varied with the exposure time; aggravated depression ($R^2 = 0.946$, $P = 0.133$), stress suffering ($R^2 = 0.014$, $P = 0.173$), public conflict ($R^2 = 0.946$, $P = 0.133$), irritation and annoyance ($R^2 = 0.371$, $P = 0.137$), behavioral affects ($R^2 = 0.596$, $P = 0.0616$) and speech interference ($R^2 = 0.355$, $P = 0.445$), hypertension ($R^2 = 0.96$, $P = 0.00095$) and cardiovascular issue ($R^2 = 0.775$, $P = 0.044$).

Keywords: Noise pollution, Traffic wardens, Physio-psychological effects, Exposure-effect relation

Introduction

Economical and social structure of society has been changed in Pakistan due to rapid population increase, urbanization and industrialization. Due to advancement in transportation facilities and technology the traffic loads are increasing at an alarming rate which initiates a major problem of noise pollution. Noise is an unwanted sound which has different frequencies and acoustic pressure without any regular pattern. The major causes of noise pollution are roads, railway and air traffic. Characteristics of noise (frequency and acoustic pressure) depend upon the characteristics of the traffic and road i.e. road gradient, road surface type, surrounding topography, grade of road, number of vehicles, type of vehicles, ages of vehicle passing, speed of vehicles, type of goods transported, packing of goods in vehicles, horns sound pressure, meteorological conditions, type of breaks and behavior of the drivers. Most effective way of controlling noise is to reduce it at source i.e. vehicle and road. Noise contribution from different parts of vehicles is different; Air intake system 9%, exhaust system 27%, tires 30% and engine 34%. At 70 km/h speed the tire noise dominate on other sources within the vehicle. Therefore, limiting the vehicle speed on busy roads substantially reduces noise the level [1]. Only 5% reduction in the gradient of road reduces 1.5 dB of Leq noise. Road surface macro-texture wavelength range should be 2-10-20 mm to reduce noise levels. Similarly, Green belts, plants and

* Correspondence: shamastabraiz28@yahoo.com
[1]Department of Environmental Engineering, University of Engineering & Technology, Taxila, Pakistan
Full list of author information is available at the end of the article

trees along the roads are effective in the reduction of noise levels. For 10 dB reduction, mounds and barriers along the roads are suggested to construct. Building treatment is also effective for the reduction of internal noise i.e. single glazed and double glazed windows reduce 10 and 25 dB noise level, respectively [2]. Management of noise is important to shun its perilous effects. High noise levels deteriorate the health of people and thus subjected to economy loss.

Access to pollution free environment is the central theme of human rights. Therefore, it should be everybody's task to keep ambient environment natural and clean as much as possible [3]. Noise is discarded, unlikable and annoying sound of such duration, intensity or other quality that can cause any kind of physiological harm to human health or other living things [4]. Traffic noise is a source of annoyance [5-7]. It is detrimental to the health of every third person. Every one out of five persons in Europe is exposed to health deteriorating noise levels even at night (*WHO*). Noise has identified as one of the main environmental problem in Europe but lower priority is given to it as compared to water and air pollution [8]. A study was carried out on healthy individuals who migrated to the high and low noise levels residential areas of Bonn city. The blood pressure got to higher level in the individuals who moved to higher noise level areas as compared to those who moved in quitter areas [9]. Another study revealed that blood pressure quickly returned to normal when elevated noise came to lower level. Relationship between the noise levels and sleep quality is complex. It is recommended by World Health Organization that noise level in the bed room should be kept below the average level of 30 dB, or maximum of 45 dB for a single event. Higher levels of sound are related to reduced sleeping quality or awakening.

Cardiovascular effects due to traffic noise exposure are an important area of research in environmental epidemiology. Adverse health effects of noise like blood pressure and hypertension were studied. Noise level and blood pressure relationship found in both children and adults was inconsistent [10,11]. Rather, more consistent results have been obtained in case of hypertension, with aircraft noise as well as road traffic noise [12-22]. Relationship between exposure to traffic noise and hearing impairment, ischemic heart disease, hypertension, annoyance and sleep disturbance was established [23]. In the recent years the cause relationship between traffic noise and cardiovascular risk has been increased [24]. Impacts of noise pollution include but not limited to the impairment of mind-peace, internal peace, impairment of hearing and the physiological impacts [25]. Noise pollution could impair physiological and psychological aspects of human health [26]. The effects of long term exposure are not well understood. Long term effects are linked to hypertension and heart diseases [8]. Traffic wardens in the cities are mostly appointed on the traffic rush junctions of the roads to regulate the traffic. Therefore, they are exposed to elevated noise levels. Long term physio-psychological effects of traffic noise on traffic wardens and effects relation with exposure time has not been evaluated yet. Therefore, this study was undertaken with these objectives. Beside that Noise levels were also measured at three wardens check posts situated along the Grand Truck Road (Taxila, Islamabad).

Materials and methods

The study was divided in two phases. In the first phase, noise levels were measured. In second phase, physio-psychological effects were evaluated.

Exposure assessment

Three wardens check posts along the Grand Truck Road were selected for noise survey in Taxila and Islamabad. The check posts selected were Taxila underpass check post, Tarnol check post and Golra mor check post. These check posts were selected keeping in view the movement of crush and other goods carrying heavy vehicles. The measurements were taken at those points where the traffic wardens were found to be performing their duties. Usually they perform their duties within 1-5-10 m range from the road side. Noise monitoring was conducted for one month. The noise level meter during measurements was held at a height of 1.5 meter above ground level at an angle of 30° from monitoring persons' body, to measure noise levels which can be perceived by the human ears. A one minute reading time was set at each location in order to obtain a stable maximum reading of noise at that particular location. In this way, 60 readings were noted at every location for an hour, each one after 1 minute then the Leq (8 hrs) for the day was calculated.

Physio- psychological effects assessment

In the second phase, a wide-ranging standardized questionnaire was developed. It took 30-40 minute to interview a warden. Wardens appointed at the aforementioned three and nearby check posts were interviewed. The questionnaire covered housing (socio-economic conditions), exposure, heritage physio-psychological diseases, any accidental disability or psycho effect, age, working exposure, medication of any physio-psycho disease. The psychological effects evaluated were; aggravated depression, stress, public conflict, irritation and annoyance, behavioral effects and speech interference. The psychological effects were judged and evaluated on the basis of questions asked verbally: "Did you have aggravated depression? (Yes, no)", "Did any one of your parents have aggravated depression?

(Yes, no)", "Did you have aggravated depression before joining the job as traffic warden? (Yes, no)", "Did you ever take anti-depressive medicine? (Yes, No)". Same types of questions were asked to assess other psychological effects. The physiological effects evaluated were; hypertension, muscle tension, exhaustion, exasperated fatigue, low performance levels, concentration loss, earache, hearing impairment, headache, cardiovascular issue, tinnitus and insomnia. The effects like muscle tension, exhaustion, exasperated fatigue, low performance levels, concentration loss, earache, headache, tinnitus and insomnia were evaluated with 7- grade verbal question; "Did you have hypertension? (Yes, no)", "Did one of your parents have hypertension? (Yes, no)", "Did you have hypertension before joining the job as traffic warden? (Yes, no)", "Did you ever take anti-hypertensive medicine? (Yes, no)", "Do you feel muscle tension during or after job hours? (Often, sometime, seldom or never)", "Is there any reason of muscle tension other than noise at your job place? (Yes, no)", if "Yes" then what? Same types of questions were asked to evaluate the effects like exhaustion, exasperated fatigue, low performance levels and concentration loss. For earache and headache; "Do you feel earache and headache due to high noise levels? (Often, sometime, seldom or never)", "Do you ever got injury in head or ear? (Yes, no)". For hearing impairment; "Do you have hearing impairment? (Yes, no)", "Was there an accident which impaired your hearing? (Yes, no)". For insomnia and tinnitus; "Do you have any sleeping problem (early wakeup, falling asleep, awakening)?: (often, sometime, seldom or never)" [27], "Do you have tinnitus problem (ear ringing)? (often, sometime, seldom or never)". For the cardiovascular problem; "Do you have any cardiovascular issue (valve problem, ischemic stroke, heart failure)? (Yes, no)". If yes, do you have heart problem before the joining of job? (Yes, no), "Are you taking any medication? (Yes, no)". All the wardens were divided into different groups based on the job experience (exposure time); group A (1-5 years), Group B (5-10 years), Group C (10-15 years) and Group D (15-20 years). After the assessment of exposure and physio-psychological effects, the simple linear regression was applied by using Microsoft excel to estimate the association of exposure time with physio-psychological effects. Two hundred and fifty wardens were interviewed to evaluate the effects.

Results and discussions

Noise levels

Noise level measurements were taken at selected locations near the check posts. Noise levels (Leq for 8 hours) of three check posts at 1m, 5 m and 10 m distance from road are given below in Figures 1, 2 and 3. Noise levels (Leq) varied from 86-106 dB. Such high levels were due to the heavy traffic and high pressure horns. Impulsive noise levels measured were up to 120 dB. Even at the distance of 10 m noise levels (Leq) did not dropped below 85 dB. At the distance of 10 m, noise levels range was 85-95 dB, at all the selected posts. While at the distance of 5 m from road, noise levels (Leq) measured were in the range 88-102 dB, at all the selected point. At 1m away from road, noise levels (Leq) range was 91-106 dB. Hearing impairment could be possible of such high noise pressure level (120 dB), even for short period (seconds) exposure. From the measurements, it was assessed that not only the wardens performing duty but people living near roads are under severe threat of possible physiological as well psychological effects.

Psychological effects

The common psychological effects i.e. hypertension, aggravated depression, stress, public conflict, irritation and annoyance, behavioral effects and speech interference were taken into account for this study. All the results of

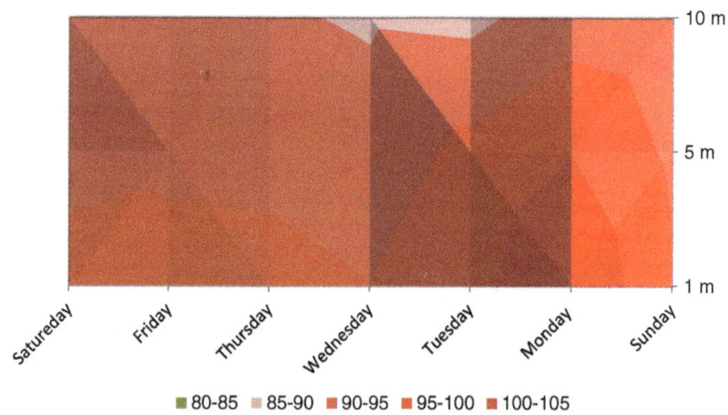

Figure 1 Noise levels (Leq 8hrs) at Taxila Underpass check post.

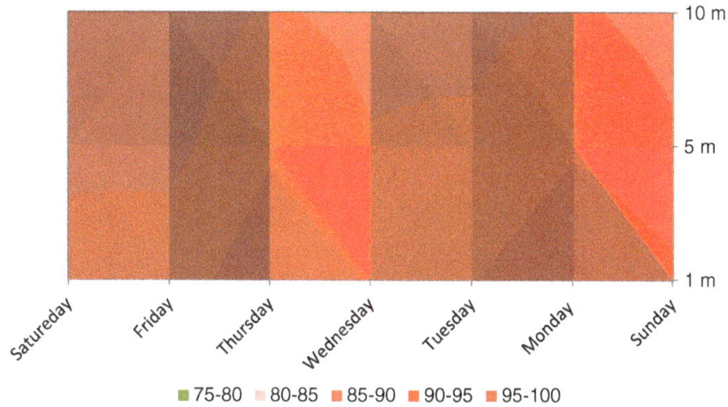

Figure 2 Noise levels (Leq 8hrs) Tarnol check post.

psychological effects are explained below and shown in Figures 4 and 5.

Aggravated depression

In aggravated depression one feels sadness, hopelessness and loss of interest in life. The survey study statistics showed that 58% of the wardens were suffering from aggravated depression. Exposure time wise, there was variance in effected warden's percentage. Aggravated depression found; 35%, 54%, 62% and 80% in group A, B, C and D, respectively. A linear relation was found between exposure time and the percentage of aggravated depression suffering wardens ($R^2 = 0.978$, $P = 0.033$).

Stress

In case of stress, 65% of the wardens of all the groups were suffering from stress. Exposure wise trend was random. Group A, B, C & D reported 60%, 70%, 50% and 80% stress, respectively. A non-linear relation was found between the exposure time and wardens percentage of stress suffering ($R^2 = 0.16$, $P = 0.093$).

Public conflict

Public conflict evaluated was 71%. It showed random trend as the exposure time increased. 70%, 69%, 45% and 100% public conflict problem found for group A, B, C and D group, respectively. A non-linear relation was found between exposure time and the percentage of wardens reported public conflict often ($R^2 = 0.143$, $P = 0.223$).

Irritation and annoyance

Irritation and annoyance was found in 54% wardens. Exposure time wise percentage suffering from irritation and annoyance was; 40%, 65%, 50% and 62% in group A, B, C and D, respectively. A non linear relationship was found between exposure time and percentage of wardens suffered irritation and annoyance ($R^2 = 0.327$, $P = 0.099$).

Behavioral effects

Behavior changes were reported by 59% wardens. Exposure time wise behavioral effects reported were 70%, 81%, 59% and 25% in group A, B, C and D, respectively. Percentage effected showed a linear relation with exposure

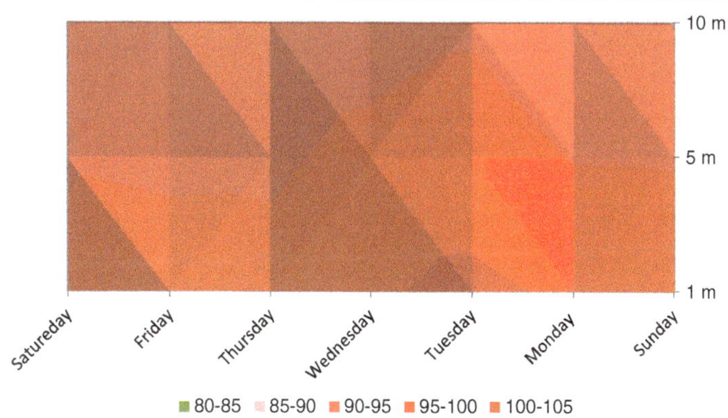

Figure 3 Noise levels (Leq 8hrs) Golra mor check post.

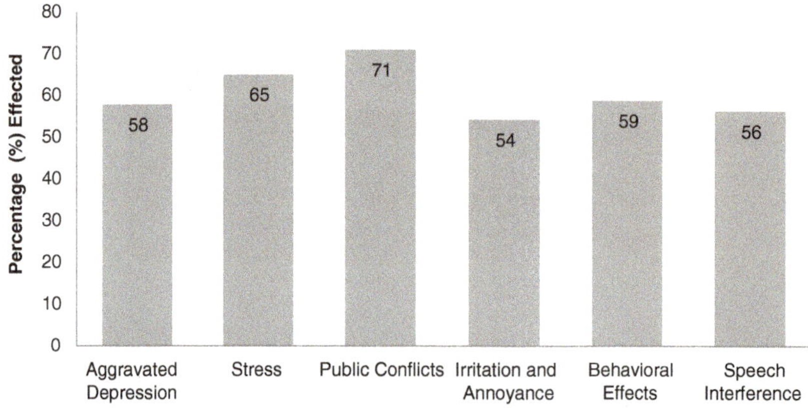

Figure 4 Psychological effects of noise on traffic wardens.

time ($R^2 = 0.7$, $P = 0.0388$). Results showed that percentage increased initially from 1-10 years of exposure then decreased afterward.

Speech interference

56% of the surveyed wardens reported that noise create difficulty in conversation. Exposure time wise percentage of wardens' reported speech interference was; 40%, 40%, 85% and 60% in group A, B, C and D, respectively. A non-linear relation was established between exposure time and the percentage of wardens reported speech interference ($R^2 = 0.4$, $P = 0.349$).

Physiological effects

In Figure 6, physiological effects that were investigated in this study for traffic wardens are given. Hypertension, muscle tension, exhaustion, exasperated fatigue, low performance levels, concentration loss, earache, hearing impairment, headache, cardiovascular issue, tinnitus and insomnia were probed. All the results of physiological effects are briefly discussed below and shown in Figure 6. Exposure time wise effects are shown in Figures 7 and 8.

Hypertension

In the survey 87% of the wardens were found to be suffering from hypertension (high blood pressures). The effected personnel percentile variation with the exposure time was as: in group A; 79%, group B; 82%, group C; 91% and group D; 94%. A linear trend was found between the exposure time and hypertension in the wardens ($R^2 = 0.976$, $P = 0.001$).

Muscle tension

Muscle tension reported was 64%. Exposure time wise muscle tension reported by group A, B, C and D were 52%, 55%, 71% and 80%, respectively. It increased as the exposure time increased. A linear relation was found between wardens' percentage affected and exposure time ($R^2 = 0.945$, $P = 0.0136$).

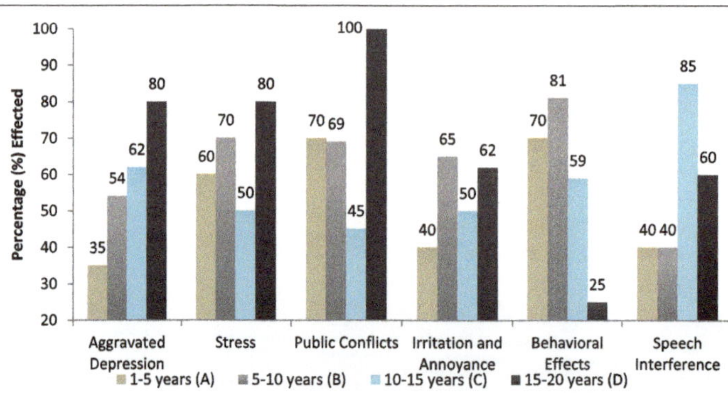

Figure 5 Exposure time wise psychological effects of noise on traffic wardens.

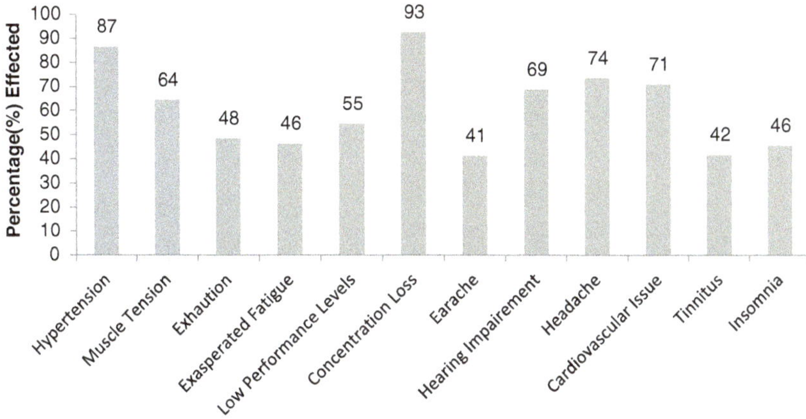

Figure 6 Physiological effects of noise on traffic wardens.

Exhaustion

Survey results showed that 48% wardens were suffering from exhaustion. Exposure time wise exhaustion in group A, B, C and D was 55%, 64%, 38% and 37%, respectively. Percentage effected showed a little linear relation with exposure time ($R^2 = 0.609$, $P = 0.0312$). Results showed that percentage reported exhaustion increased initially from 1-10 years of exposure then decreased afterward.

Exasperated fatigue

"Noise does exasperated fatigue" reported by 46% wardens. Exasperated fatigue found as 35, 65, 62 and 23% in group A, B, C and D, respectively. Percentage reported exasperated fatigue showed a non-linear relation with exposure time ($R^2 = 0.06$, $P = 0.201$). Results showed that percentage increased initially from 1-15 years of exposure then decreased afterward up to 20 years exposure.

Low performance levels

"Elevated noise levels reduce the performance levels" reported by 55% traffic wardens. Group A, B, C and D were suffering low performance level; 40%, 56%, 62% and 60%, correspondingly. With the increase in exposure time the percentage effected increased. It showed linear relation ($R^2 = 0.728$, $P = 0.0396$) between exposure time and percentage of effected wardens.

Concentration loss

Elevated noise caused concentration loss, 93% wardens reported. Exposure time wise concentration loss reported was 70%, 100%, 100% and 100% in group A, B, C and D, respectively. After 5 years of exposure 100% personnel reported concentration loss due to noise. It showed approximately constant relation after 5 years ($R^2 = 0.6$, $P = 0.0389$) between exposure time and percentage of effected wardens.

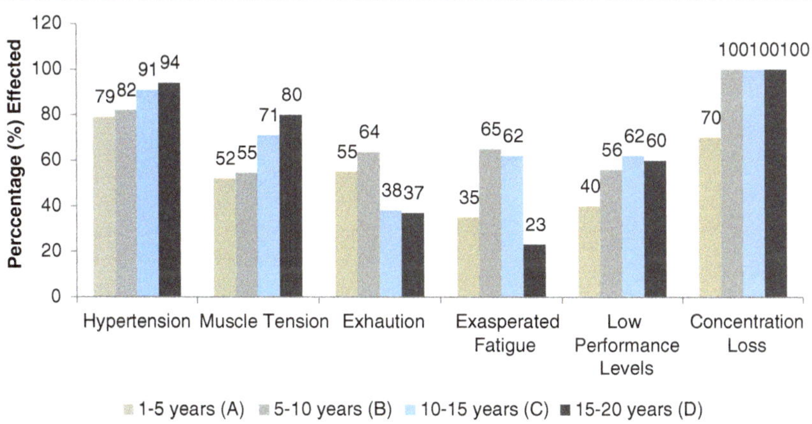

Figure 7 Exposure time wise Physiological effects of noise on traffic wardens.

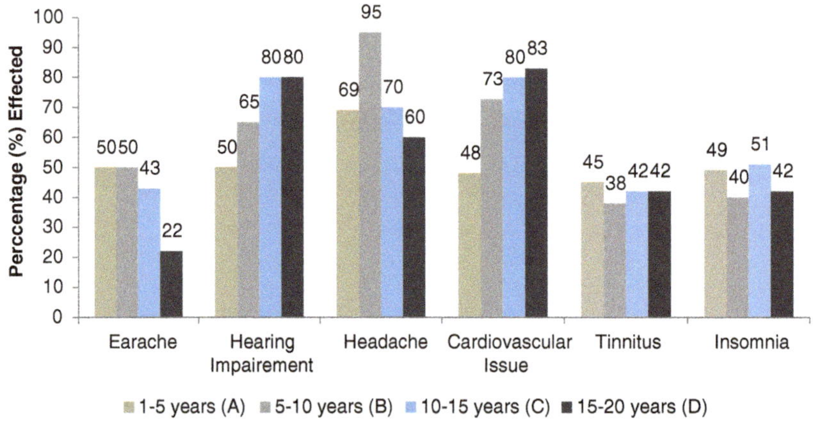

Figure 8 Exposure time wise Physiological effects of noise on traffic wardens.

Earache

Only 41% wardens were suffering from earache due to the elevated noise levels. Exposure wise wardens suffering earache; 50%, 50%, 43% and 22% in group A, B, C and D, respectively. A linear relation ($R^2 = 0.786$, $P = 0.020$) was found between warden's earache reported percentage and exposure time. Earache decreased with increase in exposure time.

Table 1 Summary of simple regression statistics (exposure-effect relation)

Physio-psychological effects	(R^2)*	P-value*	Exposure-effect relation
Aggravated Depression	0.978	0.033	Linear
Stress	0.16	0.093	Non-linear
Public Conflicts	0.143	0.233	Non-linear
Irritation and Annoyance	0.327	0.099	Non-linear
Behavioral Effects	0.7	0.0388	Non-linear
Speech Interference	0.4	0.349	Non-linear
Hypertension	0.976	0.001	Linear
Muscle Tension	0.945	0.0136	Linear
Exhaustion	0.609	0.0312	Linear
Exasperated Fatigue	0.06	0.21	Non-linear
Low Performance Levels	0.728	0.0396	Linear
Concentration Loss	0.6	0.089	Linear
Earache	0.786	0.020	Linear
Hearing Impairment	0.896	0.0269	Linear
Headache	0.199	0.052	Non-linear
Cardiovascular Issue	0.827	0.049	Linear
Tinnitus	0.05	0.0039	Constant
Insomnia	0.00588	0.025	Non-linear

*P-value is the probability of observing a test statistic more extreme than what was observed (if $p < 0.05$ then null hypothesis rejected).
*R^2 is a coefficient of determination (measures the how accurate linear model is at predicting i.e. near to unity mean more linear relation).

Hearing impairment

Hearing capability has been impaired of 69% wardens. Hearing impairment increased as the exposure time increased. The effected personnel percentile variation with the exposure time was as: in group A; 50%, group B; 65%, group C; 80% and group D; 80%. A linear trend was found between the exposure time and hearing impairment in the wardens ($R^2 = 0.896$, $P = 0.0269$).

Headache

Headache was reported by 74% traffic wardens. Headache increased and then decreased after ten years time period. Exposure time wise headache; in group A, B, C and D E was 69%, 95%, 70% and 60% respectively. Initial it increased to 10 years after that it decreased. A non linear relation ($R^2 = 0.199$, $P = 0.0502$) between exposure time and headache in warden was found.

Cardiovascular issue

On the whole, 71% wardens reported cardiovascular issues. Exposure time wise Cardiovascular issues; in group A, B, C and D was 48%, 73%, 80% and 83%, respectively. A linear relation ($R^2 = 0.827$, $P = 0.049$) between exposure time and cardiovascular issues in wardens was found. Heart problem increased as the exposure to noise increased.

Tinnitus

Ringing ear reported by 42% wardens only. This percentage was approximately constant ($R^2 = 0.05$, $P = 0.0039$) in each group A, B, C and D.

Insomnia

About 46% wardens feel difficulty in sleeping (early wakeup, falling asleep, awakening). Exposure time wise insomnia found in group A, B, C and D was 49%, 40%, 51% and 42% respectively. A non- linear relation (R^2 = 0.00588, P = 0.025) between exposure and insomnia was established. Summary of the regression test for exposure effect relation is given below in Table 1.

Conclusions

Noise pollution at selected locations has increased up to dangerous levels. Noise pollution is one of major cause of physiological and psychological deterioration but never gets due attention. Some of the physio-psychological effects showed high percentage of effected traffic wardens initially but decreased as the exposure time increased i.e. behavioral effects, speech interference, exhaustion, exasperated fatigue, ear each and head each. Most of the physio-psychological effects increased as the exposure time increased i.e. hypertension, aggravated depression, muscle tension, low performance levels, concentration loss, hearing impairment and cardiovascular issues. Few effects remained constant as the exposure time increased tinnitus and insomnia, while some effects showed random trends with increase in exposure time i.e. irritation and annoyance, public conflicts and stress. Further investigation of the effects is recommended which showed the random relation with exposure time to reveal reasons.

It can be concluded from the results that physiological and psychological deterioration lead to disturbed social life of traffic wardens. Therefore, efforts should be done to provide a framework in order to reduce duty hours for traffic wardens based on their physiological deterioration due to prolong exposure. Traffic wardens should be equipped with personal protective equipments such as ear plugs to protect their hearing impairment. Most importantly noise reduction techniques should be applied while the road planning, design and construction is in process. Legislation for the standard of vehicles should be in Pakistan. Enforcement of present rules regarding noise could help in noise pollution reduction.

Competing interests

The authors declare that they have no competing interests.

Authors' contributions

ST: Data collection, manuscript writing. SA: manuscript proof reading, improvements of results description, presentation of results. IS: psychological effects description/trends evaluation. MBA: manuscript proof reading, improvements of results description, presentation of results. All authors read and approved the final manuscript.

Acknowledgements

This study has not been completed without the assistance of my undergraduate students; Tayyab Ishtiaq, Sibtain Ali, M. Usman Saleem, Shahzeb Imtiaz, M. Naeem Zaman, Junaid Qaisar, Fawad Humayoon. They assisted author in measuring noise levels and interviewing the wardens.

Author details

[1]Department of Environmental Engineering, University of Engineering & Technology, Taxila, Pakistan. [2]Department of Psychology, University of Gujrat, Pakistan.

References

1. ARRB Transport Research. Road surface characteristics and condition: effects on road users, research report ARR314. 1998.
2. FHWA. Highway Traffic Noise Analysis and Abatement. 1995.
3. Pettersson B. Indoor noise and high sound levels –a transcription of the Swedish national board of health and welfare's guidelines. J Sound Vibration. 1997;205(4):475–80.
4. Kupchella, Charles E, Hyland, Margaret C. Living within the system of nature. Environmental Science. 3rd ed. Prentice Hall. 1993: pp. 438.
5. Trombetta Zannin PH, Bunn F. Noise annoyance through railway traffic – a case study. J Environ Health Sci Eng. 2014;12:14.
6. Golmohammadi R, Ghorbani F, Mahjub H, Daneshmehr Z. Study of school noise in the capital city of Tehran-Iran. Iran J Environ Health Sci Eng. 2010;7(4):365–70.
7. Mohammadi G. An investigation of community response to urban traffic noise. Iran J Environ Health Sci Eng. 2009;6(2):137–42.
8. The European Community Green Paper on Future Noise Policy 1996.
9. Berglund B, Institutet K. Proceedings of the 5th International Congress on Noise as a Public Health Problem: 1990; Stockholm, Sweden: Swedish Council for Building Research: Distribution Svensk Byggtjänst. 1990.
10. Van Kempen E, van Kamp I, Fischer P, Davies H, Houthuijs D, Stellato R. Noise exposure and children's blood pressure and heart rate. Occup Environ Med. 2006;63:632–9.
11. Babisch W. Transportation noise and cardiovascular risk: updated review and synthesis of epidemiological studies indicate that the evidence has increased. Noise Health. 2006;8:1–24.
12. Leon Bluhm G, Berglind N, Nordling E, Rosenlund M. Road traffic noise and hypertension. Occup Environ Med. 2007;64:122–6.
13. De Kluizenaar Y, Gansevoort RT, Miedema HM, de Jong PE. Hypertension and road traffic noise exposure. J Occup Environ Med. 2007;49:484–92.
14. Eriksson C, Rosenlund M, Pershagen G, Hilding A, Ostenson CG, Bluhm G. Aircraft noise and incidence of hypertension. Epidemiology. 2007;18:716–21.
15. Eriksson C, Bluhm G, Hilding A, Ostenson CG, Pershagen G. Aircraft noise and incidence of hypertension – gender specific effects. Environ Res. 2010;110:764–72.
16. Belojević GA, Jakovljević BD, Stojanov VJ, Slepcević VZ, Paunović KZ. Night time road-traffic noise and arterial hypertension in an urban population. Hypertens Res. 2008;31:775–81.
17. Jarup L, Babisch W, Houthuijs D, Pershagen G, Katsouyanni K, Cadum E. Hypertension and exposure to noise near airports: the HYENA study. Environ Health Perspect. 2008;116:329–33.
18. Bodin T, Albin M, Ardö J, Stroh E, Ostergren PO, Björk J. Road traffic noise and hypertension: results from a cross-sectional public health survey in southern Sweden. Environ Health. 2009;8:38.
19. Barregard L, Bonde E, Ohrström E. Risk of hypertension from exposure to road traffic noise in a population-based sample. Occup Environ Med. 2009;66:410–5.
20. Barregard L. Traffic noise and hypertension. Environ Res. 2011;111:186–7.
21. Bluhm G, Eriksson C. Cardiovascular effects of environmental noise: research in Sweden. Noise Health. 2011;13:212–6.
22. Lercher P, Botteldooren D, Widmann U, Uhrner U, Kammeringer E. Cardiovascular effects of environmental noise: research in Austria. Noise Health. 2011;13:234–50.
23. Passchier-Vermeer W, Passchier WF. Noise exposure and public health. Environ Health Perspect. 2000;108(1):123–31.
24. Babisch W. Cardiovascular effects of noise. In: Nriagu JO, editor. Encyclopedia of Environmental Health. Burlington: Elsevier; 2011. p. 532–42.
25. Akan Z, Yilmaz A, Özdemir O, Selvi Y, Korpinar MA. Noise pollution, psychiatric symptomms and quality of life: Noise problem in the east region of Turkey. Eur Psychiatry. 2012;27(1):436.
26. Gümgüm HB. Noise pollution in the Tigris river basin provinces (Turkey). J Environ. 2012;01(04):142–5.
27. Lercher P, Widdmann U. Association and moderation of self-reported hypotension with traffic noise exposure: a neglected relationship. Noise Health. 2013;15(65):205–16.

Laccase-catalyzed decolorization and detoxification of Acid Blue 92: statistical optimization, microtoxicity, kinetics, and energetics

Shahla Rezaei[1], Hamed Tahmasbi[1], Mehdi Mogharabi[1,2], Alieh Ameri[3], Hamid Forootanfar[4], Mohammad Reza Khoshayand[5] and Mohammad Ali Faramarzi[1,2*]

Abstract

Background: In recent years, enzymatic-assisted removal of hazardous dyes has been considered as an alternative and eco-friendly method compared to those of physicochemical techniques. The present study was designed in order to obtain the optimal condition for laccase-mediated (purified from the ascomycete *Paraconiothyrium variabile*) decolorization of Acid Blue 92; a monoazo dye, using response surface methodology (RSM). So, a D-optimal design with three variables, including pH, enzyme activity, and dye concentration, was applied to optimize the decolorization process. In addition, the kinetic and energetic parameters of the above mentioned enzymatic removal of Acid Blue 92 was investigated.

Results: Decolorization of Acid Blue 92 was maximally (94.1% ± 2.61) occurred at pH 8.0, laccase activity of 2.5 U/mL, and dye concentration of 75 mg/mL. The obtained results of kinetic and energetic studies introduced the laccase-catalyzed decolorization of Acid Blue 92 as an endothermic reaction (Ea, 39 kJ/mol; ΔS, 131 J/mol K; and ΔH, 40 kJ/mol) with K_m and V_{max} values of 0.48 mM and 227 mM/min mg, respectively. Furthermore, the results of microtoxicity study revealed that the toxicity of laccase-treated dye was significantly reduced compared to the untreated dye.

Conclusions: To sum up, the present investigation introduced the *Paraconiothyrium variabile* laccase as an efficient biocatalyst for decolorization of synthetic dye Acid Blue 92.

Keywords: Enzyme Biocatalysis, Optimization, Waste Treatment, Bioremediation, Laccase, Decolorization

Introduction

The wide usage of synthetic dyes in industries such as textile, paper, plastics, printing, leather, cosmetics, and pharmaceuticals has led to releasing dye containing effluents, rich in complex aromatic structures into the environment [1,2]. Almost all synthetic colorants especially the azo dyes – the most common dye group in textile dyeing processes – and/or their degradation products have been reported to be toxic, mutagenic, and carcinogenic [3]. Resistance of the environmentally hazardous dyes to light, biological treatment procedures, ozone, or other degradative environmental procedures is the main problem in the elimination of dyes discharged from wastewaters [4,5]. So, development of efficient and economical processes for treatment of the synthetic dyes still remains as a major challenge [1]. Among various physicochemical and biotechnological techniques, the enzymatic removal of synthetic dyes is the most preferred method due to its simplicity, efficiency at high and low pollutant concentration over a wide range of pH and temperature, low energy required, minimal impact on ecosystem, and less sludge production in the decolorization process [6-9].

Laccases (benzenediol:oxygen oxidoreductase, EC 1.10.3.2) are multi-copper containing oxidase mainly found in fungi, plants, and some bacterial strains [5,10]. The ability of laccases to oxidize a broad range of aromatic compounds such as benzenethiols, substituted phenols,

* Correspondence: faramarz@tums.ac.ir
[1]Department of Pharmaceutical Biotechnology, Faculty of Pharmacy and Biotechnology Research Center, Tehran University of Medical Sciences, P.O. Box 14155–6451, Tehran 1417614411, Iran
[2]Pharmaceutical Sciences Research Center, Tehran University of Medical Sciences, Tehran 1417614411, Iran
Full list of author information is available at the end of the article

and polyaromatic hydrocarbons (PAHs) in the presence of molecular oxygen as a co-substrate introduces this biotechnologically important enzyme as the first choice for xenobiotic removal experiments [11,12]. Decolorization and detoxification of synthetic dyes assisted by laccase and laccase-mediated system (LMS) have received great attention during two last decades [13,14]. However, high cost of enzymatic removal (due to low production yield) limits extensive application of laccases in xenobiotic elimination [15,16]. This constrain could be overcome via optimization of removal reaction conditions using statistical approaches [17,18].

The response surface methodology (RSM) which is utilized broadly in biotechnological processes, involves a collection of useful statistical and mathematical techniques for analyzing the causal relationship between independent variables, responses, and their interactions through the construction of polynomial mathematical models which leads to time and cost saving [19]. Several studies have been recently published on the potential applications of RSM in the enzymatic decolorization of reactive dyes such as Reactive Black 5, Reactive Red 239, Reactive Yellow 15, and Reactive Blue 114 [16,17].

In the present study, a D-optimal model for RSM as a very useful design method was applied to obtain maximal removal of Acid Blue 92 assisted by laccase. Furthermore, the kinetic and thermodynamic parameters of laccase-mediated dye removal reaction were investigated. The microtoxicity experiments were also performed in order to evaluate the toxicity of untreated and laccase-treated dye.

Materials and methods

Chemicals

2,2'-Azino-bis(3-ethylbenzthiazoline-6-sulphonate) (ABTS) was purchased from Sigma-Aldrich (St. Louis, MO, USA). Acid Blue 92 (Figure 1) was kindly donated by Alvan Sabet Co. (Tehran, Iran). All other chemicals and reagents were of the highest purity available. The extracellular laccase of *Paraconiothyrium variabile* (*Pv*L) was purified using the method previously described by Forootanfar et al. [20].

Laccase assay

Laccase activity was determined using ABTS as the substrate [21]. The reaction mixture was prepared by adding 0.5 mL ABTS (5 mM) dissolved in 100 mM citrate buffer (pH 4.5) and 0.5 mL of enzyme solution followed by incubation at 40°C and 120 rpm. Oxidation of ABTS was monitored by increase in absorbance at 420 nm (ε_{420} = 36000 M^{-1} cm^{-1}) using a UV/vis Spectrophotometer (UVD 2950, Labomed, Culver City, USA). One unit of laccase activity was defined as amount of the enzyme required to oxidize 1 μmol of ABTS per min [22].

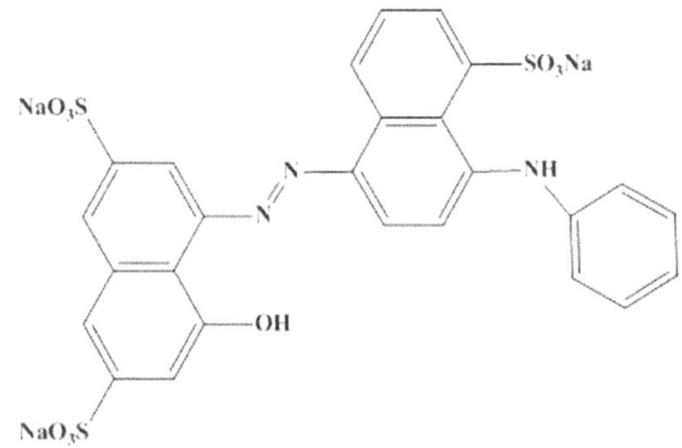

Figure 1 Chemical structure of Acid Blue 92.

Dye decolorization experiments

After preparation of dye solution (concentration range of 100–200 mg/L) in citrate-phosphate buffer (0.1 M, pH range of 3.0 – 8.0), the purified *Pv*L (final activity of 1–2.5 U/L) was added to the reaction mixture and incubated at desired temperature (30–70°C) for 30–90 min followed by measuring the absorbance of the taken samples using a UV/visible spectrophotometer at maximum absorbance of applied dye (571 nm). Decolorization percentage was then calculated using the following equation: decolorization (%) = $[A_i - A_t/A_i] \times 100$; where A_i is the initial absorbance of the reaction mixture and A_t is the absorbance after incubation time [1]. The negative control (reaction mixture containing the heat-inactivated enzyme) was prepared and incubated at the same conditions. All experiments were performed in triplicate and the means of decolorization percentages were reported.

Experimental design and statistical analysis

Screening study

The screening experiment was designed based on the fractional factorial design method, which is a definite part of full factorial design matrix with two-level factor variations containing 2^{k-p} runs ($1/2^p$ fraction of the 2^k design); where k and p are the number of independent variables and size of the fraction, respectively. Different factors including pH (A), temperature (B), enzyme activity (C), dye concentration (D), and incubation time (E) were implemented using 2^{5-1} fractional factorial design with resolution V. The design matrix was built using the statistical software package, Design-Expert (version 7.0.0; Stat-Ease, Inc., Minneapolis, Minnesota, USA). Factors and corresponding response presented in Table 1. All of the experiments were accomplished in triplicate and the averages considered as responses.

Optimization study

The D-optimal design developed to select the design points in a method that minimizes the variance associated

Table 1 Level of independent variables in fractional factorial design

Variables	Symbol	Unit	Low level (−1)	High level (+1)
pH	A	-	3	8
Temperature	B	°C	30	70
Enzyme	C	U/mL	1	2.5
Dye	D	mg/L	100	200
Incubation time	E	min	30	90

with the estimates of coefficient in a specified model. The number of runs in a D-optimal design are less and do not rise as fast as the classical design with an increasing number of factors. The D-optimal design try to minimize the determinant of the $(X'X)^{-1}$ matrix which lead to minimize the volume of the confidence for the coefficients and maximizes the determinant 'information' matrix (X'X); where "X" defined as a matrix containing the designed points generated by the computer to fulfill the D-optimally. The D-optimal designs, as an optimization method based on a chosen optimality criterion, used the determinant of the information matrix X'X. Maximizing the determinant of the information matrix (X'X) lead to minimizing the determinant of the matrix $(X'X)^{-1}$ which beneficially keep the later calculations as short as possible. Finally, the Fisher's test with P-values below 0.05 employed to evaluate the statistical significance of the effect of independent variables on the response using analysis of variance (ANOVA). While the multiple correlation coefficient (R^2) and adjusted R^2 used as quality indicators for the fit of second-order polynomial model equation, contour plots, and three-dimensional surface plots were used to graphically show the relationship and interactions between the coded variables and responses. The optimal condition for enzymatic decolorization process determined by solving the equation derived from the final quadratic model as well as grid search of the three-dimensional surface plots.

Dye removal kinetics and energetics

Kinetics of decolorization

After performing of decolorization reaction in the presence of the dye concentrations (10–500 μM) at optimal pH and temperature, the velocity for different concentrations of dye was determined. Michaelis-Menten curve was then drawn by plotting the obtained initial velocity against dye concentrations. Calculation of K_m and V_{max} values were performed by fitting the data to the Lineweaver-Burk plot, resulting of the Michaelis-Menten plot conversion [1,2].

Thermodynamics of decolorization

In order to evaluate the effect of temperature on removal of Acid Blue 92, decolorization experiment was performed at temperature range of 10–50°C and the obtained velocities (for each temperature) were plotted against initial dye concentrations. Thereafter, apparent first-order rate constant (K) for each temperature were calculated from the slope of each straight plot. The value of activation energy (E_a) (kJ/mol) was then estimated from the slope of the linearized Arrhenius curve achieved by drawing the ln (K) versus 1/T ($\times 10^3$ K^{-1}): slope of Arrhenius plot = $(-E_a/R)$; where R is the gas constant (8.3145 J/mol K), and T is the absolute temperature (K). The enthalpy (ΔH) and entropy (ΔS) of decolorization reaction were estimated by using of Van't Hoff plot which was drawn by plotting the ln (K_{eq}) against 1/T ($\times 10^3$ K^{-1}). K_{eq} is the apparent equilibrium constant and was calculated from difference of initial dye concentration and residual dye concentration at equilibrium state, when the decolorization percentage become constant and remained dye have no longer changes with passing the time [23]. Finally, the Gibbs free energy (ΔG) was calculated using the equation of $\Delta G = \Delta H - T\Delta S$.

Microtoxicity studies

In order to evaluate the toxicity of produced metabolite(s) following laccase treatment, a series of microtoxicity study was performed. Firstly, a preculture of three Gram-positive bacterial strains (*Micrococcus luteus* ATCC 10240, *Staphylococcus aureus* ATCC 6538, and *Bacillus subtilis* ATCC 6633) and three Gram-negative bacterial strains (*E. coli* ATCC 25922, *Pseudomonas aeruginosa* ATCC 9027, and *Salmonella typhi* ATCC 19430) was prepared by incubating of each bacterial strain in Muller-Hinton broth at 37°C and 150 rpm to reach the OD_{600} of 0.2. Consequently, the untreated dye solution (final concentration of 100 mg/L) and the sample obtained from enzymatic treatment of applied dye (performed at the optimized condition) was separately added to the prepared bacterial culture media and incubated at 37°C. Changes in the OD_{600} of each bacterial strain were then recorded every 2 h for 10 h. A negative control (cultivated bacterial strain in the absence of dye) was designed for each experiment. The percentage of growth inhibition (GI%) was defined as $[(1 - D_{600S}/OD_{600C}) \times 100]$; where OD_{600S} is the OD_{600} of sample and OD_{600C} is the OD_{600} of control. All experiments were performed in triplicate and mean of the obtained results was reported.

Results and discussion

Fractional factorial design for screening of important variables

In a preliminary study, fractional factorial design was used to evaluate the influence of five factors including pH (A), temperature (B), the enzyme activity (C), dye concentration (D), and incubation time (E) on the decolorization yield in order to select the most effective variables on laccase-catalyzed decolorization of Acid Blue 92. The respective high (+1) and low (−1) levels for each coded factor defined in Table 1. A total of 20 runs and the related responses with combination of five mentioned variables were presented in Table 2. The half-normal probability plot employed as a graphical tool for estimation the effect of each factor alone and in combination with other factors on decolorization percentage is illustrated in Figure 2. The large distance of pH (A), enzyme activity (C), and dye concentration (D) from zero in half-normal plot, indicated that these factors significantly affected the decolorization process (Figure 2). In addition, considerable interaction between enzyme activity (C) and dye concentration (D) emerged from half-normal probability plot (Figure 2). In the study of Khouni et al. [24], who applied central composite design (CCD) matrix and RSM for optimization of laccase-mediated decolorization of Black Novacron R and Blue Bezaktiv S-GLD 150, two variables including pH and temperature positively affected decolorization process while the third parameter (laccase activity) did not influence dye removal. Daassi et al. [25] reported the significant effect of four factors (laccase activity,1-hydroxybenzotriazole concentration, dye concentration, and reaction time) on decolorization of Acid Orange 51 assisted by crude laccase from *Trametes trogii* using a three-level Box-Behnken factorial design combined with response surface methodology to reach 87.87 ± 1.27 decolorization percent.

D-optimal design response surface methodology

Based on the results of the fractional factorial design, three variables, including pH (A), enzyme activity (C), and dye concentration (D) were found to have a greater influence on the decolorization of Acid Blue 92 assisted by laccase. Therefore, the D-optimal RSM exploited to optimize the affecting decolorization factors. The responses generated from performing a total of 20 runs are presented at Table 3. Table 4 demonstrates the quadratic model as the most suitable model for the enzymatic decolorization of Acid Blue 92. The analysis of variance for the quadratic model is shown in Table 5. The model F-value (16.08) shows the significance of the model. In addition, Table 5 indicated pH (A), dye concentration (D), and their interactions; AC, CD, A^2 and, C^2 as significant model terms (P-value < 0.05). The quadratic model explained the statistical relationship between the selected variables and the response in terms of coded factors was best fitted with the following equation.

$$\begin{aligned} Y = {} & 18.27 + 3.99A + 21.55C - 13.5D \\ & + 8.59\,AC - 5.16\,AD - 12.37CD - 6.17A^2 \\ & + 11.42C^2 + 1.56D^2 \end{aligned} \tag{1}$$

where Y is the response (yield of decolorization) and A, C, and D are pH, enzyme activity, and dye concentration, respectively. In order to investigate the relation

Table 2 Fractional factorial design matrix and their observed responses for laccase-assisted decolorization of Acid Blue 92

Run no.	pH	Temperature (°C)	Enzyme activity (U/mL)	Dye concentration (mg/L)	Incubation time (min)	Response (decolorization%)
1	5.5	50	1.00	100	60	13.62
2	5.5	50	1.00	100	60	16.60
3	8.0	30	2.50	50	90	16.28
4	8.0	70	2.50	200	90	4.64
5	8.0	30	0.25	200	90	6.94
6	3.0	70	2.50	200	30	10.55
7	3.0	30	2.50	200	90	6.65
8	3.0	30	2.50	50	30	92.66
9	8.0	30	2.50	200	30	5.27
10	3.0	70	0.25	50	30	22.90
11	8.0	30	0.25	50	30	13.31
12	3.0	70	2.50	50	90	92.45
13	3.0	30	0.25	50	90	14.57
14	3.0	70	0.25	200	90	5.89
15	3.0	30	0.25	200	30	7.87
16	8.0	70	0.25	50	90	3.81
17	8.0	70	2.50	50	30	32.40
18	8.0	70	0.25	200	30	3.65
19	5.5	50	1.00	100	60	22.25

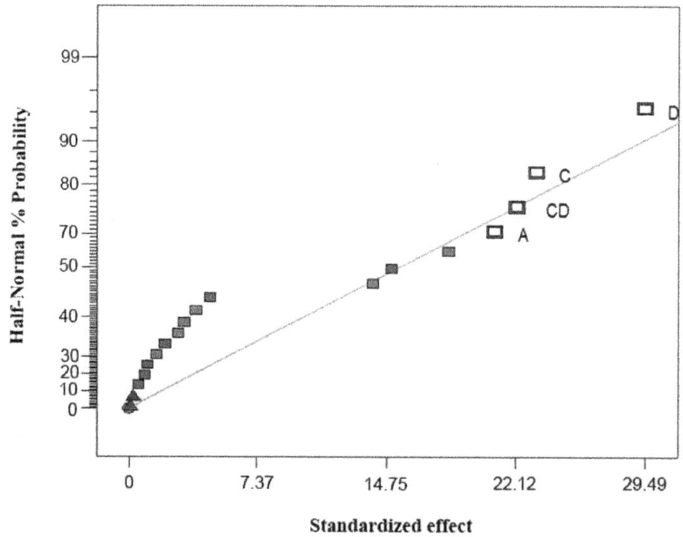

Figure 2 Half-normal probability plot for statistical analysis of fractional factorial design. A: pH; C: Enzyme; D: Dye.

between the independent variables and the responses, contour plots generated by RSM using the Design-Expert software. The response surface plots (Figure 3) present the decolorization of Acid Blue 92 as function of two variables, while the third one kept at a constant level. Surface plots demonstrate an increase in dye decolorization with raising the pH value. Figure 3a and Figure 3c clearly show that the decolorization percentage influenced by small alterations of the enzyme activity.

Table 3 D-optimal design matrix containing various conditions and related responses

Run no.	pH	Enzyme activity (U/mL)	Dye concentration (mg/L)	Response (decolorization%)
1	8.00	0.75	75.00	4.4
2	1.30	1.63	137.50	2.1
3	5.50	3.10	137.50	99.3
4	3.00	0.75	200.00	5.3
5	5.50	1.63	137.50	19.2
6	5.50	1.63	137.50	14.4
7	3.00	0.75	75.00	11.8
8	5.50	1.63	137.50	15.5
9	3.00	2.50	200.00	10.3
10	3.00	2.50	75.00	42.2
11	5.50	1.63	32.39	58.1
12	5.50	1.63	137.50	16.3
13	5.50	1.63	137.50	17.3
14	8.00	2.50	200.00	14.4
15	9.70	3.25	137.50	6.1
16	8.00	2.50	75.00	92.1
17	5.50	1.63	242.61	4.3
18	5.50	1.63	137.50	18.7
19	8.00	0.75	200.00	0.8
20	5.50	0.15	137.50	4.4

Table 4 Sequential model sum of squares for D-optimal design

Source	Sum of squares	D_f	Mean squares	F-value	P-value
Mean vs Total	10503.78	1	10503.78	-	-
Linear vs Mean	9046.07	3	3015.36	8.54	0.0013
2FI vs Linear	2028.02	3	676.01	2.43	0.1121
Quadratic vs 2FI	2671.26	3	890.42	9.38	0.003
Cubic vs Quadratic	820.89	4	205.22	9.57	0.0089
Residual	128.66	6	21.44	-	-
Total	25198.68	20	1259.93	-	-

Optimization

By solving the equation (1) and analyzing three dimensional response surface graphs (Figure 3), the optimal levels for pH, enzyme activity, and dye concentration were found to be 8.0, 2.5 U/L, and 75 mg/L, respectively. The yield of laccase-catalyzed decolorization of Acid Blue 92 in optimal condition was 94.1% ± 2.61. The obtained results introduced pH as the most relevant factor for the enzymatic decolorization of Acid Blue 92. Low yield of decolorization achieved at acidic pH, while pH 8.0 was the optimal pH where maximum decolorization occurred. In general, most of fungal laccases optimally work at acidic pH levels [26]. For example, the crude laccase of *Ganoderma lucidum* preferred the acidic range for dye elimination and its decolorization activity decreased significantly at pH levels above 6.0 [16]. However, Tavares et al. [17] reported that more than 80% decolorization of Reactive Blue 114 occurred at pH 7.0–7.5. The maximum removal of Azure B using the purified laccase of *Trametes trogii* BAFC 463 mediated by 1-hydroxybenzotriazole was achieved at pH 7.0 [18]. In contrast, the laccases originated from bacterial strains maximally work at alkaline pH values. For instance, maximum decolorization activity of the purified laccase of *Bacillus pumilus* strain W3 was achieved at pH 9.0 [27]. The effect of the enzyme activity and dye concentration on decolorization of Acid Blue 92 at pH 8.0 (Figure 2c) revealed that decolorization decreased with increasing dye concentration, but increased with raising the activity of laccase. The results of the present study showed that more than 75% of the decolorization obtained at dye concentration of 75 mg/L and laccase activity of 2.5 U/L. Roriz et al. [15] reported that increasing of laccase activity (the crude laccase from *Trametes pubescens*) positively affected the decolorization of Reactive Black 5. Ashrafi et al. [1] demonstrated significant increase in removal of thirteen synthetic dyes when laccase activity increased in the range of 0.025–0.1 U/L.

Table 5 Analysis of variance for D-optimal design

Source	Sum of squares	D_f	Mean squares	F-value	P-value
Model	13745.35	9	1527.26	16.08	<0.0001
A-A	217.65	1	217.65	2.29	0.1610
C-C	6340.77	1	6340.77	66.78	<0.0001
D-D	2487.64	1	2487.64	26.20	0.0005
AC	590.82	1	590.82	6.22	0.0317
AD	213.31	1	213.31	2.25	0.1648
CD	1223.89	1	1223.89	12.89	0.0049
A^2	548.32	1	548.32	5.77	0.0371
C^2	1878.05	1	1878.05	19.78	0.0012
D^2	34.93	1	34.93	0.37	0.5577
Residual	949.55	10	94.95	-	-
Lack of fit	883.05	5	176.61	13.28	0.0065
Pure error	66.5	5	13.3	-	-
Cor total	14694.9	19	-	-	-

Validation of model

In order to determine the correctness of the model, five verification experiments were performed using the statistically optimal condition. The results showed maximum decolorization efficiency of 90.2%, which is about 99% of the predicted value, implying a strong similarity between experimental and predicted values calculated from the model that confirms the precision and validity of the model.

Kinetics and energetics of decolorization

Kinetic study

The linear relationship between the initial velocity and dye concentration (Figure 4) indicated that the decolorization is a first order reaction. Based on the Michaelis-Menten and Lineweaver-Burk plots for enzymatic decolorization of Acid Blue 92 (Figure 4a and Figure 4b), the K_m and V_{max} values were 0.48 mM and 227 mM/min mg, respectively. In the study of Ashrafi et al. [1] who investigated on the ability of laccase for decolorization of a number of synthetic dyes, the lowest K_m (14.8 μM) found for the anthraquinone-based dye of Disperse Blue 56 followed by the monoazo dye of Acid Red 18 and the triazo dye of Direct Blue 71 with K_m value of 25 μM and 800 μM, respectively. In general, the azo dyes that are classified

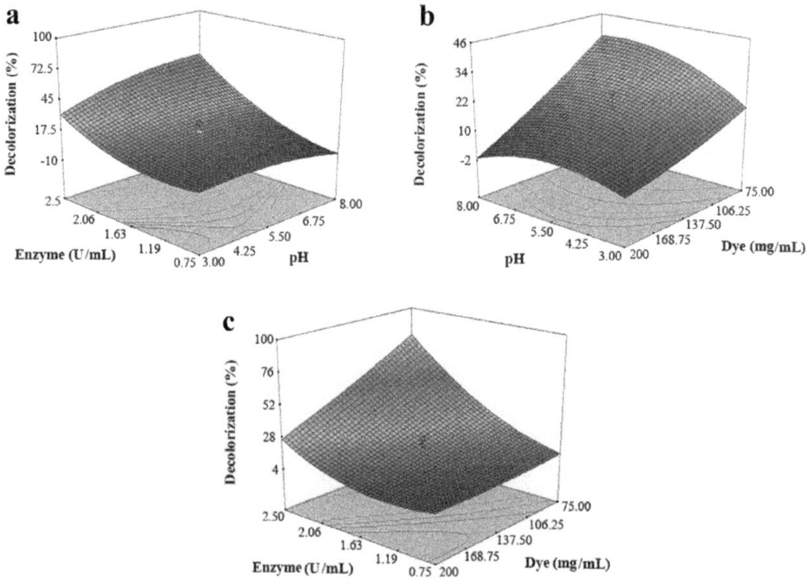

Figure 3 Response surface plot indicating the effects of interactions between **a)** pH and the enzyme activity, **b)** pH and the dye concentration, **c)** the enzyme activity and the dye concentration.

as monoazo, diazo, and triazo, have been identified as recalcitrant colorants for oxidation using laccases compared to that of anthraquinone-based dyes [1,3].

Energetic study

Energetic results are illustrated in Figure 5. A significant effect of temperature on decolorization of Acid Blue 92 was observed, as decolorization rate increased from 10 to 50°C. The activation energy (obtained from the slope of Arrhenius plot, Figure 5a) was 39 kJ/mol, which is typical for the decolorization reaction. The calculated values for ΔH and ΔS were 40 kJ/mol and 131 J/mol K, respectively. The positive sign of ΔH (Figure 5b) or the negative slope of van't Hoff plot (Figure 5b) implies that the reaction is endothermic.

Dye toxicity

The obtained results of toxicity evaluation of untreated and laccase-treated dye solution showed that when Acid Blue 92 was used, the calculated GI% in the presence of *M. luteus*, *S. aureus*, *B. subtilis*, *E. coli*, *P. aeruginosa*, and *S. typhi* was found to be 60 ± 1.9, 49 ± 1.4, 51 ± 1.3, 42 ± 1.1, 30 ± 1.1, and 37 ± 0.8, respectively. However, the GI% for laccase treated dye solution was 31 ± 3.9, 34 ± 2.8, 27 ± 2.5, 20 ± 2.3, 17 ± 0.4, and 14 ± 0.9 in in the presence of *M. luteus*, *S. aureus*, *B. subtilis*, *E. coli*, *P. aeruginosa*, and *S. typhi*, respectively, which was significantly

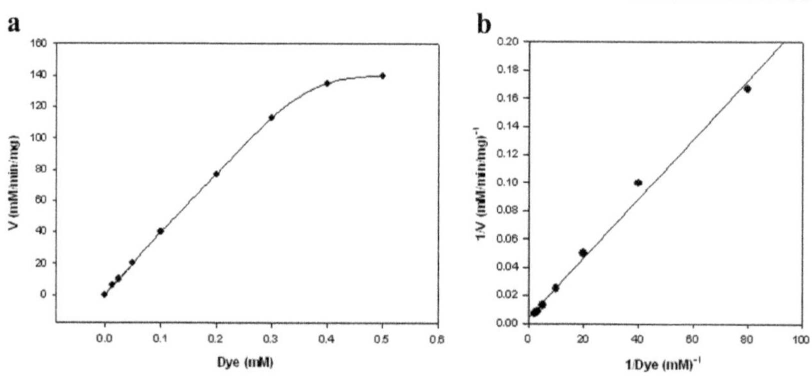

Figure 4 Kinetic study. **a)** Michaelis-Menten plot, **b)** Lineweaver-Burk plot.

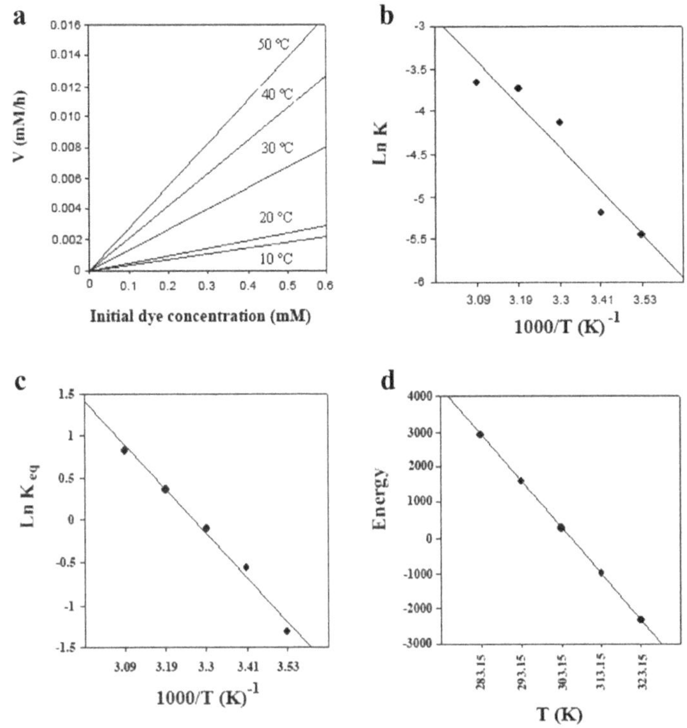

Figure 5 Energetic Study. **a)** dependence of decolorization rate on temperature (10–50°C), **b)** Arrhenius plot, **c)** van't Hoff plot and, **d)** Gibbs free energy changes plot.

lower for all bacterial strains. Due to the probable toxicity of produced metabolite following physicochemical or enzymatic treatment of synthetic dyes, many protocols have been developed for determination the toxicity of dyes and related compounds based on the growth inhibition of bacterial (*B. megaterium* and *E. coli*), yeast (*Saccaromyces cerevisiae*), or animal cells (human cervix cells; HeLa), and phytotoxicity studies on plant seeds (*Oryza sativa* or *Triticum aestivum*) [1,3]. In the present work, the toxicity of laccase-treated dye sample was significantly reduced compared to that of parent dye. Same results was reported by Palmieri et al. [28] who determined 95% viability for *B. cereus* in the presence of sample obtained from enzymatic elimination of Remazol Brilliant Blue R (an anthraquinone dye) using the purified laccase of *Pleurotus ostreatus*. The growth of *B. megaterium* and *E. coli* were reported to be 99% and 94%, respectively, in the presence of laccase-treated Malachite Green (a triphenylmethane dye). In general, azo dyes have been shown to be more resistant to enzymatic removal compared to other classes of synthetic dyes (like anthraquinone dyes) and provide higher toxicity after enzymatic removal [3,29]. However, Champagne and Ramsay [30] reported lower toxicity for azo dyes (Acid red 27 and Reactive black 5) compared to anthraquinone dyes (Reactive blue 19 and Dispersed blue 3) after treatment using free and immobilized laccases.

Conclusion

Response surface methodology is an important tool to optimize the conditions for textile dye wastewater treatment. Such statistical approach reduces the number of runs and provides valuable information on possible interactions between the variables and response. The present study was designed to optimize the reaction condition for laccase- catalyzed decolorization of Acid Blue 92. The results indicated pH, enzyme activity, and dye concentration as the most important variables in the enzymatic decolorization of Acid Blue 92. Further studies should be performed on the decolorization of dyes from different families such as triphenylmethan, indigo, and anthraquinone dyes, extensively employing in the textile and chemical industries.

Competing interests
The authors declare that they have no competing interests.

Author's information
Correspondence for statistical experimental design; Mohammad Reza Khoshayand.

Authors' contributions

SR and HT carried out the decolorization studies of Acid Blue 92. Production and purification of laccase from *P. variabile* culture and detoxification studies was performed by HF and AA. MM contributed in writing of the manuscript. MRK involved in design of removal experiments, analyzing of data and reviewing of the manuscript. MAF involved in purchasing of required materials and instruments, designing of removal experiments, analyzing of data and reviewing of the manuscript. All authors read and approved the final manuscript.

Acknowledgement

This work was supported financially by the grant No. 92-03-33-24540 from Tehran University of Medical Sciences, Tehran, Iran to M.A.F.

Author details

[1]Department of Pharmaceutical Biotechnology, Faculty of Pharmacy and Biotechnology Research Center, Tehran University of Medical Sciences, P.O. Box 14155–6451, Tehran 1417614411, Iran. [2]Pharmaceutical Sciences Research Center, Tehran University of Medical Sciences, Tehran 1417614411, Iran. [3]Department of Medicinal Chemistry, Faculty of Pharmacy, Kerman University of Medical Sciences, Kerman, Iran. [4]Department of Pharmaceutical Biotechnology, Faculty of Pharmacy, Kerman University of Medical Sciences, Kerman, Iran. [5]Department of Drug and Food Control, Faculty of Pharmacy and Pharmaceuticals Quality Assurance Research Center, Tehran University of Medical Sciences, Tehran 1417614411, Iran.

References

1. Ashrafi SD, Rezaei S, Forootanfar H, Mahvi AH, Faramarzi MA. The enzymatic decolorization and detoxification of synthetic dyes by the laccase from a soil-isolated ascomycete, *Paraconiothyrium variabile*. Int Biodeter Biodegr. 2013;85:173–81.
2. Mirzadeh SS, Khezri SM, Rezaei S, Forootanfar H, Mahvi AH, Faramarzi MA. Decolorization of two synthetic dyes using the purified laccase of *Paraconiothyrium variabile* immobilized on porous silica beads. J Environ Health Sci Eng. 2014;12:9.
3. Saratale RG, Saratale GD, Chang JS, Govindwar SP. Bacterial decolorization and degradation of azo dyes: a review. J Taiwan Ins Chem Eng. 2011;42:138–57.
4. Khlifi R, Belbahri L, Woodward S, Ellouz M, Dhouib A, Sayadi S. Decolourization and detoxification of textile industry wastewater by the laccase-mediator system. J Hazard Mater. 2010;175:802–8.
5. Aghaie-Khouzani M, Forootanfar H, Moshfegh M, Khoshayand MR, Faramarzi MA. Decolourization of some synthetic dyes using optimized culture broth of laccase producing ascomycete *Paraconiothyrium variabile*. Biochem Eng J. 2012;60:9–15.
6. Gholami-Borujeni F, Mahvi AH, Nasseri S, Faramarzi MA, Nabizadeh R, Alimohammadi M. Enzymatic treatment and detoxification of acid orange 7 from textile wastewater. Appl Biochem Biotechnol. 2011;165:1274–84.
7. Gholami-Borujeni F, Faramarzi MA, Nejatzadeh-Barandozi F, Mahvi AH. Oxidative degradation and detoxification of textile azo dye by horseradish peroxidase enzyme. Fresenius Environ Bull. 2013;22:739–44.
8. Gholami-Borujeni F, Mahvi AH, Naseri S, Faramarzi MA, Nabizadeh R, Alimohammadi M. Application of immobilized horseradish peroxidase for removal and detoxification of azo dye from aqueous solution. Res J Chem Environ. 2011;15:217–22.
9. Asadgol Z, Forootanfar H, Rezaei S, Mahvi AH, Faramarzi MA. Removal of phenol and bisphenol-A catalyzed by laccase in aqueous solution. J Environ Health Sci Eng. 2014;12:93.
10. Mogharabi M, Faramarzi MA. Laccase and laccase-mediated systems in the synthesis of organic compounds. Adv Synt Catal. 2014;356:897–927.
11. Forootanfar H, Movahednia MM, Yaghmaei S, Tabatabaei-Sameni M, Rastegar H, Sadighi A, et al. Removal of chlorophenolic derivatives by soil isolated ascomycete of *Paraconiothyrium variabile* and studying the role of its extracellular laccase. J Hazard Mater. 2012;209-210:199–203.
12. Ostadhadi-Dehkordi S, Tabatabaei-Sameni M, Forootanfar H, Kolahdouz S, Ghazi-Khansari M, Faramarzi MA. Degradation of some benzodiazepines by a laccase-mediated system in aqueous solution. Bioresour Technol. 2012;125:344–7.
13. Mogharabi M, Nassiri-Koopaei N, Bozorgi-Koushalshahi M, Nafissi-Varcheh N, Bagherzadeh G, Faramarzi MA. Immobilization of laccase in alginate-gelatin mixed gel and decolorization of synthetic dyes. Bioinorg Chem Appl. 2012;2012:6.
14. Jiang M, Ten Z, Ding S. Decolorization of synthetic dyes by crude and purified laccases from *Coprinus comatus* grown under different cultures: The role of major isoenzyme in dyes decolorization. Appl Biochem Biotechnol. 2013;169:660–72.
15. Roriz MS, Osma JF, Teixeira JA, Rodríguez-Couto S. Application of response surface methodological approach to optimise Reactive Black 5 decolouration by crude laccase from *Trametes pubescens*. J Hazard Mater. 2009;169:691–6.
16. Murugesan K, Dhamija A, Nam IH, Kim YM, Chang YS. Decolourization of reactive black 5 by laccase: optimization by response surface methodology. Dyes Pigm. 2007;75:176–84.
17. Tavares APM, Cristóvão RO, Loureiro JM, Boaventura RAR, Macedo EA. Application of statistical experimental methodology to optimize reactive dye decolourization by commercial laccase. J Hazard Mater. 2009;162:1255–60.
18. Grassi E, Scodeller P, Filiel N, Carballo R, Levin L. Potential of *Trametes trogii* culture fluids and its purified laccase for the decolorization of different types of recalcitrant dyes without the addition of redox mediators. Int Biodeter Biodegr. 2011;65:635–43.
19. Myers RH, Montgomery D. Response surface methodology: Process and product optimization using designed experiments. 2nd edition, Wiley, 2002.
20. Forootanfar H, Faramarzi MA, Shahverdi AR, Tabatabaei Yazdi M. Purification and biochemical characterization of extracellular laccase from the ascomycete *Paraconiothyrium variabile*. Bioresour Technol. 2011;102:1808–14.
21. Faramarzi MA, Forootanfar H. Biosynthesis and characterization of goldnanoparticles produced by laccase from *Paraconiothyrium variabile*. Colloids Surf, B. 2011;87:23–7.
22. Alberts JF, Gelderblom WCA, Botha A, Vanzyl WH. Degradation of aflatoxin B1 by fungal laccase enzymes. Int J Food Microbiol. 2009;135:47–52.
23. Annuar MSM, Adnan S, Vikineswary S, Chisti Y. Kinetics and energetics of azo dye decolorization by *Pycnoporus sanguineus*. Water Air Soil Pollut. 2009;202:179–88.
24. Khouni I, Marrot B, Amar RB. Decolourization of the reconstituted dye bath effluent by commercial laccase treatment: Optimization through response surface methodology. Chem Eng J. 2010;156:121–33.
25. Daassi D, Zouari-Mechichi H, Frikha F, Martinez JM, Nasri M, Mechichi T. Decolorization of the azo dye Acid Orange 51 by laccase produced in solid culture of a newly isolated *Trametes trogii* strain. 3. Biotech. 2013;3:115–25.
26. Baldrian P. Fungal laccases – occurrence and properties. FEMS Microbiol Rev. 2006;30:215–42.
27. Guan ZB, Song CM, Zhang N, Zhou W, Xu CW, Zhou LX, et al. Overexpression, characterization, and dye-decolorizing ability of a thermostable, pH-stable, and organic solvent-tolerant laccase from *Bacillus pumilus* W3. J Mol Catal B: Enzym. 2013;101:1–6.
28. Palmieri G, Cennamo G, Sannia G. Remazol Brilliant Blue R decolourisation by the fungus *Pleurotus ostreatus* and its oxidative enzymatic system. Enzyme Microb Technol. 2005;36:17–24.
29. Couto SR. Decolouration of industrial azo dyes by crude laccase from *Trametes hirsute*. J Hazard Mater. 2007;148:768–70.
30. Champagne P-P, Ramsay JA. Dye decolorization and detoxification by laccase immobilized on porous glass beads. Bioresour Technol. 2010;101:2230–5.

Experimental design approach to the optimization of PAHs bioremediation from artificially contaminated soil: application of variables screening development

Masoumeh Ravanipour[1], Roshanak Rezaei Kalantary[2], Anoushiravan Mohseni-Bandpi[3*], Ali Esrafili[2], Mahdi Farzadkia[2] and Samireh Hashemi-Najafabadi[4]

Abstract

Background: The effectiveness of bioremediation systems for PAH-contaminated soil may be constrained by physicochemical properties of contaminants and environmental factors. Information on what is the most effective factor in bioremediation process is essential in the decision of what stimulations can be taken to assist the biodegradation efficacy.

Methods: In this study, four factors of surfactant (Tween 80), humic acid (HA), salinity and nutrients in a 2^4 full factorial design were screened in bioremediation of phenanthrene contaminated soil by using a consortium of bacteria.

Results: Between the employed levels of the factors only salinity had not significant effect. Optimal concentrations of surfactant, HA and nutrient were obtained by a response surface design. For phenanthrene biodegradation, a central composite face centred design (CCFD) showed that nutrient, surfactant and HA concentrations had highly significant, significant and insignificant effects, respectively. The best conditions with 87.1% phenanthrene biodegradation were 150 mg HA/Kg soil, 12.68 μg/L surfactant, and nutrients as K_2HPO_4, 0.8; KH_2PO_4, 0.2 and KNO_3, 1 g/L. A high similarity was between the model prediction and experimental results.

Conclusions: This study showed that nutrient with 81.27% efficiency could be considered as the most effective factor for practical implications of bioremediation process for PAHs contaminated soil cleanup strategies.

Keywords: Bioremediation, Polycyclic Aromatic Hydrocarbons, Response Surface Method, Nutrient, Tween80

Background

Petroleum derivatives are introduced into the environment through different ways such as anthropogenic activities, incomplete combustion of petroleum products, wood and coal, undesirable discharging of oil tankers, spills around petroleum refineries and gas plant facilities [1]. Polycyclic Aromatic Hydrocarbons (PAHs) are a group of these compounds with carcinogenic and toxic potentially [2,3] which contribute to environmental contamination and health hazards. Soil and sediments are the most important environmental reservoir for PAHs.

At present, employing biological treatment is the most popular and cost – effective strategy among the different methods to remove these pollutants from the soil [4]. However, successful application of bioremediation is often limited by environmental, physical and chemical factors [5] such as availability of pollutants to undergo biological transformations [6] toxicity and complex structures of PAHs derivatives, limitation for nitrogen, phosphorus or other nutrients, pH and temperature [7].

One of the main factors limiting bioavailability is the low aqueous solubility of PAHs. Addition of surfactant

* Correspondence: a.mohseni8@yahoo.com
[3]Department of Environmental Health Engineering, School of Public Health, Shahid Beheshti University of Medical Sciences, Tehran, Iran
Full list of author information is available at the end of the article

can be used for increasing the phase partitioning of organic compounds and their bioavailability. On the other hand bioavailability may be decreased by up taking of contaminants into the surfactant micelle [8]. Thus, the concentration of surfactant plays an important role in biodegradation of PAHs.

Many studies have been conducted to overcome problems related to the poor bioavailability of PAHs by using organic matter. Borresen and Rike [9] have shown that humic substances (HS) can increase the solubility of PAHs leads in increasing the bioavailability of PAHs in soil.

Nutrient such as nitrogen and phosphorus are the other important factors on biodegradation. Microorganisms need for nutrients similar to their composition of cells [10,11]. Betancur-Galvis LA et al. [12] used biosolid and inorganic fertilizer in bioremediation of phenanthrene.They found that the removal of phenanthrene in the soil with fertilization was 25 fold more than the other soils. There is a relation between mineralization rates of phenanthrene and the initial concentrations of nitrogen and phosphorus in the soils [13].

The marine soil and sediment is one of the most petroleum contaminants sites, so the salinity is an important factor which must be investigated. Minai-Tehrani D. et al. [14] showed that increasing salinity content of soil had decreasing effect on the biodegradation of total crude oil and PAHs.

The bioremediation strategy is dependent on the optimizing the factors which affect on the microbial growth and biodegradation of pollutant [15]. In combination of several factors, the effect of anyone may be influenced by the others and interactions among them may be occurred too.

The experimental design can be used for optimizing operational conditions for the multivariable system [16] and the interaction between variables would be considered too [17]. The number of experiment's runs would be reduced by statistical design of experiments [17,18]. In this study numerous factors, involving surfactant (Tween 80 (Su.)), salinity (Sal.), soil nutrients (N, P (Nu.)) and organic matters (humic acid (HA)), individually and in combination, have been studied to remedy PAH artificially contaminated soils. The aim of this study was to investigate the effect of physicochemical factors; HA, surfactant (Tween 80), salinity and nutrient (N, P) together on the bioremediation of phenanthrene a three-ring PAH, an appropriate model compound, in soil slurry. In order to find out the most effective factor and the sequence importance of them in biodegradation of phenanthrene a three-ring PAH in the soil slurry, the study was conducted in two phase: 1) Screening the factors by using full factorial experimental design, and 2) Optimization of the phenanthrene-contaminated soil biodegradation by using a central composite face centered design (CCFD) under response surface methodology (RSM). Then the optimized condition was examined in PAHs real contaminated soil.

Methods

Chemicals

Acetone in HPLC grade was purchased from ROMIL Company. Phenanthrene (Purity > 98%), NaCl and chemical materials for mineral salt medium (MSM) were purchased from Merck Company. HA and Tween80 were supplied by Sigma Aldridge and Fluka, respectively. The formula of HA was $C_{17}H_{17}BrN_2O_5$. Nutrient Broth and R2A Agar were purchased from BIOMARK Company.

Phenanthrene biodegradation

Clean soil was collected from a depth of 5–20 cm of ground's surface, Tehran, Iran. It was air dried and passed through a 2-mm sieve. The soil was classified as sand (consisted of 89% sand, 11.9% silt and 5% clay) by the use of standard sieves. Total nitrogen and phosphorus were 0.025% and 0.0012%, respectively. Total organic carbon was 0.18%. The pH and electrical conductivity (EC) were 7.4 and 3.2 ds/m, respectively.

Two grams of dry soil was placed into 50 mL Erlenmeyer flask as non-continuous bioreactors. The bioreactors containing clean soil were autoclaved. Then, it was spiked with dissolved phenanthrene in acetone to have 500 mg phenanthrene/kg dry soil. The bioreactors containing spiked soil were placed in a shaker (Heidolph, ProMax 2020 model) at the velocity of 180 rpm in room temperature and dark condition to have a uniform dispersion of phenanthrene and evaporation of acetone.

The soil was inoculated with a consortium of bacteria in different concentration of MSM with an optical density of 1 at 630 nm [5] using CECIL UV/Vis spectrophotometer (model 7100). The bacterial consortium was consisted of *Bacillus sporogenes*, *Bacillus licheniformis*, *Capnocytophaga ochracea* (presumably), *Acinetobacter sporogenes* and *Staphylococcus xylosus*. Enrichment of the consortium and the potential of it in bioremediation of phenanthrene contaminated soils had been proved in our previous study [6]. The base of MSM was contained of the following (per liter): 0.2 g $MgSO_4.7H_2O$, 0.1 g $CaCl_2.2H_2O$, 0.1 g NaCl, 0.01 g $FeCl_3.6H_2O$ and 1 mL trace element solution. The trace element solution contained the following (per liter): 23 mg $MnCl_2.2H_2O$, 30 mg $MnCl_4$, 32 mg H_3BO_3, 39 mg $CoCl_2.2H_2O$, 50 mg $ZnCl_2$, 30 mg $NaMnO_4.2H_2O$ and 20 mg $NiCl_2$ [5]. Then the amount of K_2HPO_4, KH_2PO_4, KNO_3, HA, Surfactant and NaCl were added according to Tables 1 and 2 for phase 1 and 2 respectively. The pH was adjusted to 6.8 ± 0.2 using a pH meter (HACH 40d model). At the end, the slurry of soil liquid ratio was 10% w: v [6]. All the samples and their similar blanks were put in the

Table 1 Actual values coded and of variables used in the full factorial (2^4) design

Run	Actual value/coded Levels					
	Salinity (Sal.) %W/V	Surfactant (Su.)L/Lμ	Humic Acid (HA)mg/Kg	Nutrient(N, P)(Nu.) g/L		
				KNO_3	KH_2PO_4	K_2HPO_4
R1	0 (−1)	0 (−1)	0 (−1)	0.000132 (−1)	0.000103 (−1)	0.0017 (−1)
R2	200 (1)	0 (−1)	0 (−1)	0.000132 (−1)	0.000103 (−1)	0.0017 (−1)
R3	0 (−1)	13 (1)	0 (−1)	0.000132 (−1)	0.000103 (−1)	0.0017 (−1)
R4	200 (1)	13 (1)	0 (−1)	0.000132 (−1)	0.000103 (−1)	0.0017 (−1)
R5	0 (−1)	0 (−1)	2 (1)	0.000132 (−1)	0.000103 (−1)	0.0017 (−1)
R6	200 (1)	0 (−1)	2 (1)	0.000132 (−1)	0.000103 (−1)	0.0017 (−1)
R7	0 (−1)	13 (1)	2 (1)	0.000132 (−1)	0.000103 (−1)	0.0017 (−1)
R8	200 (1)	13 (1)	2 (1)	0.000132 (−1)	0.000103 (−1)	0.0017 (−1)
R9	0 (−1)	0 (−1)	0 (−1)	0.8 (1)	0.2 (1)	1 (1)
R10	200 (1)	0 (−1)	0 (−1)	0.8 (1)	0.2 (1)	1 (1)
R11	0 (−1)	13 (1)	0 (−1)	0.8 (1)	0.2 (1)	1 (1)
R12	200 (1)	13 (1)	0 (−1)	0.8 (1)	0.2 (1)	1 (1)
R13	0 (−1)	0 (−1)	2 (1)	0.8 (1)	0.2 (1)	1 (1)
R14	200 (1)	0 (−1)	2 (1)	0.8 (1)	0.2 (1)	1 (1)
R15	0 (−1)	13 (1)	2 (1)	0.8 (1)	0.2 (1)	1 (1)
R16	200 (1)	13 (1)	2 (1)	0.8 (1)	0.2 (1)	1 (1)

Table 2 Experimental matrix for central composite design for optimization

Run	Actual value/coded levels					Removed amount of phenanthrene(mg/Kg)	
	HA(mg/Kg)	Su.(μg/L)	Nu. (N , P) (g/L)				
			K2HPO4	KH2PO4	KNO3	Experimented value	Predicted value
R1	0(−1)*	5(−1)	0.4(−1)	0.1(−1)	0.5(−1)	196.4	190.67
R2	150(+1)*	5(−1)	0.4(−1)	0.1(−1)	0.5(−1)	222	213.63
R3	0(−1)	13(+1)	0.4(−1)	0.1(−1)	0.5(−1)	208.4	214.23
R4	150(+1)	13(+1)	0.4(−1)	0.1(−1)	0.5(−1)	275.7	274.69
R5	0(−1)	5(−1)	0.8(+1)	0.2(+1)	1(+1)	332.1	331.61
R6	150(+1)	5(−1)	0.8(+1)	0.2(+1)	1(+1)	342.8	335.47
R7	0(−1)	13(+1)	0.8(+1)	0.2(+1)	1(+1)	380.8	387.67
R8	150(+1)	13(+1)	0.8(+1)	0.2(+1)	1(+1)	424.8	429.03
R9	0(−1)	9(0)	0.6(0)	0.15(0)	0.75(0)	247.6	281.05
R10	150(+1)	9(0)	0.6(0)	0.15(0)	0.75(0)	260.8	313.21
R11	75(0)*	5(−1)	0.6(0)	0.15(0)	0.75(0)	283.6	267.85
R12	75(0)	13(+1)	0.6(0)	0.15(0)	0.75(0)	380	326.41
R13	75(0)	9(0)	0.4(−1)	0.1(−1)	0.5(−1)	234.2	223.31
R14	75(0)	9(0)	0.8(+1)	0.2(+1)	1(+1)	394.41	370.95
R15	75(0)	9(0)	0.6(0)	0.15(0)	0.75(0)	340	297.13
R16	75(0)	9(0)	0.6(0)	0.15(0)	0.75(0)	279	297.13
R17	75(0)	9(0)	0.6(0)	0.15(0)	0.75(0)	285.2	297.13
R18	75(0)	9(0)	0.6(0)	0.15(0)	0.75(0)	260.5	297.13

*Low Level: (−1); Middle: (0); High Level: (−1).

shaker at the velocity of 180 rpm in room temperature (22 ± 3°C) for 8 weeks.

Experimental designs

The experiment was accomplished in two phases; screening of important variables and the levels of them that significantly influenced phenanthrene degradation, followed by optimization of variables levels by using response surface methodology.

Screening of variables

Screening step was used for identifying the important of four factors based on full factorial design (2^4). These relevant factors were Tween 80, as a non-ionic surfactant (Su), HA, Nutrient and Salinity in two levels of high (+1) and low (−1). The importance of the factors was on the base of the largest effect on the biodegradation of phenanthrene in contaminated soil. In this phase 16-run was applied to evaluate factors (variables). Table 1 illustrates the variables and their corresponding levels. The levels of the factors were on the base of previous studies in literature for PAHs bioremediation [14,19-21].

All the experiment runs were performed in triplicates and the average of them was taken as the result. Each of experiment run had the similar chemical control without any inoculation. The statistical software Design Expert V.7, (Stat-Ease, USA) was used to evaluate the analysis of variance ($P < 0.05$) to determine the significance of each term.

Optimization of variables by RSM

The eventual objective of RSM is optimization of the variables by considering the interactive effects of independent factors. With the aim of increasing phenanthrene biodegradation, a central composite face centered design (CCFD) was applied with three factors, which were screened in the first phase. These factors were surfactant, HA and nutrient in 3 levels: High (+1), middle (0) and low (−1) with 4 center points with a CCFD under response surface methodology (RSM) (Table 2). In the first phase salinity had not significant effect on the results, so it was fixed on the 1% W/V.

After running the 18 trials, a second-order polynomial model corresponding to CCFD was fitted to correlate the relationship between the independent variables and the highest percent removal of phenanthrene as a response. The linear computer-generated model is given as [14] (Eq. 1):

$$Y = \beta_0 + \beta_1 x_1 + \beta_2 x_2 + \ldots + \beta_k x_k + \epsilon \qquad (1)$$

Where y is the measured response; β_0 is an intercept, β_1 and β_2 are the linear effect (coefficients computed from the observed experimental values of y) regression terms; x_1 and x_2 is the coded level of independent variable; and k indicates the number of independent variables and ϵ is the error. The significance of each coefficient in the equation was determined by Student t-test and P-values. Also, F-test indicated all of the factors and interactions considered in the experimental design that are statistically significant ($P < 0.05$) at the 95 per cent confidence level.

Determination of microbial population

Most Probable Number (MPN) was used for identification of bacterial population at different time intervals. The bacterial suspension was diluted tenfold serially in a sterile ringer solution (8.5 g NaCl per 1 L DW) and added to the sterile Nutrient Broth in the ratio of 10% of volume in triplicates in five series. After 48 hours of incubation in 30°C, the turbidity of positive growth was seen with direct observation. The population of bacterial consortium was estimated according to statistical table of MPN [6].

Extraction and analysis

The residual phenanthrene in the soil was extracted with acetone by ultrasonic homogenizer (Bandelin Sonoplus HD 2070) according to EPA 3550B (EPA) for two minutes [22]. The extracted sample was then centrifuged (Hettich D7200) for 15 minutes at 6000 rpm and filtered by Whatman cellulose filter papers N.42 through 2–3 cm of glass wool. A portion of the filtered solution was used for analysis.

The extract was quantified by gas chromatography (GC; Chrompack CP 9001) using a flame ionization detector (FID) with an HP5 capillary column (length of 30 m, inside diameter of 0.32 mm and a coated-film thickness of 0.2 μm). α-Naphthol was used as an internal standard. Initial temperature of the oven was maintained at 100°C for one minute, and then it was increased at a rate of 10°C/min until 250°C. The injector and detector temperatures were set to 250°C and 270°C, respectively. The concentration of phenanthrene was determined after the calibration of the method with standard phenanthrene samples.

Results and discussion

Complete factorial design and bioremediation results

This stage has attempted to investigate the effect of four independed variables on reducing the phenanthrene level by biodegradation. The phenanthrene removal efficiency in 16 tests using biodegradation compared with their similar control (without inoculation) in 8 weeks was shown in Figure 1 which is about the effectiveness of variables in comparison with each other and Bonferroni line (significant). In these 16 experiments the decrease of phenanthrene concentration in the soil was in

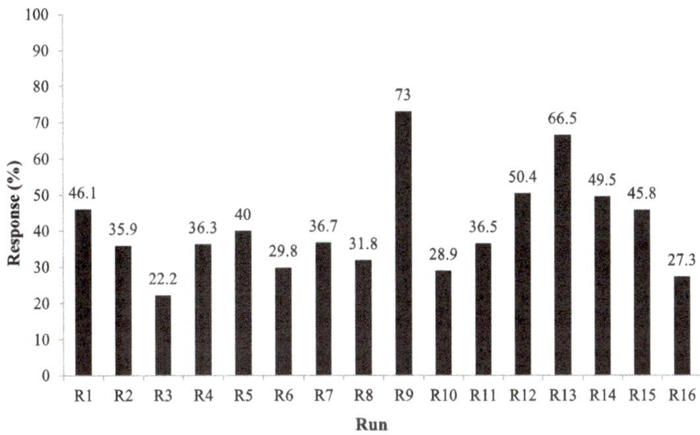

Figure 1 The phenanthrene removal percentage from Soil in 16 run of complete factorial design samples in comparison to similar controls (Response).

the range of 22–73 percent. These results indicate that the addition of microorganisms increased the rate of biodegradation. Our previous study showed that the presence of bacteria in soil has the critical role in the removal of phenanthrene [4,6]. Kästner et al. [23] reported that the biodegradation of pyrene and anthracene had been enhanced by using pure bacterial culture. This increasing was about sixfold in pyrene and tenfold in anthracene removal. The promotion of PAHs mineralization by using mix bacterial culture was reported by Jacques et al. too [24].

After performing 16 runs of complete factorial design, the statistical analysis of the responses was done with regard to the coded design matrix. The responses showed that linear term of HA, Surfactant and Nutrient have remarkable effects on phenanthrene removal efficiency, but salinity was not significant variable in these concentrations. The idea of less effect of salinity is not agree with Chen et al. [7] who reported that salinity was the most significant factor on phenanthrene biodegradation.

Analysis of variance (ANOVA) indicated that nutrient with the largest effect was the most significant variable ($p < 0.0001$), and HA and Su were possibly significant. The statistically relevant effects were sorted from the largest to the smallest in the Pareto chart presented in Figure 2 that shows the main effect of variables in relation with the slope of the lines. In the Pareto chart all

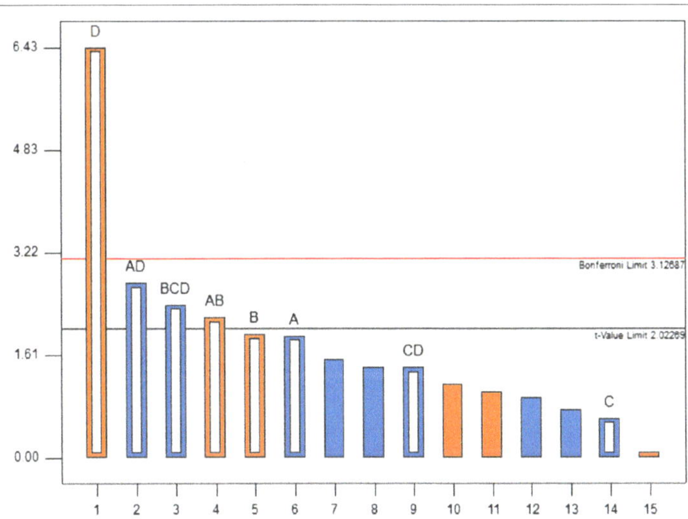

Figure 2 Standardization (P = 95%) main effects Pareto chart for the complete factorial design: A, HA— humic acid (mg/kg); B, Su—Tween 80 (μL/L); C, Sal— salinity (%W/V); D, Nut— nutrients (g/L), Red line is Bonferroni limitation line.

the effect which are in the right the t-value line are significant and each effect which are larger than Bonferroni limitation line have high significant. Hence HA, Su and nutrient were selected as important factors for bioremediation of phenanthrene.

Figure 3 shows the statistical fitness between actual and predicted values. The main effect plot (Figure 3) shows the comparative effects of all parameters on biodegradation of phenanthrene. The slope of nutrient effect shows that the response of phenanthrene removal was sensitive to this factor. The relatively flat line of salinity shows insensitivity of the responses to change in this parameter. The slope of other two parameters confirmed the possibly significant role in the process. Betancur-Galvis et al. [12] reported that the concentration of phenanthrene was 1.7-times and 2.9-times lower in the soil amended with NP and biosolid respectively compared to sterilized soil. In our study application of high level of N and P in run 9 had been increased the phenanthrene removal by more than 1.5 fold compared to low level of N and P in run 1. Da Silva et al. [25] showed that bioremediation of coastal sand contaminated crude oil would be improved through biostimulation by using commercial mineral NPK fertilizer.

Optimization and verification

The structural matrix using three factors CCFD with 4 center points, experimental and predicted responses were presented in Table 2.

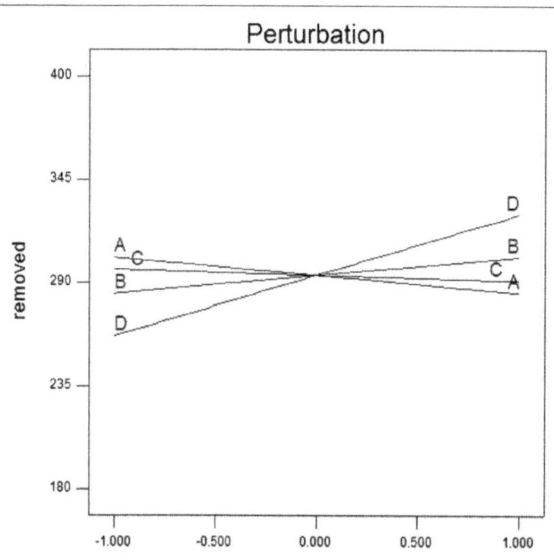

Figure 3 Perturbation plots for phenanthrene removal (mg/kg) in screening phase: A, HA— humic acid (mg/kg); B, Su—Tween 80 (μL/L); C, Sal— salinity (%W/V); D, Nut— nutrients (g/L).

In order to explain the effect of each variable and their interactions on the phenanthrene removal, the first-order model was given by the following equation (Eq.2):

$$Removed\ phenanthrene(mg/kg) = 297.13 + 16.08A + 29.28B + 73.82C + 9.38AB - 4.77AC + 8.13BC \quad (2)$$

Where A is HA concentration, B is Surfactant concentration and C is the concentrations of Nutrient (as N and P). The model was examined by the regression equation and determination coefficient ($R^2 = 0.883$) which suggests that more than 88.3 percent of the variance is attributable to the variables (Figure 4). The F-value of the regression model was 10.63, which implied that the application of linear model was satisfactory for the assessment of phenanthrene biodegradation in the soil. Also the low probability value (<0.0005) indicated a high significance of the model. Among the independed variables the linear term of surfactant had significant effect ($P < 0.05$), while the linear term of nutrient had highly significant effect ($P < 0.0001$) on phenanthrene removal. All the parameters and their interaction had positive effect except the interaction of HA and nutrient had negative effect, which indicated that increasing their levels would decrease the biodegradation of phenanthrene. Figure 5 represent three-dimensional plot of the variables. The parallel contours in Figure 5a,b and c, respectively suggested that interactions of HA to nutrient, HA to surfactant and surfactant to nutrient could be negligible that confirmed by the percent of their effects which was <1%.

With emphasis on nutrient effect, lower HA values together with higher nutrient levels or higher HA values together with lower nutrient levels could result in higher biodegradation values. Very slight variations in phenanthrene removal efficiency (Figure 5a and b) were obtained when the HA was used up to concentration of 150 mg/kg soil corresponding to surfactant and nutrient from low level to high level. The F-ratio of 3.85% had confirmed it too. In spite of increasing the solubility of PAH in the presence of humic substance [26,27], HA had no significant effect on phenanthrene biodegradation and it is agree with Heywood et al. [28] who reported that there was not any relationship between PAHs biodegradation and the organic matter content of the soil. The bond between the contaminants and HA can make the contaminants to be more resistant against desorption and biodegradation by microorganisms [29]. Macleod and Semple and Semple et al. reported that in the soil containing humic compounds the mineralization of pyrene occurred with a retardation which was attributed to the slow desorption of HOCs (Hydrophobic Organic Contaminants) as a rate-limiting factor in biodegradation [15,30]. In our previous study in presence of humic

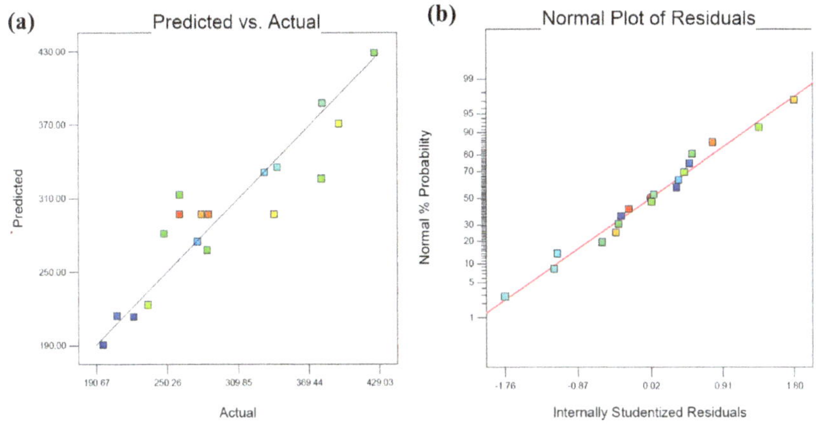

Figure 4 (a) **Predicted versus actual and** (b) **Normal plot of phenanthrene in optimization phase.**

compounds, retardation in phenanthrene biodegradation was seen too [6].

The analysis of variance shows that the effect of surfactant on phenanthrene biodegradation was more than HA and less than nutrient. Its middle effect was shown in 3D plot too (Figure 5a and b). Mixed results have been reported concerning the effect on the biodegradation of PAHs by the addition of surfactant. The positive effect of surfactant is enhancement of biodegradation via increasing in solubility of phenanthrene in water phase which may lead to be more bioavailable for degrading microorganisms [8,31]. The negative effect of surfactant has been attributed to be used by microorganisms as a preferential growth substrate or may be toxic for microorganisms because of high amount of contaminant which have been soluble [8]. The lower amount of MPN may be related to high concentration of phenanthrene in liquid phase (data was not shown). The bacterial cell lyses was reported by Avramova et al. [8] in prevention of phenanthrene mineralization by *Pseudomonas sp.* In their research addition of Triton X-100 inhibited the phenanthrene mineralization too. Piskonen and Itävaara [21] showed that using four types of surfactant at lower concentrations of CMC could not enhance the PHAs bioremediation in contaminated soils.

In Figure 3, a steep slope of in nutrient curve shows that the response of phenanthrene removal was very sensitive to this factor. The effect of this factor was the highest in second phase of experiment too, which was confirmed by ANOVA. Figure 5a and c. show that nutrient is a crucial factor for the biodegradation of phenanthrene. A highly variation of hydrocarbon biodegradation in the same aquifer samples with nutrient addition was reported in the other research too. Breedveld and Sparrevik [32] observed a strong relation between biodegradation of fluoranthene with nitrogen and phosphorous content. They reported that the transformation rates of hydrocarbons have been increased by addition of inorganic nitrogen and phosphorous. Similar conclusion was also drawn from Coulon et al. [33] who showed that the addition of a commercial oleophilic fertilizer containing N and P had a positive effect on degradation of TPH in contaminated sub-Antarctic soil. But it is not agree with Chen et al. [7] who reported that nutrient was the insignificant factor on phenanthrene biodegradation [8].

Enhancement in biodegradation rates of oil-contaminated sediments with addition of inorganic nutrients had been reported before [34]. Nutrient comprises nitrogen, phosphorous and other inorganic elements are essential for bacterial growth [11] and addition of them can enhance bacterial activity, which causes in biodegradation of pollutant [12]. Similar to our research, Mohajeri et al. [35] reported that the higher dose of nutrient amendment can accelerate the phenanthrene biodegradation rate.

The predicted values of phenanthrene removal were calculated from the first-order model and the actual values were determined for particular runs. The predicted versus actual plot of phenanthrene biodegradation and normal percent probability versus the studentized residuals plot were shown in Figure 4a and b. the similarity between the observed values of experiment and the fitted values under the suggested model indicate that model prediction is accurate. The studentized residuals illustrate the normal distribution, the goodness of fit and linearity of the fitted model.

According to the results, the highest biodegradation (85% = 424.8 mg/kg) was obtained from run8 (150 mg HA/Kg soil, 13 μg/L surfactant, and nutrients as K_2HPO_4: 0.8; KH_2PO_4: 0.2; KNO_3: 1 g/L). In this phase the important sequence of the factors was Nutrient, Surfactant and HA with the effect of 81.27%, 12.78% and 3.58%, respectively.

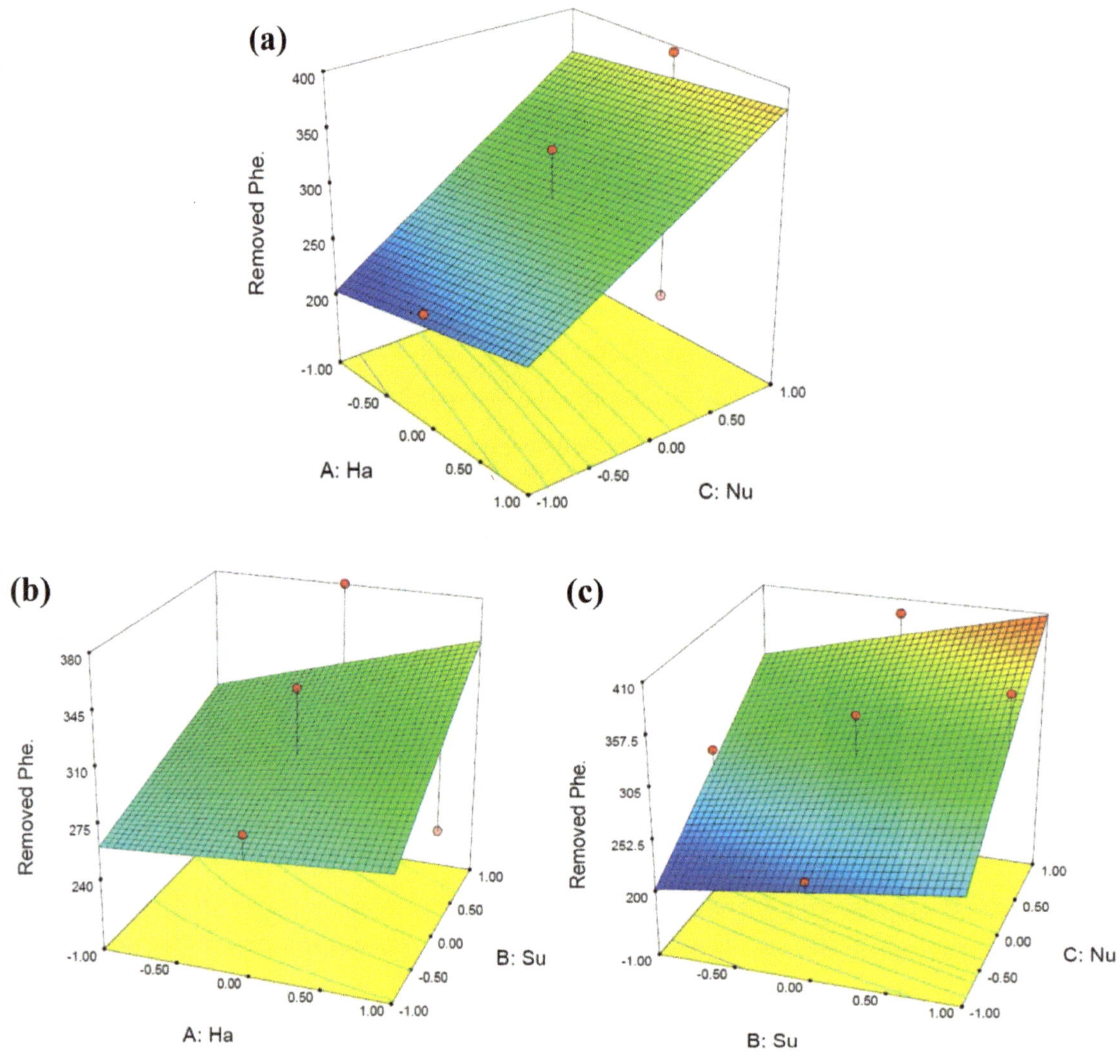

Figure 5 Response surface plot for phenanthrene removal as function of; (a) HA and nutrient, (b) HA and surfactant, (c) Nutrient and surfactant interactions.

CCFD method predicted that the removal of phenanthrene under the optimum conditions (150 mg HA/Kg soil, 12.68 μg/L surfactant, and nutrients as K_2HPO_4: 0.8; KH_2PO_4: 0.2; KNO_3: 1 g/L) will be 87.1%. The average (n = 5) of the obtained percent removal of phenanthrene in synthetic contaminated soil under the optimum conditions was 85.3 ± 1.5%. The potential application of optimum conditions resulted from the model was tested by determining the phenanthrene removal from the real contaminated soil samples. The contaminated soil samples were obtained near the Tehran Petroleum Refinery. The compositions of soil samples were consisted of 69% sand, 10.0% silt and 21% clay and the phenanthrene concentration was in the range of 32–45 mg/kg. The average (n = 5) removal of phenanthrene in contaminated soil samples under the optimum conditions was 72.2 ± 5.8%. Based on the established results, there are satisfactory agreements between the results for the estimated responses and those obtained under the optimum conditions.

The highest biodegradation rate of phenanthrene in optimum condition was 7.78 mg phenanthrene per kg soil per day; whereas the phenanthrene biodegradation rate in lowest amount of these factors were only 3.5078 mg phenanthrene per kg soil per day. Data of this research showed that optimization of effective factors in bioremediation of PAHs contaminated soil could increase the biodegradation rate by two fold of magnitude.

Conclusions

The screening experiments showed that significant factors in phenanthrene biodegradation were HA, surfactant and nutrient contents. Therefore, these variables were used for RSM. The obtained results from RSM, point out the importance of nutrients for phenanthrene biodegradation. Surfactant displayed significant positive effect and HA had an insignificant effect on phenanthrene removal. A high similarity was between the model prediction and experimental results. According to the effective ratio, the sequence importance of the variables on phenanthrene biodegradation in contaminated soils were as Nutrient > Surfactant > Humic acid > Salinity. The biodegradation rate at optimum condition was 7.78 mg phenanthrene/kg soil/day.

Competing interests

The authors declare that they have no competing interests.

Authors' contributions

MR, RRK, AM, AE, MF and SHN carried out the article with the title of Experimental design approach to the optimization of PAHs bioremediation from artificially contaminated soil: Application of variables screening development participated in the sequence alignment and drafted the manuscript. All authors read and approved the final manuscript.

Acknowledgment

The authors sincerely thank the Department of Environmental Health Engineering, Tehran University of medical Sciences for financial support of this project.

Author details

[1]Department of Environmental Health Engineering, School of Public Health, Tehran University of Medical Sciences, Tehran, Iran. [2]Department of Environmental Health Engineering, School of Public Health, Iran University of Medical Sciences, Tehran, Iran. [3]Department of Environmental Health Engineering, School of Public Health, Shahid Beheshti University of Medical Sciences, Tehran, Iran. [4]Department of Chemical Engineering, Biotechnology Group, Tarbiat Modares University, Tehran, Iran.

References

1. Mahvi AH, Maleki A, Alimohamadi M, Ghasri A. Photo-oxidation of phenol in aqueous solution: toxicity of intermediates. Korean J Chem Eng. 2007;24:79–82.
2. Mohseni Bandpi A, Rezaei Kalantary R, Ameli AEA, Zinatizadeh AA, Jonidi Jafari A. Application of response surface methodology for optimization of Fenton process for phenanthrene removal from soil. Environ Eng Manag J. 2013, in press.
3. Nadim F, Hoag GE, Liu S, Carley RJ, Zack P. Detection and remediation of soil and aquifer systems contaminated with petroleum products: an overview. J Pet Sci Eng. 2000;26:169–78.
4. Nasseri S, Kalantary R, Nourieh N, Naddafi K, Mahvi A, Baradaran N. Influence of bioaugmentation in biodegradation of PAHs-contaminated soil in bio-slurry phase reactor. Iranian J Environ Health Sci Eng. 2010;7:199–208.
5. Ressler BP, Kneifel H, Winter J. Bioavailability of polycyclic aromatic hydrocarbons and formation of humic acid-like residues during bacterial PAH degradation. Appl Microbiol Biotechnol. 1999;53:85–91.
6. Rezaei KR, Badkoubi A, Mohseni-Bandpi A, Esrafili A, Jorfi S, Dehghanifard E, et al. Modification of PAHs biodegradation with humic compounds, soil and sediment contamination. An Int J. 2013;22:185–98.
7. Chen J, Wong MH, Wong YS, Tam NFY. Multi-factors on biodegradation kinetics of polycyclic aromatic hydrocarbons (PAHs) by Sphingomonas sp. a bacterial strain isolated from mangrove sediment. Mar Pollut Bull. 2008;57:695–702.
8. Avramova T, Sotirova A, Galabova D, Karpenko E. Effect of Triton X-100 and rhamnolipid PS-17 on the mineralization of phenanthrene by Pseudomonas sp. cells. Int Biodeterior Biodegrad. 2008;62:415–20.
9. Borresen MH, Rike AG. Effects of nutrient content, moisture content and salinity on mineralization of hexadecane in an Arctic soil. Cold Reg Sci Technol. 2007;48:129–38.
10. Devi MP, Reddy MV, Juwarkar A, Sarma PN, Mohan SRV. Effect of co-culture and nutrients supplementation on bioremediation of crude petroleum sludge. CLEAN Soil, Air Water. 2011;39:900–7.
11. Liebeg EW, Cutright TJ. The investigation of enhanced bioremediation through the addition of macro and micro nutrients in a PAH contaminated soil. Int Biodeterior Biodegrad. 1999;44:55–64.
12. Betancur-Galvis LA, Alvarez-Bernal D, Ramos-Valdivia AC, Dendooven L. Bioremediation of polycyclic aromatic hydrocarbon-contaminated saline–alkaline soils of the former Lake Texcoco. Chemosphere. 2006;62:1749–60.
13. Teng Y, Luo Y, Ping L, Zou D, Li Z, Christie P. Effects of soil amendment with different carbon sources and other factors on the bioremediation of an aged PAH-contaminated soil. Biodegradation. 2010;21:167–78.
14. Minai-Tehrani D, Minoui S, Herfatmanesh A. Effect of salinity on biodegradation of Polycyclic Aromatic Hydrocarbons (PAHs) of heavy crude oil in soil. Bull Environ Contam Toxicol. 2009;82:179–84.
15. Semple KT, Reid BJ, Fermor TR. Impact of composting strategies on the treatment of soils contaminated with organic pollutants. Environ Pollut. 2001;112:269–83.
16. Montgomery DC, Wiley J. Design and Analysis of Engineering Experiments. In: Book Design and Analysis of Engineering Experiments. New York: Wiley; 2001.
17. Khajvand T, Chaichi MJ, Nazari OL, Golchoubian H. Application of Box–Behnken design in the optimization of catalytic behavior of a new mixed chelate of copper (II) complex in chemiluminescence reaction of luminol. J Lumin. 2011;131:838–42.
18. Lu M, Yuan D, Li Q, Ouyang T. Application of response surface methodology to analyze the effects of soil/liquid ratio, pH, and incubation time on the bioaccessibility of PAHs from soil in in vitro method. Water Air Soil Pollut. 2009;200:387–97.
19. Liang Y, Sorensen DL, McLean JE, Sims RC. Pyrene fate affected by humic acid amendment in soil slurry systems. J Biol Eng. 2008;2:11.
20. Maier RM, Pepper IL, Gerba CP. Environmental Microbiology. Canada: Academic Press; 2009.
21. Piskonen R, Itaevaara M. Evaluation of chemical pretreatment of contaminated soil for improved PAH bioremediation. Appl Microbiol Biotechnol. 2004;65:627–34.
22. USEPA. Ultrasonic extraction. Center for environmental research information. EPA, 3550B. 1996;2:1–14.
23. Kästner M, Breuer-Jammali M, Mahro B. Impact of inoculation protocols, salinity, and pH on the degradation of polycyclic aromatic hydrocarbons (PAHs) and survival of PAH-degrading bacteria introduced into soil. Appl Environ Microbiol. 1998;64:359–62.
24. Jacques RJS, Okeke BC, Bento FM, Teixeira AS, Peralba MCR, Camargo FAO. Microbial consortium bioaugmentation of a polycyclic aromatic hydrocarbons contaminated soil. Bioresour Technol. 2008;99:2637–43.
25. da Silva AC, de Oliveira FJS, Bernardes DS, de França FP. Bioremediation of marine sediments impacted by petroleum. Appl Biochem Biotechnol. 2009;153:58–66.
26. Grasso D, Subramaniam K, Pignatello J, Yang Y, Ratte D. Micellar desorption of polynuclear aromatic hydrocarbons from contaminated soil. Colloids Surf A Physicochem Eng Asp. 2001;194:65–74.
27. Van Stempvoort DR, Lesage S, Novakowski KS, Millar K, Brown S, Lawrence JR. Humic acid enhanced remediation of an emplaced diesel source in groundwater.: 1. Laboratory-based pilot scale test. J Contam Hydrol. 2002;54:249–76.
28. Heywood E, Wright J, Wienburg CL, Black HIJ, Sara M, Osborn D, et al. Factors influencing the national distribution of polycyclic aromatic hydrocarbons and polychlorinated biphenyls in British soils. Environ Sci Technol. 2006;40:7629–35.
29. Plaza C, Xing B, Fernández JM, Senesi N, Polo A. Binding of polycyclic aromatic hydrocarbons by humic acids formed during composting. Environ Pollut. 2009;157:257–63.
30. Macleod CJA, Semple KT. The adaptation of two similar soils to pyrene catabolism. Environ Pollut. 2002;119:357–64.
31. Ahn CK, Woo SH, Park JM. Enhanced sorption of phenanthrene on activated carbon in surfactant solution. Carbon. 2008;46:1401–10.

32. Breedveld GD, Sparrevik M. Nutrient-limited biodegradation of PAH in various soil strata at a creosote contaminated site. Biodegradation. 2000;11:391–9.
33. Coulon F, Pelletier E, Gourhant L, Delille D. Effects of nutrient and temperature on degradation of petroleum hydrocarbons in contaminated sub-Antarctic soil. Chemosphere. 2005;58:1439–48.
34. Yu KSH, Wong AHY, Yau KWY, Wong YS, Tam NFY. Natural attenuation, biostimulation and bioaugmentation on biodegradation of polycyclic aromatic hydrocarbons (PAHs) in mangrove sediments. Mar Pollut Bull. 2005;51:1071–7.
35. Mohajeri L, Aziz HA, Isa MH, Zahed MA. A statistical experiment design approach for optimizing biodegradation of weathered crude oil in coastal sediments. Bioresour Technol. 2010;101:893–900.

8

Performance evaluation and modeling of a submerged membrane bioreactor treating combined municipal and industrial wastewater using radial basis function artificial neural networks

Seyed Ahmad Mirbagheri[1], Majid Bagheri[1*], Siamak Boudaghpour[1], Majid Ehteshami[1] and Zahra Bagheri[2]

Abstract

Treatment process models are efficient tools to assure proper operation and better control of wastewater treatment systems. The current research was an effort to evaluate performance of a submerged membrane bioreactor (SMBR) treating combined municipal and industrial wastewater and to simulate effluent quality parameters of the SMBR using a radial basis function artificial neural network (RBFANN). The results showed that the treatment efficiencies increase and hydraulic retention time (HRT) decreases for combined wastewater compared with municipal and industrial wastewaters. The BOD, COD, NH_4^+−N and total phosphorous (TP) removal efficiencies for combined wastewater at HRT of 7 hours were 96.9%, 96%, 96.7% and 92%, respectively. As desirable criteria for treating wastewater, the TBOD/TP ratio increased, the BOD and COD concentrations decreased to 700 and 1000 mg/L, respectively and the BOD/COD ratio was about 0.5 for combined wastewater. The training procedures of the RBFANN models were successful for all predicted components. The train and test models showed an almost perfect match between the experimental and predicted values of effluent BOD, COD, NH_4^+−N and TP. The coefficient of determination (R^2) values were higher than 0.98 and root mean squared error (RMSE) values did not exceed 7% for train and test models.

Keywords: Combined wastewater, Submerged membrane bioreactor, Treatment efficiency, Artificial neural network, Radial basis function

Introduction

The membrane bioreactor (MBR), especially the submerged membrane bioreactor (SMBR), has been extensively investigated and applied for municipal and industrial wastewater treatment. There are more than 2200 MBR installations in operations or under construction worldwide and most of them are for municipal wastewater treatment [1,2]. Earlier studies have already shown that MBRs can be operated at much higher efficiency than of what is needed for municipal wastewater [3,4]. Treatment performances were generally good, and deterioration of the performance was not observed [5]. Rosenberger et al. [3] studied aerobic treatment of municipal wastewater in an MBR for 535 day. The pilot plant comprised an anoxic zone to enable denitrification. The hydraulic retention time (HRT) varied between 10.4 and 15.6 hours. Treatment performance was very stable and on a high level. The chemical oxygen demand (COD) was reduced by 95%. Nitrification was complete and up to 82% of the total nitrogen could be denitrified. The excellent capability of SMBRs in the treatment of municipal wastewater has decreased the HRT to the minimum possible amount compared with conventional activated sludge processes. In other words, it seems that SMBRs are over designed for the treatment of municipal wastewater.

* Correspondence: bagherimajead@yahoo.com
[1]Department of Civil Engineering, K.N. Toosi University of Technology, Vanak square, Tehran, Iran
Full list of author information is available at the end of the article

The interest of using MBR instead of classical activated sludge system for the treatment of industrial wastewater was demonstrated [6,7]. Zhao et al. [8] used a laboratory-scale anaerobic/anoxic/oxic membrane bioreactor system to treat heavily loaded and toxic coke plant wastewater and operated for more than 500 days. Treatment performance, acute toxicity assessment, and dissolved organic characteristics of the system were investigated. When the HRT of the system was 40 hours, the removal efficiencies of COD, phenol, $NH_3 - N$, total nitrogen (TN) and acute toxicity were 89.8%, 99.9%, 99.5%, 71.5% and 98.3%, respectively. A desirable treated wastewater is water that is not only low in organic or mineral components, and free from biological entities such as bacteria, pathogens, and viruses but also cost efficient and reliable [9,10]. HRT plays an important role in the removal of pollutants in activated sludge processes coupled with membranes [11]. The amount of HRT for most of industrial wastewaters is higher than 2 days. As a result, the treatment of industrial wastewater is more expensive than treatment of municipal wastewater by considering the important role of HRT in efficiency and the cost of wastewater treatment.

The components in the industrial wastewater are in huge amount; for instance, high amount of COD and biochemical oxygen demand (BOD) [12], ammonia, suspended solid or heavy metal [13] and sometimes shock loading will happen. High strength of industrial wastewater results in the low biodegradability characteristic of wastewater [14-16]. A solution used to increase the biodegradation of the slowly biodegradable compounds is the adsorbent addition in the bioreactor [17-19]. Generally, BOD/COD equal to 0.5 is considered as readily biodegradable or easily treatable [20-24]. The BOD/COD ratio for the industrial wastewaters is not equal to 0.5 and varies from 0.117 to 0.773. The idea of combined municipal and industrial wastewater is an approach to set this ratio to 0.5 and improve the other criteria for a desirable wastewater treatment such as the total BOD to total phosphorous (TBOD/TP) ratio and the influent BOD and COD concentrations.

Treatment process models are essential tools to assure proper operation and better control of wastewater treatment plants. Considerable effort has been devoted to the modeling of activated sludge processes (ASPs) since early 1970s [25]. Some deterministic models have been developed basing on the fundamental biokinetics such as activated sludge model number one (ASM1) [26,27]. Following ASM1, ASM2, ASM2d and ASM3 models were developed. The ASM2 [28] models extended the capabilities of ASM1 to involve the biological phosphorus and nitrogen removals. Whereas, ASM3 [29,30] introduced an alternative concept to the previous ASM biokinetics and aimed at simplifying the model application. Despite the availability of ASM models, the diagnosis of the process interactions and modeling of ASP in an SMBR is still difficult [30,31]. Parameter estimation and calibration of ASM models require expertise and significant effort. Moreover, calibration has to be performed for each specific treatment system. Therefore, application of ASM models to real systems can be cumbersome and problematic [25,32].

In recent years, artificial neural networks (ANNs) have been used for monitoring, controlling, classification and simulation of ASPs. ANN is a non-parametric model which utilizes interconnected mathematical nodes or neurons to form a network that can model complex functional relationships [33]. So far, different types of neural network architectures and their performances have been studied for the purpose of neuroidentification [34-37]. It includes radial basis functions (RBFs), multilayer perceptrons (MLPs), recurrent neural networks (RNNs), and echo-state networks (ESNs). In the literature to date, a limited number of applications of ANNs have been made to SMBRs for modeling of a plant operation [31,38]. Geissler et al. [31] used an ANN model to predict the filtration performance in a submerged capillary hollow fiber membrane treating municipal wastewater. The training procedure for the ANN was conducted based upon pilot-studies with an MBR system using a novel submerged capillary module. Good correlations were found between the predicted and measured permeability using ANN. Cinar et al. [38] have also proposed an ANN model for a SMBR treating cheese whey and evaluated its performance at different sludge residence time. The results of the training procedure for effluent total phosphate, COD, ammonia, nitrate were successful. However, the results of the testing procedure for effluent total phosphate were not as good as for the effluent ammonia and nitrate, although they were better than the results of effluent COD. Up till now, there have not been any investigations on treating combined municipal and industrial wastewater by SMBRs for the purpose of optimizing HRT or performance improvement. Furthermore, no attempt has been made on the modeling of the combined municipal and industrial treatment systems.

In order to achieve the objective of this study, it was decided to employ a type of RBF, which is most commonly used in classification problems [39]. RBFs have been successfully applied for solving dynamic system problems, because they can predict the behavior directly from input/output data [40-42]. The radial basis function artificial neural network (RBFANN) was applied to model the effluent quality parameters of an SMBR treating combined municipal and industrial wastewater. The influent concentration of parameters, HRT, mixed liquor volatile suspended solids (MLVSS), total dissolved solids (TDS) and pH were inputs of the RBFANN models.

Sensitivity analyses were performed to determine the effect and importance order of each input parameter on the changes of effluent concentrations.

Material and methods

Pilot plant configuration

An SMBR was used in order to treat combined municipal and industrial wastewater in this study. Figure 1 shows the schematic diagram of the hollow fiber SMBR. The SMBR consisted of a storage tank, an anaerobic reactor, an anoxic reactor and an oxic reactor as simultaneous aeration/filtration reactor. The storage tank was made of plastic measuring 0.7 by 0.7 meter and total volume of 0.49 m^3. The influent pump established a continuous influent wastewater flow from feeding tank to the anaerobic reactor. The anaerobic, anoxic and oxic reactors were made of Plexiglas with total volume of 0.06 m^3, 0.1 m^3 and 0.24 m^3, respectively. The anaerobic reactor measuring 0.4 by 0.3 meter was located 1.2 meter above the ground level to establish a continuous flow to anoxic reactor. The anoxic reactor measuring 0.5 by 0.4 meter was located 0.9 meter above the ground level to establish a continuous flow to oxic reactor. The oxic reactor measuring 0.8 by 0.5 meter performed a simultaneous aeration/filtration treating role in the SMBR system. A small portion of the sludge in the oxic reactor was recirculated back into the anoxic reactor using a recirculation pump, where it was mixed with the effluent of anaerobic reactor. This recirculation is a key feature of the activated sludge process. The temperature was kept about 25°C and the HRT of the SMBR system varied during the experiments. The SMBR consisted of a polypropylene hollow fiber membrane with a nominal pore size of 0.04 μm. The overall membrane surface area was 8 m^2 per module. The maximum permitted pressure for the hollow fiber membrane was about 30 kPa. A pressure gauge was installed on the suction path to turn off the suction pump and open the backwash path when the trans-membrane pressure exceeded permitted limit and membrane foiling occurred. Table 1 shows the detailed specification of hollow fiber membrane.

Municipal wastewater characteristics

The pilot plant was located in Ekbatan wastewater treatment plant in Tehran, Iran, which has been operating since 1988. Influent wastewater analysis for the wastewater treatment plant was carried out for a four month period. According to the results obtained from raw wastewater analysis, the maximum values were selected as critical design parameters. Table 2 shows the critical values of the influent wastewater characteristics to Ekbatan wastewater treatment plant.

Industrial wastewater characteristics

The industrial wastewaters are defined by high strength wastewaters because of the high concentration of their components. Table 3 shows the characteristics of high strength wastewater for different industries. The COD, BOD and total suspended solids (TSS) are three most high concentration components of industrial wastewaters

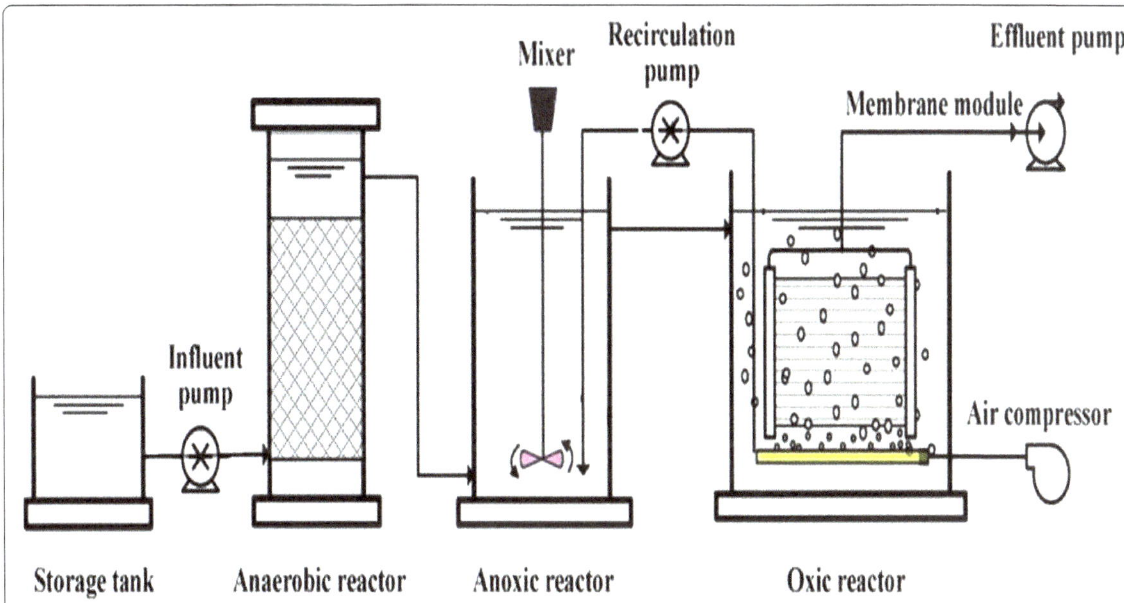

Figure 1 Schematic flow diagram of the experimental apparatus.

Table 1 Specifications of the hollow fiber membrane used in this study

Description	Value
Material	Polypropylene
Capillary Thickness	40 ~ 50 μm
Capillary Outer Diameter	450 μm
Capillary Pore Diameter	0.01 ~ 0.2 μm
Gas permeation	$7.0 * 10^{-2}$ $cm^3/cm^2 \cdot S \cdot cm$ Hg
Porosity	40 ~ 50%
Lengthways strength	120,000 kPa
Designed flux	6 ~ 9 $L/M^2/H$
Area of membrane module	8 m^2/module
Operating Pressure	−10 ~ −30 kPa
Flow rate	1.0 ~ 1.2 m^3/ day

compared with municipal wastewater. In the current research, the high strength wastewater was simulated by increasing the influent COD, BOD, and TSS concentration to 2000, 1300 and 5000 mg/L, respectively. The simulated wastewater had the characteristics close to Beverage wastewater [12], and its TSS concentration was increased greatly. The TDS concentration was increased to 4500 mg/L and pH was different for each experiment.

Combined municipal and industrial wastewater characteristics

Much work has been performed to study the performance of SMBRs in the treatment of municipal and industrial wastewater, including influences of biological processes and membrane fouling. A factor which influences membrane performance in an optimum treatment of wastewater is decreasing HRT while keeping effluent components lower than discharge limits. Moreover, wastewater with BOD/COD ratio equal to 0.5 is considered as readily biodegradable. It has been shown biodegradability greater than 0.5 for spent caustic wastewater after treatment by using wet air oxidation [43]. If the ratio value is less than 0.5, the wastewater needs to have physical or chemical treatment before a biological treatment takes place [43,44]. Table 3 shows that for the most of industries this ratio is not equal to 0.5 and HRT of industrial wastewater is noticeably higher than HRT of municipal wastewater. Therefore, the municipal and industrial wastewater were combined in proportions that the BOD/COD ratio approached 0.5 for the produced combined wastewater.

Table 2 Municipal wastewater characteristics in the critical conditions

Parameter	Value	Parameter	Value
Temperature (°C)	25.8	Org-N (mg/L)	16.8
DO (mg/L)	0	TKN (mg/L)	39.9
BOD_5 (mg/L)	180	TS (mg/L)	810
COD (mg/L)	380	TDS (mg/L)	630
NO_3^-−N (mg/L)	0.96	TSS (mg/L)	180
NH_4^+−N (mg/L)	23.1	TP (mg/L)	16.54

Analytical methods

Temperature, pH, dissolved oxygen (DO), BOD, COD, TSS, TDS, mixed liquor suspended solids (MLSS) and MLVSS concentration, TP, NH_4^+−N and NO_3^-−N were measured in this study. The pH and temperature were measured using a digital pH meter. A dissolved oxygen meter (YSI 5000) was utilized to determine DO. Biodegradability was measured by 5-day biochemical oxygen demand test according to the standard methods [45]. The seed for BOD test was obtained from the Ekbatan wastewater treatment plant [46]. The COD was determined according to the standard methods [45]. Weekly analyses included mixed liquor volatile and total suspended solids (MLVSS, MLSS) in the oxic reactor [47]. MLSS and MLVSS were determined at the Ekbatan wastewater treatment plant laboratory at the temperature of 550°C [20]. The TP and NH_4^+−N were measured with aid of a spectrophotometer (The Hach DR 5000 UV–vis Laboratory Spectrophotometer) at the wastewater treatment plant laboratory.

RBFANN; background and methodology

ANNs can be used for monitoring, controlling, classification and simulation of experimental data. ANNs are mathematical models simulating important parameters based on past observations in complex systems. There are many types of ANNs for modeling and function approximation of the engineering problems [35]. The two well-known ANN models, namely RBFANN and multilayer perceptron artificial neural network (MLPANN) have been used in engineering applications to model or approximate properties [48]. An RBFANN is the most commonly used neural network for pattern recognition problems, and it is also widely used for fault diagnosis. The RBFANN has the advantages of a fast learning process, a learning stage without any iteration of updating weights, robust ability, and adaptation capability compared with other ANNs such as MLP, RNN and ESN [41,49]. RBFANNs have a very strong mathematical foundation rooted in regularization theory for solving ill-conditioned problems [48]. Tomenko et al. [50] reported that RBFANNs produced better results than multiple regression in a wetland treatment system. Madaeni et al. [51] modeled O_2 separation from air in a hollow fiber membrane module with glassy membrane using ANN. They found RBFANN as the best network with minimum training error and checked generalization capability of their network with some unseen data (test data).

Table 3 Characteristics of high strength wastewater for different industries

Industry	HRT (day)	COD (mg/L)	BOD (mg/L)	BOD/COD	NH_4 -N (mg/L)	TSS (mg/L)	So_4^{2-} (mg/L)	po_4^{3-} (mg/L)	Oil (mg/L)	Phenol (mg/L)
Tannery [14]		2000	-	-	-	-	400	-	-	-
Tannery [12]		16000	5000	0.313	450	-	-	-	-	-
Textile [15]	2	6000	700	0.117	20	-	-	120	-	-
Textile [16]	0.7-4	4000	500	0.125	4.8	-	200	2	-	-
Dyeing [21]		1300	250	0.192	100	200	-	-	40	-
Textile [22]	0.58	1500	500	0.333	50	140	-	7	-	-
Wheat starch [10]		35000	16000	0.457	-	13300	-	-	-	-
Dairy [12]		3500	2200	0.629	120	-	-	-	-	-
Beverage [12]		1800	1000	0.556	-	-	-	-	-	-
Palm oil [23]	0.8	67000	34000	0.507	50	24000	-	-	100000	-
Pet food [24]	2.9	21000	10000	0.476	110	54000	-	200	-	-
Dairy product [18]		880	680	0.773	-	2480	-	-	-	-
Phenolic [19]	0.42	797	-	-	131	-	-	-	-	37.3
Pharmaceutical [11]	1	6300	3225	0.51	-	-	-	-	-	-

Shahsavand and Pourafshari Chenar [52] found RBFANN a better choice for the prediction of gas separation performance in a membrane system because of its powerful noise filtering capability. The RBFANNs could offer more successful solutions operating on limited data (less time consuming, supplying additional information on existing relationships), thus indicating that the use of hyperspheres to divide up the pattern space into various classes is more advantageous when dealing with data described by a small number of variables [39].

The structure of the basic RBFANN consisted of one input layer, one output layer, and one hidden layer. A single-output RBFANN with N hidden layer nodes can be described by Eq. (1) and (2).

$$Y = \sum_{n=1}^{N} w_n \theta_n(X) \tag{1}$$

where X and Y are the input and output of the network, X = (x1, $x2$,..., xm)T, wn is the connecting weights between nth hidden node and the output layer, θn is the output value of the nth hidden node, and

$$\theta_n(X) = e^{\left(-x-\mu_n/\sigma_n^2\right)} \tag{2}$$

where μn is the center vector of nth hidden node, $x - \mu_n$ is the Euclidean distance between x and μn, and σn is the radius of the nth hidden node.

As its name implies, radially symmetric basis function is used as activation functions of hidden nodes [39,48]. The transformation from the input nodes to the hidden nodes is a non-linear one, and training of this portion of the network is generally accomplished by an unsupervised fashion. The training of the network parameters (weight) between the hidden and output layers occurs in a supervised fashion based on target outputs [48]. For a RBFANN, the capabilities of the final network are determined by the parameter optimization algorithms and the structure size. However, the number of hidden nodes in these RBFANNs is often assumed to be constant [53]. The transfer function of the RBFANN hidden in layer unit is symmetrical RBF (such as Gaussian function). RBF network has approximation to nonlinear continuous function, and it can learn at high speed and undertake a wide range of data fusion and high-speed processing data in parallel [54]. In this study, the RBFANN applies different network functions such as newrbe and newrb to the input data. The newrb function designs a radial basis neural network and the newrbe function designs an exact radial basis ANN.

Model development

Simulation model of operational parameters was established based on the theory of feed forward artificial neural networks, namely RBFANN using the mathematical software program MATLAB. The operating parameters, including influent BOD, COD, NH_4^+-N or TP as well as HRT, MLVSS, TDS and pH were the input variables of the RBFANN models. These variables were used to train RBFANN in order to simulate effluent BOD, COD, NH_4^+-N and TP. Experimental data over 90 days (30 total data points) were used in RBFANN modeling process. Table 4 shows the statistical characteristics of the measured

Table 4 Characteristics of measured variables used for modeling by RBFANN

Input variable no.	Input variable	Value	Output variable	Value
1	Influent conc.		Effluent conc.	
	BOD (mg/L)	500–600	BOD (mg/L)	5.5–172.3
	COD (mg/L)	1000–12000	COD (mg/L)	11–396.5
	NH_4^+–N (mg/L)	21–27	NH_4^+–N (mg/L)	0.2–3.1
	TP (mg/L)	15–16.4	TP (mg/L)	1.4–6.4
2	HRT (h)	3–11		
3	MLVSS (mg/L)	4120–5990		
4	TDS (mg/L)	500–4900		
5	pH	6.2–7.6		

process variables used in this study to model effluent BOD, COD, NH_4^+–N and TP by RBFANN.

In order to obtain convergence within a reasonable number of cycles, the input and output data should be normalized and scaled to the range of 0–1 by Eq. (3) [55,56]:

$$x_{ni} = \frac{x_i - x_{min}}{x_{max} - x_{min}} \tag{3}$$

where xi is the initial value, xmax and xmin are the maximum and minimum of the initial values, and xni is the scaled value. After the training and testing of the ANN, the output data were scaled to the real-world values through the Eq. (4).

$$x_i = x_{ni}(x_{max} - x_{min}) + x_{min} \tag{4}$$

In the current research, the developed networks consisted of three layers including one input layer that comprised five neurons (including influent BOD, COD, NH_4^+–N or TP, HRT, MLVSS, TDS and pH), one hidden layer consisting of five neurons (which were constant due to the application of newrbe function) and the output layer that had one output neuron (which was effluent BOD or COD, NH_4^+–N or TP). The RBFANN applied network function of newrbe (design exact radial basis network) to the input data. The newrbe function creates a two-layer network with biases for the both layers (Figure 2). The first layer has radial basis transfer function (radbas). Consequently, it calculates its weighted inputs with Euclidean distance weight function (dist) and its net input with product net input function (netprod). The second layer has linear transfer function (purelin). Consequently, it calculates its weighted inputs with Dot product weight function (dotprod) and its net input with sum net input function (netsum). For the newrbe function, the center is selected by using the fixed centers selected at random method [57]. This function can produce a network with zero error on training vectors. The stopping criterion for the newrbe function is when the outputs are exactly the matrix of target class vectors when the inputs are the matrix of input vectors [58]. The RBFANN was designed in an innovative loop that can apply newrbe function to the input data for user defined number of times in order to minimize error.

The network was trained until the network average root mean squared error (RMSE) was minimum and coefficient of determination (R^2) approached 1. Other parameters for network were chosen as the default values of the software. The performances of the ANN models were measured by R^2 and RMSE between the predicted values of the network and the experimental values, which were calculated by Eq. (5) and (6), respectively [59].

$$R^2 = 1 - \frac{\sum_{i=1}^{n}\left(y_i^* - y_P^{(i)}\right)^2}{\sum_{i=1}^{n}(y_i^* - \bar{y})^2} \tag{5}$$

$$\text{RMSE} = \sqrt{\frac{1}{n}\sum_{i=1}^{n}\left(y_P^{(i)} - y_i^*\right)^2} \tag{6}$$

where y is the average of y over the n data, and y_i^* and $y_P^{(i)}$ are the ith target and predicted responses, respectively.

Results and discussion

BOD removal efficiency and modeling outcomes

The experiments were executed from HRT of 1 to 7 h in order to optimize HRT for the municipal wastewater. The influent BOD and MLVSS concentration varied from 140 to 180 mg/L and 3380 to 5470 mg/L, respectively. The removal performance of the SMBR increased from the HRT of 1 to 5 h so the effluent decreased to 5 mg/L with removal efficiency of 97% at HRT of 5 h. The effluent BOD concentration increased and removal performance decreased after 5 h due to self-degradation of microorganisms [60]. Consequently, the kinetic constants including the half saturation coefficient (K_s), the maximum substrate utilization rate (k) and endogenous decay coefficient (K_d) were calculated as 113 mg L^{-1}, 2.05 d^{-1} and 0.036 d^{-1} different HRTs. The influent BOD concentration decreased from 175 mg/L to 25 mg/L for the effluent concentration at HRT of 1 h. It means that the effluent BOD concentration at HRT of 1 h meets discharge limits of municipal wastewater (BOD < 30 mg/L).

The HRT varied from 5 to 20 h in order to optimize HRT for the industrial wastewater. The influent BOD concentration varied from 1265 to 1360 mg/L and MLVSS concentration changed from 4350 to 7890 mg/L. The BOD removal efficiency increased from the HRT of 5 to 17 h in a trend like municipal wastewater. The effluent BOD decreased to 8.7 mg/L with removal

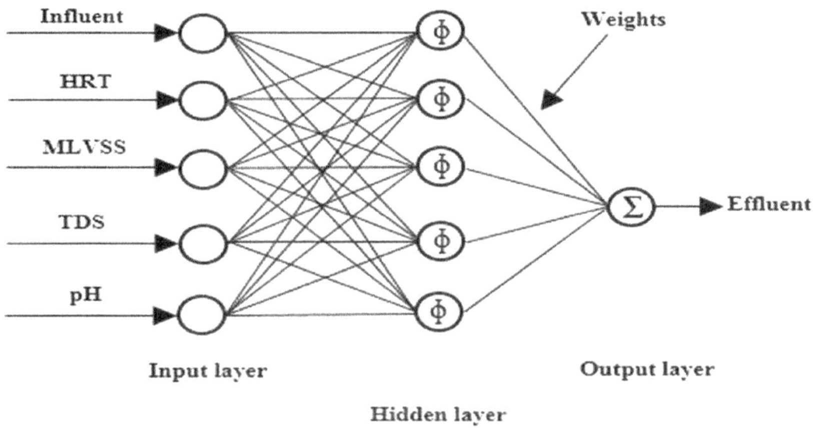

Figure 2 Topological architecture of the RBF artificial neural network used in this study.

efficiency of 99.34% at HRT of 17 h. Then efficiency decreased because of self-degradation [61]. Consequently, the kinetic constants were calculated as K_s equal to 163.55 mg L^{-1} and k equal to 3.56 d^{-1}, K_d equal to 0.013 d^{-1}. At HRT of 13 h the effluent BOD concentration decreased to 42.1 mg/L with removal efficiency of 96.9%. It was concluded that the HRT of 13 h cannot meet the discharge limits and the HRT of 17 h was the optimal result for the treatment of industrial wastewater. The difference between the required HRT for the treatment of municipal and industrial wastewater denotes the noticeable difference in the cost of municipal and industrial wastewater treatment. Therefore, the idea of combined municipal and industrial wastewater was followed as a key to improve the efficiency and decrease the cost of wastewater treatment.

The concentration of components of combined municipal and industrial wastewater was between municipal and industrial wastewaters. The BOD/COD ratio for the combined wastewater was about 0.5 compared with municipal and industrial wastewaters, which changed from 0.38 to 0.5 and from 0.6 to 0.7, respectively [9]. The HRT varied from 3 to 11 h in order to decrease effluent BOD concentration to discharge limits. The influent BOD and MLVSS concentration for the combined wastewater varied from 500 to 600 and from 4120 to 5990 mg/L, respectively. The influent BOD concentration decreased from 557 mg/L to 5.5 mg/L for the effluent of BOD with removal efficiency of 99% at HRT of 9 h. The removal efficiency decreased by increasing HRT from 9 to 11 h because of self-degradation [60,61]. Consequently, the kinetic constants were calculated as K_s equal to 177.84 mg L^{-1} and k equal to 5.29 d^{-1}, K_d equal to 0.011 d^{-1}. At HRT of 7 h for the combined wastewater, influent BOD concentration decreased from 600 mg/L to 19 mg/L for the effluent of BOD with removal efficiency of 96.9%. Therefore, the effluent BOD concentration met discharge limits at HRT of 7 h. It was observed that the performance of the SMBR increases when the influent BOD concentration is lower than 600 to 700 mg/L, which is in a good agreement with the findings of previous studies [3,24]. We concluded that by combining municipal and industrial wastewaters, the treatability of wastewater could be improved by setting BOD/COD ratio to 0.5 and reduction of wastewater strength by decreasing influent BOD concentration to lower than 700 mg/L.

In order to model the effluent BOD concentration by RBFANN, the influent BOD, HRT, MLVSS, TDS and pH were input variables of the neural network. The RBFANN applied network function of newrbe to the input data and the spread of RBF was considered equal to its default value, 1. A large spread results in a smooth function approximation, but, by contrast, a large spread can cause numerical problems [48]. The network function of newrbe selected 70% of normalized data to train and 30% to test RBFANN models [38]. The RBFANN was designed in an innovative loop that applied newrbe to the data for more than 30 times in order to minimize error. Optimal network was chosen on the basis of the minimum average error. Figure 3 shows the denormalized results of the effluent BOD modeling using the RBFANN according to training and testing data sets. The results of the training procedure by RBFANN were successful for the effluent BOD concentration. Figure 3 shows that the train and test models by RBFANN indicated an almost perfect match between the experimental and the predicted effluent BOD values compared with the results of models introduced by Cinar et al. [38]. The RMSE values for train and test (verification process) models were 8.67 and 4.56 mg/L, respectively, and the R^2 values were greater than 0.99 for both models. The

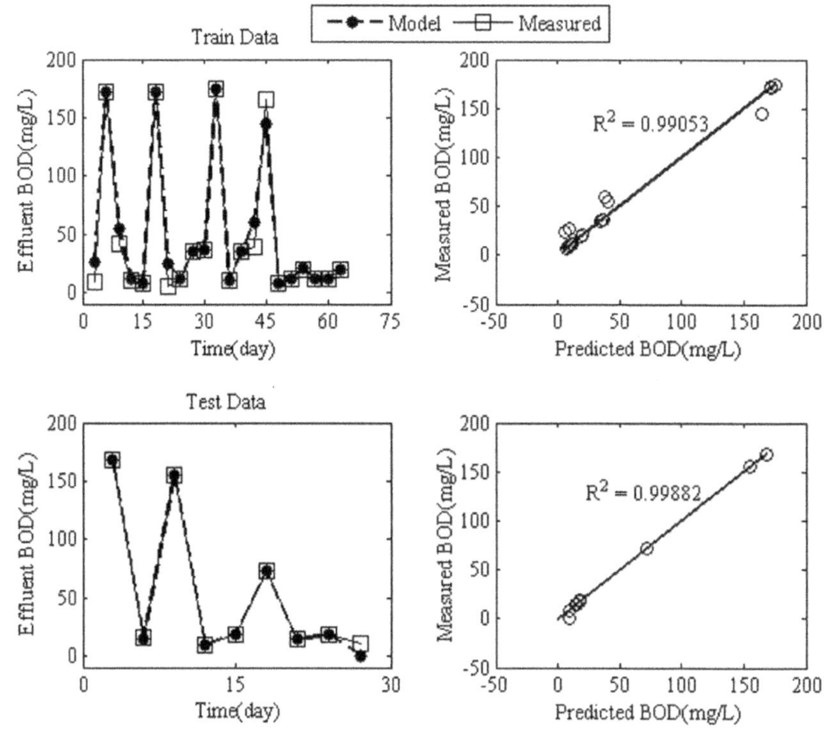

Figure 3 Simulated effluent concentration of BOD by RBFANN model for train and test data.

RBFANN predicted effluent BOD so accurate that the mean average error for train and test models did not exceed 5% and 3%, respectively.

The effluent BOD was modeled individually by considering different single and joint variables as inputs of neural network to examine the effect of each variable on the changes of effluent BOD concentration. The joint inputs to train the neural network were groups of two, three and four variables. Table 5 shows HRT among single input variables, and HRT and MLVSS among groups of two variables had the most considerable effect on the effluent BOD concentration. Furthermore, HRT, MLVSS and TDS among groups of three variables, and HRT, MLVSS, TDS and influent BOD between groups of four variables were determined to have the greatest effects on the effluent BOD concentration.

The sensitivity [62] of effluent BOD concentration to changes of an input variable such as HRT determines the influence and importance of HRT on the effluent BOD models. The effect of each variable on the RBFANN models to simulate effluent of BOD compared with other variables was determined by its importance order. Table 5 shows the importance order of each input variable and the joint variables for effluent BOD. The variable with higher rank of importance denoted not only a favorable match between experimental and simulated values by RBFANN

Table 5 Effect of different single and joint variables on the effluent BOD models

Input variable no.	R^2		RMSE (mg/L)		Importance order
	Train	Test	Train	Test	
1	0.804	0.863	35.62	32.80	5
2	0.991	0.999	8.15	2.79	1
3	0.907	0.995	24.76	6.43	2
4	0.705	0.674	48.71	42.27	4
5	0.804	0.863	35.62	32.8	3
2-1	0.973	0.961	17.39	16.28	4
2-3	0.998	0.999	2.98	2.15	1
2-4	0.996	0.978	6.82	2.46	3
2-5	0.995	0.999	6.65	1.76	2
2-3-1	0.992	0.999	6.17	2.41	2
2-3-4	0.998	1	4.08	0.12	1
2-3-5	0.996	0.998	5.62	4.23	3
2-3-4-1	0.998	0.998	3.69	3.07	1
2-3-4-5	0.997	0999	4.29	3.25	2
2-3-4-1-5	0.990	0.998	8.67	4.56	1

The numbers 1 to 5 refers to input variables identified in Table 4.

models but also low RMSE and high R^2 values. The variation of effluent BOD concentration was influenced by HRT, MLVSS, TDS, influent BOD concentration and pH, respectively. The results of this study show that the HRT and MLVSS have the greatest influence on the effluent BOD, which are in a good agreement with earlier experimental studies [63,64].

COD removal efficiency and modeling outcomes

To perform COD experiments for the municipal wastewater the influent COD concentration varied from 310 to 360 mg/L. By increasing HRT from 1 to 5 h the removal efficiencies increased and the best results obtained at HRT of 5 h. At this point, the effluent COD concentration reached to 8 mg/L with removal efficiency of 97.9%. The effluent COD concentration increased and removal efficiency decreased after 5 h [60]. Consequently, the kinetic constants were calculated as K_s equal to 96 mg L^{-1}, k equal to 2.31 d^{-1} and K_d equal to 0.043 d^{-1}. The influent COD concentration decreased from 354 mg/L to 53 mg/L for the effluent COD at HRT of 1 h. It means that the effluent COD concentration at HRT of 1 h meets discharge limits of the municipal wastewater (COD < 60 mg/L).

For the industrial wastewater the influent COD concentration varied from 2050 to 2120 mg/L. From the HRT of 5 to 17 h the removal efficiency increased and the best results obtained at HRT of 17 h. At this point, the influent COD concentration decreased from 2100 mg/L to 14.8 mg/L for the effluent COD. Consequently, the kinetic constants were calculated as K_s equal to 308 mg L^{-1}, k equal to 2.81 d^{-1}, K_d equal to 0.019 d^{-1}. The influent COD concentration decreased from 2055 mg/L to 71.2 mg/L for the effluent with removal efficiency of 96.5% at HRT of 13 h. Therefore, it was concluded that the HRT of 13 h cannot meet the discharge limits and HRT of 17 h was considered as the optimal result.

The COD experiments were performed for the combined municipal/industrial wastewater with the HRT varying from 3 to 11 h. The influent COD concentration for the combined wastewater varied from 1000 to 1200 mg/L. The influent COD concentration at HRT of 9 h decreased from 1130 mg/L to 10.3 mg/L for the effluent of COD with removal efficiency of 99.1%. By increasing HRT from 9 to 11 h the removal efficiency decreased [60,61]. Consequently, the kinetic constants were calculated as K_s equal to 113.2 mg L^{-1} k equal to 2.72 d^{-1}, K_d equal to 0.022 d^{-1}. The results indicated that at HRT of 7 h the influent COD concentration decreases from 1080 mg/L to 50 mg/L for the effluent with removal efficiency of 96%. Therefore, the effluent COD concentration at HRT of 7 h met discharge limits. It was observed that the performance of the SMBR increases due to reduction of wastewater strength [9,24] when the influent COD concentration is lower than 1000 to 1100 mg/L.

In order to model the effluent COD by RBFANN, the influent COD, HRT, MLVSS, TDS and pH were the input variables of the neural network. The results of the effluent COD modeling using the RBFANN for train and test data were denormalized to compare the observed values of effluent COD concentration with simulated values. The results of the training procedure by RBFANN were successful for the effluent COD concentration. Figure 4 shows the train and test models by RBFANN indicated an almost perfect match between the experimental and the simulated effluent COD values compared with the results of models introduced by Cinar et al. [38]. The RMSE values for train and test (verification process) models were 25.62 and 9.12 mg/L, respectively. The values of R^2 for train and test models were 0.99 and 0.98, respectively. The RBF models predicted effluent COD concentration with a mean average error for train and test models, which did not exceed 7% and 3%, respectively.

Table 6 shows HRT among single input variables, and HRT and MLVSS among groups of two variables had the most considerable effects on the effluent COD. Furthermore, HRT, MLVSS and TDS among groups of three variables, and HRT, MLVSS, TDS and influent COD between groups of four variables were determined to have the greatest effect on the effluent COD. Table 6 also shows the importance order of each input and joint variable for effluent COD according to sensitivity analysis procedure. The variation of effluent COD concentration was influenced by HRT, MLVSS, TDS, influent COD concentration and pH, respectively. The results showed that the HRT and MLVSS have the greatest influence on the effluent BOD, which are in a good agreement with earlier experimental studies [63,64]. The results of sensitivity analyses were highly collaborated for both BOD and COD models performed by RBFANN. The variation of both effluent BOD and COD concentration to all input variables were the same in their simulation processes, which can be justified by similarities of their natures. As a result, to control the changes of effluent BOD and COD concentration the most effective variables are HRT and MLVSS.

NH_4^+–N removal efficiency and modeling outcomes

The influent NH_4^+–N concentration for the municipal wastewater varied from 18 to 24 mg/L. The results showed that the removal efficiency of NH_4^+–N is improved with increase of the HRT. The influent NH_4^+–N concentration decreased from 24 mg/L to 0.4 mg/L for the effluent at HRT of 7 h. The influent NH_4^+–N

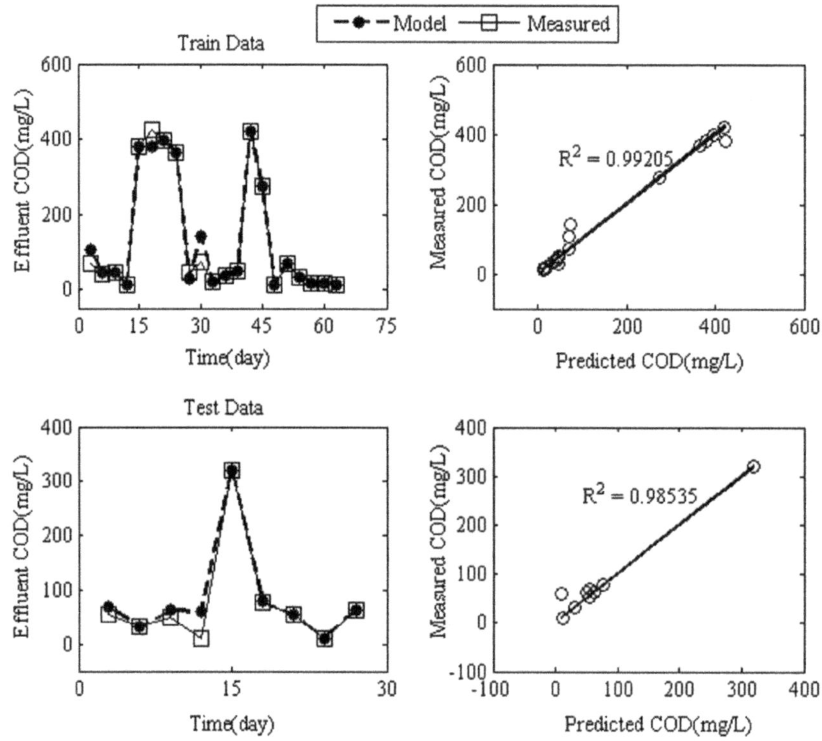

Figure 4 Simulated effluent concentration of COD by RBFANN model for train and test data.

concentration decreased from 23 mg/L to 0.8 mg/L at HRT of 5 h. As the optimal result, the effluent NH_4^+-N concentration with removal efficiency of 96.5% at HRT of 5 h met the discharge limit (NH_4^+-N <1 mg/L). Increasing the cell retention time increases Azotobacters and Organotrophic bacteria and then followed by rapid removal of dissolved carbonaceous biochemical oxygen demand (CBOD) [20]. Rapid removal of dissolved CBOD increases the aeration time for the nitrification process [5,20]. Increases of the Azotobacter populations in activated sludge process occurs at relatively high hydraulic retention time [5,64]. Hence, the required HRT of municipal wastewater treatment was increased to 5 h.

For the industrial wastewater the influent NH_4^+-N concentration varied from 24 to 31 mg/L. The NH_4^+-N concentration decreased with increase of HRT for the industrial wastewater. And at HRT of 17 h the effluent N H_4^+-N concentration was equal to 0.2 mg/L with the removal efficiency of 99.35%. The effluent NH_4^+-N concentration at HRT of 13 h decreased from 30 mg/L to 0.7 mg/L for the effluent with the removal efficiency of 97.7%.

For the combined municipal and industrial wastewater the influent NH_4^+-N concentration varied from 21 to 27 mg/L. The effluent concentration at HRT of 7 h met

Table 6 Effect of different single and joint variables on the effluent COD models

Input variable no.	R^2		RMSE (mg/L)		Importance order
	Train	Test	Train	Test	
1	0.179	0.599	153.49	91.66	5
2	0.947	0.997	44.36	14.1	1
3	0.941	0.994	48.23	35.67	2
4	0.6	0.8	123	63.41	3
5	0.862	0.715	70	109.03	4
2-1	0.951	0.999	47	4.3	4
2-3	0.995	0.999	15.43	6.19	1
2-4	0.998	0.994	11.23	13.58	3
2-5	0.987	0.998	22.45	7.66	2
2-3-1	0.984	0.996	26.66	13.34	3
2-3-4	0.994	0.996	16.48	8.48	1
2-3-5	0.974	0.997	33.13	11.56	2
2-3-4-1	0.992	0.999	19.58	2.61	1
2-3-4-5	0.987	0.991	22.08	14.93	2
2-3-4-1-5	0.992	0.985	25.62	9.12	1

The numbers 1 to 5 refers to input variables identified in Table 4.

the discharge limits. The influent concentration decreased from 24 mg/L to 0.8 mg/L for the effluent with removal efficiency of 96.7%. The results of NH_4^+-N experiments for three systems showed that the required HRT for NH_4^+-N removal increases the required HRT for the treatment of municipal wastewater and does not have any effect on the required HRT for industrial and combined municipal and industrial wastewaters. The idea of combined wastewater is more useful because the treatment system is able to receive sewage with higher influent BOD and COD concentration without any increase in the HRT.

In order to model the effluent NH_4^+-N by RBFANN, the influent NH_4^+-N concentration, HRT, MLVSS, TDS and pH were the input variables of network. The results of the effluent NH_4^+-N modeling using the RBFANN for train and test data sets were denormalized to compare the observed values of effluent NH_4^+-N concentration with simulated values. The results of the training procedure by RBFANN were successful for the effluent N H_4^+-N. Figure 5 shows that the train and test models by RBFANN indicated an almost perfect match between the experimental and the simulated values of effluent NH_4^+-N concentration compared with the results of models introduced by Cinar et al. [38]. The RMSE values for train and test (verification process) models were 0.06 and 0.14 mg/L, respectively. The R^2 values for train and test models were 0.98 and 0.99, respectively. The RBF models simulated effluent NH_4^+-N with a mean average error for train and test models, which did not exceed 2% and 5%, respectively.

Table 7 shows HRT among single input variables, and HRT and MLVSS among groups of two variables had the most considerable effect on the COD effluent. Furthermore, HRT, MLVSS and pH among groups of three variables, and HRT, MLVSS, pH and influent NH_4^+-N between groups of four variables were determined to have the greatest effect on the COD effluent. Table 7 also shows the importance order of each input variable and the joint variables for effluent NH_4^+-N according to sensitivity analysis procedure. The variation of effluent NH_4^+-N concentration was influenced by HRT, MLVSS, pH, influent NH_4^+-N and TDS, respectively. The current research shows that the HRT and MLVSS have the greatest influence on the changes of effluent NH_4^+-N, which are in a good agreement with earlier experimental studies [63,64]. The results of sensitivity analysis for the effluent NH_4^+-N models indicated that to control the changes of effluent NH_4^+-N concentration the most effective variables are HRT and MLVSS.

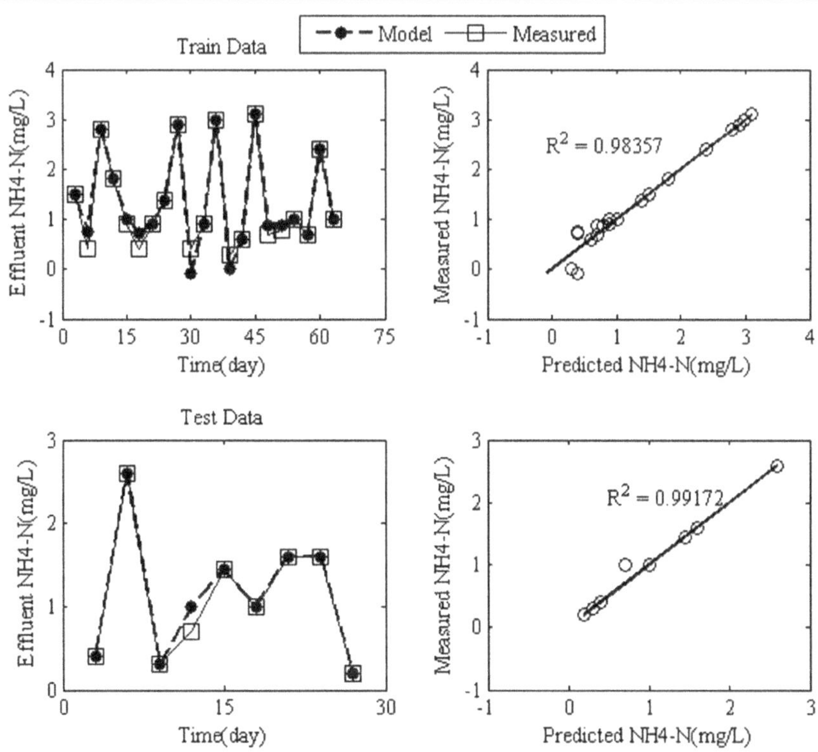

Figure 5 Simulated effluent concentration of NH_4^+-N by RBFANN model for train and test data.

Table 7 Effect of different single and joint variables on the effluent NH_4^+–N models

Input variable no.	R^2		RMSE (mg/L)		Importance order
	Train	Test	Train	Test	
1	0.59	0.44	0.78	0.62	5
2	0.983	0.998	0.16	0.12	1
3	0.923	0.985	0.35	0.17	2
4	0.54	0.6	0.81	0.52	4
5	0.78	0.94	0.56	0.29	3
2-1	0.99	0.993	0.13	0.13	3
2-3	0.99	0.994	0.13	0.09	1
2-4	0.99	0.964	0.12	0.23	4
2-5	0.99	0.992	0.14	0.11	2
2-3-1	0.987	0.995	0.15	0.09	3
2-3-4	0.99	0.996	0.13	0.05	2
2-3-5	0.988	1	0.14	0	1
2-3-5-1	0.99	0.998	0.14	0.06	1
2-3-5-4	0.97	0.996	0.2	0.1	2
2-3-5-1-4	0.98	0.991	0.06	0.14	1

The numbers 1 to 5 refers to input variables identified in Table 4.

Results of TP removal efficiency and modeling

The influent TP concentration for the municipal wastewater varied from 13.2 to 15.1 mg/L. The results showed that the removal efficiency of TP is improved with increase of the HRT from 1 to 5 h. Then, the removal efficiency decreased for the HRT of 5 to 7 h. The influent TP concentration decreased from 14.8 to 2 mg/L for the effluent at HRT of 5 h with removal efficiency of 86.5%. The influent TP concentration decreased from 14.9 to 5.8 mg/L for the effluent at HRT of 1 h and met discharge limit (effluent TP <6 mg/L). The effluent TP concentration depends on the TBOD/TP ratio so with a ratio less than 20, it is not possible to achieve the effluent TP concentration lower than 2 mg/L [65]. We observed that the TBOD/TP ratio for the influent of Ekbatan wastewater treatment plant was less than 20. Subsequently, the combination of municipal and industrial wastewater was an effective method to correct this problem.

The influent TP concentration for the combined wastewater varied from 15 to 16.4 mg/L. The influent TP concentration decreased from 16.2 mg/L to 5.7 mg/L for the effluent TP at HRT of 1 h. With increasing the HRT from 7 to 9 h the effluent TP concentration reached to lower than 1 mg/L with removal efficiency of 92%. The influent TP varied from 17.24 to 20.3 mg/L for the industrial wastewater. The high ratios of the TBOD/TP for the industrial wastewater allowed the effluent TP to be lower than 1 mg/L. The effluent TP concentration reached to 0.7 mg/L with the removal efficiency of 96.55% at HRT of 17 h.

In order to model the effluent TP by RBFANN, the influent TP concentration, HRT, MLVSS, TDS and pH were the input variables of the neural network. The results of the effluent TP modeling using the RBFANN for train and test data sets were denormalized to compare the observed values of effluent TP with simulated values. The results of the training procedure by RBFANN were successful for the effluent TP. Figure 6 shows that the train and test models by RBFANN indicated an almost perfect match between the experimental and the simulated effluent TP concentration compared with the results of models introduced by Cinar et al. [38]. The RMSE values for train and test (verification process) models were 0.32 and 0.17 mg/L respectively, and the value of R^2 was 0.99 for both models. The RBFANN models simulated effluent TP so accurate that the mean average error for train and test models did not exceed 5% and 3%, respectively.

Table 8 shows HRT among single input variables, and HRT and pH among groups of two variables had the most considerable effect on the effluent TP. Furthermore, HRT, pH and MLVSS among groups of three variables, and HRT, pH, MLVSS and influent TP concentration between groups of four variables were determined to have the greatest effect on the effluent TP. Table 8 also shows the importance order of each input variable and the joint variables for effluent TP according to sensitivity analysis procedure. The variation of effluent TP concentration was influenced by HRT, pH, MLVSS, influent TP concentration and TDS, respectively, which is in a good agreement with earlier experimental studies [63,64].

Conclusions

The current research was an effort to evaluate performance of an SMBR treating combined municipal and industrial wastewater compared with treating municipal and industrial wastewaters. The combined municipal and industrial wastewater showed more satisfactory treating features for wastewater treatment by an SMBR compared with the municipal and industrial wastewaters. Although the concentration of components in combined wastewater were almost half of the industrial wastewater, required HRT for combined wastewater was 7 h in comparison to 17 h for industrial wastewater. It was observed that treatment performance of the SMBR improves and HRT decreases noticeably by decreasing BOD and COD concentration to lower than 700 mg/L and 1000 mg/L, respectively. The results indicated that the combined wastewater improves treatment performance by increasing TBOD/TP ratio and setting BOD/COD ratio to 0.5. Therefore, effluent TP concentration was lower than 2 mg/L by increasing TBOD/TP ratio to more than 20. This study showed that it is possible to

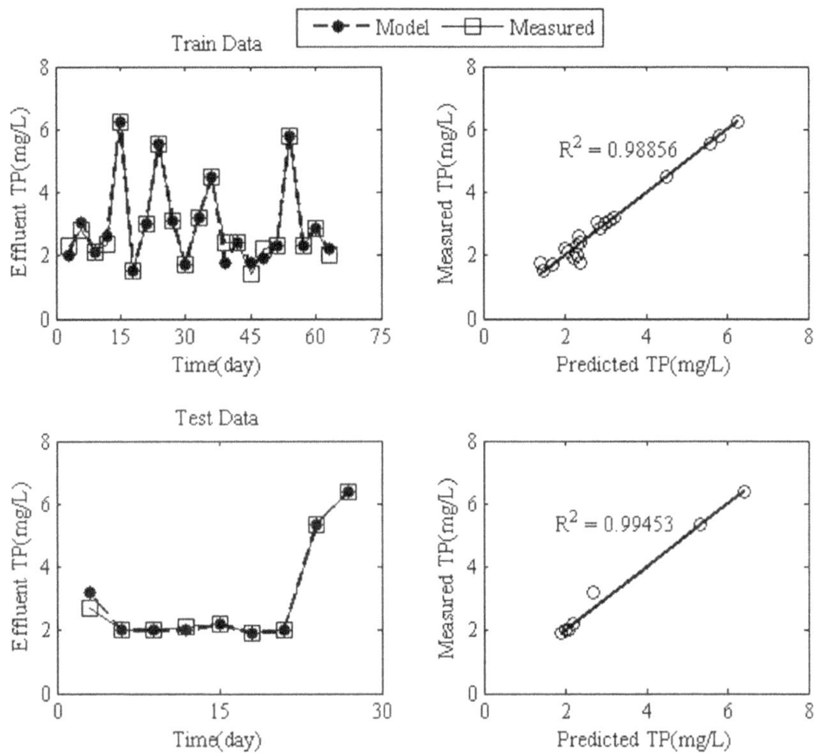

Figure 6 Simulated effluent concentration of TP by RBFANN model for train and test data.

Table 8 Effect of different single and joint variables on the effluent TP models

Input variable no.	R^2		RMSE (mg/L)		Importance order
	Train	Test	Train	Test	
1	0.34	0.3	1.15	1.88	5
2	0.972	0.992	0.35	0.14	1
3	0.878	0.978	0.7	0.29	2
4	0.51	0.8	1.39	0.82	4
5	0.77	0.96	0.99	0.4	3
2-1	0.975	0.97	0.37	0.37	4
2-3	0.99	0.99	0.22	0.16	2
2-4	0.92	0.99	0.53	0.09	3
2-5	0.995	0.997	0.11	0.11	1
2-5-1	0.99	0.998	0.2	0.13	2
2-5-3	0.991	1	0.14	0	1
2-5-4	0.96	0.999	0.35	0.09	3
2-5-3-1	0.99	0.998	0.2	0.1	1
2-5-3-4	0.99	0.993	0.23	0.11	2
2-5-3-1-4	0.988	0.994	0.32	0.17	1

The numbers 1 to 5 refers to input variables identified in Table 4.

achieve a cost efficient wastewater treatment by SMBR for the combined wastewater because required HRT was at the same range for BOD, COD and NH_4^+-N compared with the municipal and industrial wastewaters.

Treatment process models are efficient tools to assure proper operation and better control of wastewater treatment systems. ANNs have been successfully used for monitoring, controlling, classification and simulation of experimental data. In this study, an RBFANN was utilized in order to simulate effluent quality parameters of the SMBR treating combined municipal and industrial wastewater. An RBFANN is the most commonly used neural network for pattern recognition problems, it is also widely used for fault diagnosis and solving ill-conditioned problems. The RBFANN has the advantages of a fast learning process, a learning stage without any iteration of updating weights, robust ability, and adaptation capability compared with other ANNs such as MLP, RNN and ESN. The results showed that the training and testing procedure by RBFANN for effluent BOD, COD, NH_4^+-N and TP were successful. The train and test models showed an almost perfect match between the experimental and predicted values of effluent BOD, COD, NH_4^+-N and TP.

Abbreviations

MBR: Membrane bioreactor; SMBR: Submerged membrane bioreactor; HRT: Hydraulic retention time; COD: Chemical oxygen demand; TN: Total nitrogen; BOD: Biochemical oxygen demand; TBOD/TP: Total BOD to total phosphorous; ASP: Activated sludge process; ASM: Activated sludge model; ANN: Artificial neural network; RBF: Radial basis function; MLP: Multi-layer perceptron; RNN: Recurrent neural network; ESN: Echo-state network; RBFANN: Radial basis function artificial neural network; MLVSS: Mixed liquor volatile suspended solids; TDS: Total dissolved solids; TSS: Total suspended solids; DO: Dissolved oxygen; MLSS: Mixed liquor suspended solids; MLPANN: Multi-layer perceptron artificial neural network; RMSE: Root mean squared error; R^2: Coefficient of determination; K_s: Half saturation coefficient; k: Maximum substrate utilization rate.K_d, Endogenous decay coefficient; CBOD: Carbonaceous biochemical oxygen demand..

Competing interests

The authors declare that they have no competing interests.

Authors' contributions

SAM participated in design of the pilot plant and contributed in the modeling process. MB participated in the design of pilot, carried out the experiments and modeling and drafted the manuscript. SB participated in modeling process and helped to draft manuscript. ME participated in the experiments and helped to draft manuscript. ZB carried out the biological interpretation, participated in modeling process and helped to draft manuscript. All authors read and approved the final manuscript.

Acknowledgments

The authors are grateful to Ekbatan wastewater treatment plant for their technical and logistical assistance during this work which was supported by authors. And also, we wish to thank Ali Reza Jafari and Ali Morad Kamarkhani for technical help with modeling by artificial neural network.

Author details

[1]Department of Civil Engineering, K.N. Toosi University of Technology, Vanak square, Tehran, Iran. [2]Department and Faculty of Basic Sciences, PUK University, Kermanshah, Iran.

References

1. Yang W, Cicek N, Ilg J. State-of-the-art of membrane bioreactors: Worldwide research and commercial applications in North America. J Membr Sci. 2006;270:201–11.
2. Katsou E, Malamis S, Loizidou M. Performance of a membrane bioreactor used for the treatment of wastewater contaminated with heavy metals. Bioresour Technol. 2011;102:4325–32.
3. Rosenberger S, Krüger U, Witzig R, Manz W, Szewzyk U, Kraume M. Performance of a bioreactor with submerged membranes for aerobic treatment of municipal waste water. Water Res. 2002;36:413–20.
4. Ferrai M, Guglielmi G, Andreottola G. Modelling respirometric tests for the assessment of kinetic and stoichiometric parameters on MBBR biofilm for municipal wastewater treatment. Environ Model Softw. 2010;25:626–32.
5. Muller E, Stouthamer A, Van Verseveld HW, Eikelboom D. Aerobic domestic waste water treatment in a pilot plant with complete sludge retention by cross-flow filtration. Water Res. 1995;29:1179–89.
6. Zaloum R, Lessard S, Mourato D, Carriere J. Membrane bioreactor treatment of oily wastes from a metal transformation mill. Water Sci Technol. 1994;30:21–7.
7. Scholz W, Fuchs W. Treatment of oil contaminated wastewater in a membrane bioreactor. Water Res. 2000;34:3621–9.
8. W-t Z, Huang X, Lee D-j. Enhanced treatment of coke plant wastewater using an anaerobic–anoxic–oxic membrane bioreactor system. Sep Purif Technol. 2009;66:279–86.
9. Mutamim NSA, Noor ZZ, Hassan MAA, Olsson G. Application of membrane bioreactor technology in treating high strength industrial wastewater: a performance review. Desalination. 2012;305:1–11.
10. Sutton PM. Membrane bioreactors for industrial wastewater treatment: Applicability and selection of optimal system configuration. Proceedings Water Environ Federation. 2006;2006:3233–48.
11. Chang C-Y, Chang J-S, Vigneswaran S, Kandasamy J. Pharmaceutical wastewater treatment by membrane bioreactor process–a case study in southern Taiwan. Desalination. 2008;234:393–401.
12. Marrot B, Barrios-Martinez A, Moulin P, Roche N. Industrial wastewater treatment in a membrane bioreactor: a review. Environ Prog. 2004;23:59–68.
13. Barakat MA. New trends in removing heavy metals from industrial wastewater. Arab J Chem. 2011;4:361–77.
14. Artiga P, Ficara E, Malpei F, Garrido J, Mendez R. Treatment of two industrial wastewaters in a submerged membrane bioreactor. Desalination. 2005;179:161–9.
15. Badani Z, Ait-Amar H, Si-Salah A, Brik M, Fuchs W. Treatment of textile waste water by membrane bioreactor and reuse. Desalination. 2005;185:411–7.
16. Brik M, Schoeberl P, Chamam B, Braun R, Fuchs W. Advanced treatment of textile wastewater towards reuse using a membrane bioreactor. Process Biochem. 2006;41:1751–7.
17. Lesage N, Sperandio M, Cabassud C. Study of a hybrid process: Adsorption on activated carbon/membrane bioreactor for the treatment of an industrial wastewater. Chemi Eng Process: Process Intensif. 2008;47:303–7.
18. Katayon S, Megat Mohd Noor M, Ahmad J, Abdul Ghani L, Nagaoka H, Aya H. Effects of mixed liquor suspended solid concentrations on membrane bioreactor efficiency for treatment of food industry wastewater. Desalination. 2004;167:153–8.
19. Viero AF, De Melo TM, Torres APR, Ferreira NR. The effects of long-term feeding of high organic loading in a submerged membrane bioreactor treating oil refinery wastewater. J Membr Sci. 2008;319:223–30.
20. Tchobanoglous G, Burton FL, Stensel HD, Metcalf & Eddy. Wastewater Engineering; Treatment and Reuse. New York, NY: McGraw-Hill Education; 2003.
21. Feng F, Xu Z, Li X, You W, Zhen Y. Advanced treatment of dyeing wastewater towards reuse by the combined Fenton oxidation and membrane bioreactor process. J Environ Sci. 2010;22:1657–65.
22. Yigit N, Uzal N, Koseoglu H, Harman I, Yukseler H, Yetis U, et al. Treatment of a denim producing textile industry wastewater using pilot-scale membrane bioreactor. Desalination. 2009;240:143–50.
23. Yuniarto A. Ujang Z. Performance of bio-fouling reducer in submerged membrane bioreactor for palm oil mill effluent treatment. In International Conference & Exposition on Environmental Management and Technologies, PWTC, Kuala Lumpur: Noor ZZ; 2008.
24. Acharya C, Nakhla G, Bassi A. Operational optimization and mass balances in a two-stage MBR treating high strength pet food wastewater. J Environ Eng. 2006;132:810–7.
25. Moral H, Aksoy A, Gokcay CF. Modeling of the activated sludge process by using artificial neural networks with automated architecture screening. Comput Chem Eng. 2008;32:2471–8.
26. Gernaey KV, Van Loosdrecht M, Henze M, Lind M, Jørgensen SB. Activated sludge wastewater treatment plant modelling and simulation: state of the art. Environ Model Softw. 2004;19:763–83.
27. Henze M, Grady C, Gujer W, Marais G, Matsuo T. A general model for single-sludge wastewater treatment systems. Water Res. 1987;21:505–15.
28. Henze M, Gujer W, Mino T, Matsuo T, Wentzel M, Marais G. Wastewater and biomass characterization for the activated sludge model no. 2: biological phosphorus removal. Water Sci Technol. 1995;31:13–23.
29. Gujer W, Henze M, Mino T, Loosdrecht M. Activated sludge model no. 3. Water Sci Technol. 1999;39:183–93.
30. Chen L, Tian Y, Cao C, Zhang S, Zhang S. Sensitivity and uncertainty analyses of an extended ASM3-SMP model describing membrane bioreactor operation. J Membr Sci. 2012;389:99–109.
31. Geissler S, Wintgens T, Melin T, Vossenkaul K, Kullmann C. Modelling approaches for filtration processes with novel submerged capillary modules in membrane bioreactors for wastewater treatment. Desalination. 2005;178:125–34.
32. Mannina G, Cosenza A, Viviani G. Uncertainty assessment of a model for biological nitrogen and phosphorus removal: Application to a large wastewater treatment plant. Phys Chem Earth, Parts A/B/C. 2012;42:61–9.
33. Sha W, Edwards K. The use of artificial neural networks in materials science based research. Mater Des. 2007;28:1747–52.
34. Azmy AM, Erlich I, Sowa P. Artificial neural network-based dynamic equivalents for distribution systems containing active sources. IEE Proceedings-Generation, Trans Distrib. 2004;151:681–8.
35. Park J-W, Venayagamoorthy GK, Harley RG. MLP/RBF neural-networks-based online global model identification of synchronous generator. Ind Electron, IEEE Trans. 2005;52:1685–95.

36. Singh S, Venayagamoorthy G. Online identification of turbogenerators in a multimachine power system using RBF neural networks. St. Louis, Missouri, USA: Artificial Neural Networks in Engineering Conference (ANNIE) 2000; 2002. p. 485–90.
37. Venayagamoorthy GK. Online design of an echo state network based wide area monitor for a multimachine power system. Neural Netw. 2007;20:404–13.
38. Çinar Ö, Hasar H, Kinaci C. Modeling of submerged membrane bioreactor treating cheese whey wastewater by artificial neural network. J Biotechnol. 2006;123:204–9.
39. Suchacz B, Wesołowski M. The recognition of similarities in trace elements content in medicinal plants using MLP and RBF neural networks. Talanta. 2006;69:37–42.
40. Ferrari S, Bellocchio F, Piuri V, Borghese NA. A hierarchical RBF online learning algorithm for real-time 3-D scanner. Neural Netw, IEEE Trans. 2010;21:275–85.
41. Lee C-M, Ko C-N. Time series prediction using RBF neural networks with a nonlinear time-varying evolution PSO algorithm. Neurocomput. 2009;73:449–60.
42. Wang S, Yu D. Adaptive RBF network for parameter estimation and stable air–fuel ratio control. Neural Netw. 2008;21:102–12.
43. Kumfer B, Felch C, Maugans C. Wet air oxidation treatment of spent caustic in petroleum refineries. In: N.P.R.s. Association, editor. National Petroleum Refiner's Association Conference, Phoenix, AZ. 2010. p. 21–3.
44. Samudro G, Mangkoedihardjo S. Review on BOD, COD and BOD/COD ratio: a triangle zone for toxic, biodegradable and stable levels. Int J Acad Res. 2010;2:235–9.
45. Andrew D. Standard methods for the examination of water and wastewater. APHA-AWWA-WEF: Washington, D.C; 2005.
46. Elmolla ES, Chaudhuri M. Combined photo-Fenton–SBR process for antibiotic wastewater treatment. J Hazard Mater. 2011;192:1418–26.
47. L-y F, Wen X-h, Yi Qian Q-IL. Treatment of dyeing wastewater in two SBR systems. Process Biochem. 2001;36:1111–8.
48. Kashaninejad M, Dehghani A, Kashiri M. Modeling of wheat soaking using two artificial neural networks (MLP and RBF). J Food Eng. 2009;91:602–7.
49. Abyaneh HZ. Evaluation of multivariate linear regression and artificial neural networks in prediction of water quality parameters. J Environ Health Sci Eng. 2014;12:1–8.
50. Tomenko V, Ahmed S, Popov V. Modelling constructed wetland treatment system performance. Ecol Model. 2007;205:355–64.
51. Madaeni S, Zahedi G, Aminnejad M. Artificial neural network modeling of O2 separation from air in a hollow fiber membrane module. Asia-Pacific J Chem Eng. 2008;3:357–63.
52. Shahsavand A, Chenar MP. Neural networks modeling of hollow fiber membrane processes. J Membr Sci. 2007;297:59–73.
53. Eslamimanesh A, Gharagheizi F, Mohammadi AH, Richon D. Artificial neural network modeling of solubility of supercritical carbon dioxide in 24 commonly used ionic liquids. Chem Eng Sci. 2011;66:3039–44.
54. Liu S, Zhang Y, Ma P, Lu B, Su H. A Novel Spatial Interpolation Method Based on the Integrated RBF Neural Network. Procedia Environ Sci. 2011;10:568–75.
55. Sahoo GB, Ray C. Predicting flux decline in crossflow membranes using artificial neural networks and genetic algorithms. J Membr Sci. 2006;283:147–57.
56. Xi X, Cui Y, Wang Z, Qian J, Wang J, Yang L, et al. Study of dead-end microfiltration features in sequencing batch reactor (SBR) by optimized neural networks. Desalination. 2011;272:27–35.
57. Lowe D, Broomhead D. Multivariable functional interpolation and adaptive networks. Complex syst. 1988;2:321–55.
58. Demuth H, Beale M. Neural network toolbox user's guide. 2000.
59. Pendashteh AR, Fakhru'l-Razi A, Chaibakhsh N, Abdullah LC, Madaeni SS, Abidin ZZ. Modeling of membrane bioreactor treating hypersaline oily wastewater by artificial neural network. J Hazard Mater. 2011;192:568–75.
60. De la Cruz N, Gimenez J, Esplugas S, Grandjean D, De Alencastro L, Pulgarin C. Degradation of 32 emergent contaminants by UV and neutral photo-fenton in domestic wastewater effluent previously treated by activated sludge. Water Res. 2012;46:1947–57.
61. Qin L, Zhang G, Meng Q, Xu L, Lv B. Enhanced MBR by internal micro-electrolysis for degradation of anthraquinone dye wastewater. Chem Eng J. 2012;210:575–84.
62. Saltelli A, Ratto M, Andres T, Campolongo F, Cariboni J, Gatelli D, et al. Global sensitivity analysis: the primer. England: John Wiley & Sons Ltd, The Atrium, Southern Gate, Chichester, West Sussex PO19 8SQ; 2008.
63. Campos JC, Borges RMH, Oliveira Filho AM, Nobrega R, Sant'Anna G. Oilfield wastewater treatment by combined microfiltration and biological processes. Water Res. 2002;36:95–104.
64. Tay J-H, Zeng JL, Sun DD. Effects of hydraulic retention time on system performance of a submerged membrane bioreactor. Sep Sci Technol. 2003;38:851–68.
65. Kim SH, Baumann ER. Investigation of Chemical Phosphate Removal from an Oxidation Ditch by Field Evaluation. Environ Eng Res. 1997;2:207–16.

9

Immobilized laccase mediated dye decolorization and transformation pathway of azo dye acid red 27

Meenu Chhabra[1*], Saroj Mishra[2] and Trichur Ramaswamy Sreekrishnan[2]

Abstract

Background: Laccases have good potential as bioremediating agents and can be used continuously in the immobilized form like many other enzymes.

Methods: In the present study, laccase from *Cyathus bulleri* was immobilized by entrapment in Poly Vinyl Alcohol (PVA) beads cross-linked with either nitrate or boric acid. Immobilized laccase was used for dye decolorization in both batch and continuous mode employing a packed bed column. The products of degradation of dye Acid Red 27 were identified by LC MS/MS analysis.

Results: The method led to very effective (90%) laccase immobilization and also imparted significant stability to the enzyme (more than 70% after 5 months of storage at 4°C). In batch decolorization, 90-95% decolorization was achieved of the simulated dye effluent for up to 10–20 cycles. Continuous decolorization in a packed bed bioreactor led to nearly 90% decolorization for up to 5 days. The immobilized laccase was also effective in decolorization and degradation of Acid Red 27 in the presence of a mediator. Four products of degradation were identified by LC-MS/MS analysis.

Conclusions: The immobilized laccase in PVA-nitrate was concluded to be an effective agent in treatment of textile dye effluents.

Keywords: Immobilization, Polyvinylalcohol beads, Laccase-mediator, Acid Red 27, Azo dye degradation

Introduction

The effluents generated from textile industries are reported to be one of the top ten contaminating sources of water bodies. Ligninolytic enzymes, including laccases, produced by white rot fungi find immense applications in treatment of the toxic wastes generated from textile industries [1,2]. Application of laccases for dye degradation has not been reported so far on an industrial scale mainly because of the high cost of production of these enzymes. In any treatment method, development of a reusable enzyme system, such as that generated by immobilization of enzymes, is most beneficial.

Several attempts have been made to immobilize laccases [3,4]. A number of studies report on the use of immobilized laccase preparations for dye decolorization [4-12]. Immobilization of laccase has been reported to increase thermostability and resistance to inhibitors when compared to the free enzyme [6,7,12]. While most of these studies report the use of immobilized preparations in the batch mode, Palmieri et al. [13] used laccase, immobilized in copper alginate beads, to perform continuous decolorization of dye solutions. However, an additional step that involved coating immobilized enzyme beads with chitosan was added to prevent leaching of laccase during decolorization [13]. Keeping in view the benefits of continuous decolorization and to circumvent additional steps required to reduce laccase leaching, the aim of this study was to immobilize *Cyathus bulleri* laccase. Laccase from this fungus has been demonstrated to decolorize a number of reactive dyes [14,15]. The rates of dye decolorization were significantly enhanced in the presence of mediators [16,17]. This laccase was immobilized in Poly Vinyl Alcohol (PVA) based polymers crosslinked either by nitrate or boric acid. PVA is a reasonably cheap material and its polymers are reported to have high mechanical strength [18]. The

* Correspondence: meenuchhabra@iitj.ac.in
[1]Department of Biology, Indian Institute of Technology-Jodhpur, Jodhpur 342011, Rajasthan, India
Full list of author information is available at the end of the article

PVA immobilized laccase preparation was tested for its applicability in batch and continuous decolorization processes. Immobilized laccase was also used to investigate degradation of an azo dye, namely, Acid Red (AR) 27. AR 27, also known as Amaranth, is an anionic azo dye and is applied to natural and synthetic fibers, paper, leather, and phenol-formaldehyde resins. It was more commonly used as food and cosmetic colorant. However, its use has been banned in USA by food and drug administration (FDA) as it is a suspected carcinogen. The dye is known to be decolorized by treatment with laccase/laccase-mediators [10] but no work has been done to elucidate its breakdown by laccase.

The present study was thus conducted with the following major objectives: (i) to immobilize laccase in PVA based beads and to test its stability and efficacy for both batch and continuous decolorization and (ii) to determine degradation capability of the immobilized laccase + mediator on AR 27 and to characterize the degradation products

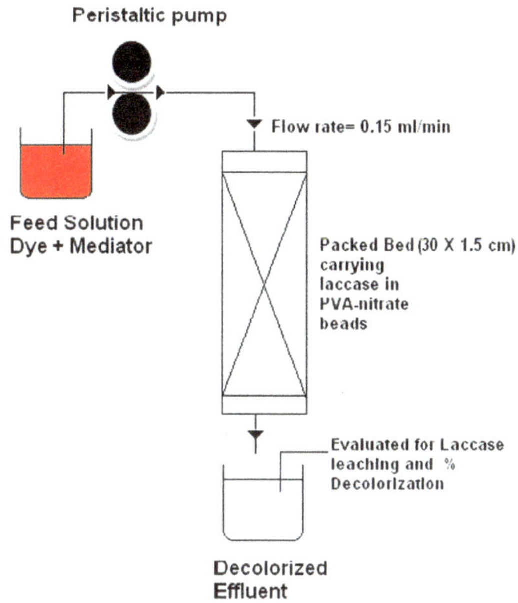

Figure 1 Schematic diagram of the continuous packed column.

Materials and methods

Laccase production

Laccase was produced in the culture supernatant of *C. bulleri* by cultivating it in basal liquid medium as described previously [14]. The culture filtrate was concentrated by ultrafiltration using 10 kDa membrane in a stirred cell (Amicon). The specific activity of laccase was 31 U/mg protein using guaiacol as a substrate.

Laccase immobilization

Laccase was entrapped in beads by two different methods. The materials and gel matrices (sodium alginate, calcium chloride, PVA, boric acid and sodium nitrate) were obtained from Himedia labs and Merck chemicals. The concentrated culture filtrate was mixed thoroughly with PVA-alginate solution. The final concentration of PVA and alginate was 8% and 1% respectively. This solution was centrifuged at 4000 rpm for 5 min to remove air bubbles and filled in a 10 ml syringe. Beads were prepared by extruding the mixture as drops into a stirred saturated boric acid solution to form beads and these were allowed to stay in the same solution for 20 min. The beads were then transferred to 1 M sodium dihydrogen phosphate solution for hardening and further incubated for 1 h. The hardened beads were then extensively washed with distilled water and stored in a minimum amount of water at 4°C. PVA-nitrate beads were prepared by extruding the PVA-alginate-enzyme mixture as drops into stirred 50% (w/v) solution of sodium nitrate containing 1% calcium chloride (w/v). The beads were allowed to stand in this solution for 6 h after which washings with distilled water were performed. The resultant beads were stored in minimum amount of water at 4°C.

The washings were collected to determine laccase activity. Percent immobilization was calculated by subtracting the activity lost in washings from the number of initially loaded units used for immobilization. In order to determine the specific activity, i.e. number of active laccase units/g of beads, the laccase assay was performed at 55°C which resulted in bead dissolution. Total protein was estimated by Lowry's method.

Determination of laccase leaching and bead stability

Leaching of laccase from beads was determined by incubating specific number of beads with a fixed initial activity in 5 ml of 50 mM sodium acetate buffer, pH 5.0. The

Table 1 Laccase immobilization and residual laccase activity in immobilized preparations

Entrapment method	% immobilization	% Laccase activity retained after 108 h incubation at 100 rpm, pH 5.5 at 30°C.	% Intact beads at 1000 rpm	% Laccase activity retained after incubating at 1000 rpm
PVA-boric acid	65 ± 5	43 ± 2	92 ± 3	0
PVA-nitrate	90 ± 2	73 ± 2	100 ± 0	20 ± 1

Table 2 Comparison of properties of immobilized and free laccase

Parameter	PVA-nitrate immobilized laccase	Free laccase
IC 50 for chloride	450 ± 22 mM	600 ± 33 mM
IC 50 for EDTA	250 ± 17 mM	250 ± 10 mM
IC 50 for sodium azide	6 ± 0.3 μM	5 ± 0.5 μM
K_m (ABTS)	65 ± 4 μM	37 ± 2.1 μM
K_{cat} (ABTS)	1675 ± 75 sec^{-1}	2996 ± 179 sec^{-1}

beads were shaken at 100 rpm at 30°C. Enzyme activity was measured in the samples withdrawn after specific time intervals. The final activity was subtracted from the initial activity to determine enzyme loss due to leaching.

Bead stability was determined by incubating 10 intact beads in 2 ml of distilled water in a 25 ml conical flask and stirring at different speeds (ranging from 50–1000 rpm) for 30 min (modified protocol Yujian et al. [19]). The % intact granules v/s rpm plot was obtained for two different kinds of beads. The amount of laccase activity retained by immobilized preparations was determined.

Characterization of immobilized laccase

Laccase stability, optimum pH, temperature, resistance to some laccase inhibitors (chloride ions, sodium azide and EDTA after 1 h pre-incubation) and kinetic parameters (K_m and K_{cat}) were determined and compared to that of the free enzyme. Laccase assays were performed using either guaiacol or 2,2'-azino-bis (3-ethylbenzothiazoline-6-sulphonic acid) or ABTS as a substrate as described earlier [14] by adding fixed amount of either free or immobilized laccase. The concentration range of laccase inhibitors used for IC50 calculation ranged from 0–2000 mM NaCl, 0–1000 mM EDTA and 0–50 μM sodium azide. The ABTS concentration range for calculating K_m and K_{cat} values was 0–1 mM. The values of catalytic constants were measured using the Lineweaver Burk plot. The value of K_{cat} was determined by dividing V_{max} with total enzyme cocentration [E_{tot}].

Decolorization experiment using immobilized enzyme

Decolorization experiments were performed using the conditions optimized previously [17]. Decolorization re-

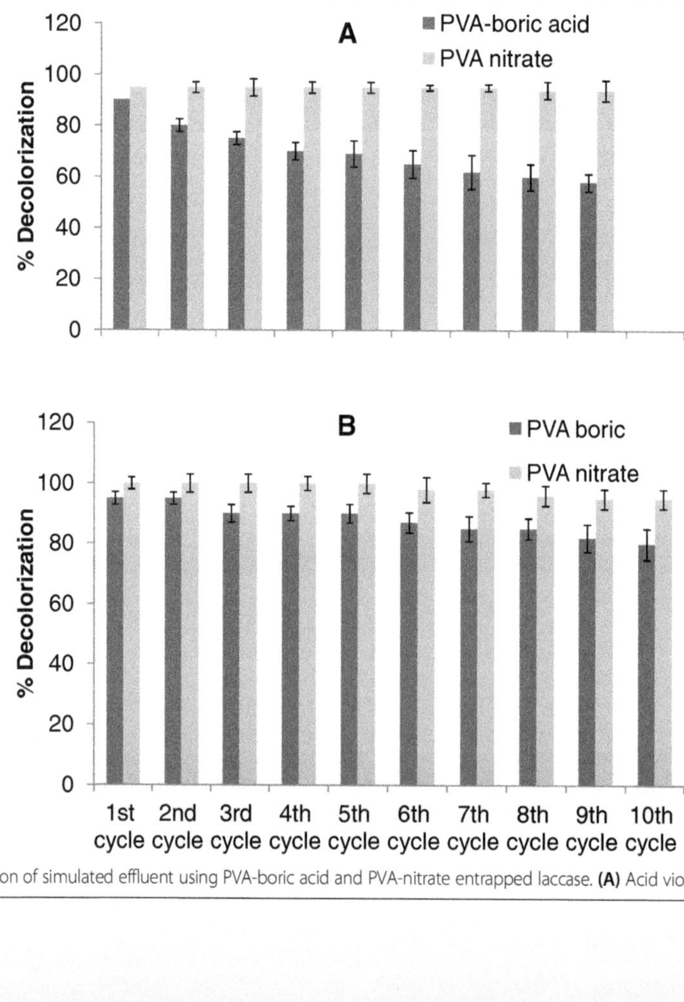

Figure 2 Batch decolorization of simulated effluent using PVA-boric acid and PVA-nitrate entrapped laccase. **(A)** Acid violet 17 or **(B)** Basic Green 4.

action was performed in a 25 ml conical flask. The reaction mixture consisted of either 100 μM of Acid Violet 17 or Basic Green 4 solution (that simulated absorbance of the textile effluent), 100 μM of ABTS and immobilized enzyme (100 mU/ml). The incubation was done at 30°C at 100 rpm. The change in absorbance at 542 nm (λ_{max} of Acid Violet 17) and 618 nm (λ_{max} of Basic Green 4) was used to compute % decolorization. After one cycle of decolorization (~90% decolorization), liquid was withdrawn and fresh dye solution was added to initiate a new cycle of decolorization.

Continuous decolorization in a packed bed column

Continuous decolorization of the simulated effluent was carried out in a column (30 cm × 1.5 cm) with approximate working volume of 50 ml. The schematic diagram of the same is shown in Figure 1. The simulated effluent contained 100 μM Acid Violet 17, as used in the local textile mills (m/z 710 and also with some unidentified peaks), 2.5 g l^{-1} sodium sulfate as per the specifications for dyeing wool, and 100 μM ABTS in 50 mM sodium acetate buffer (pH 5.5). This feed solution was saturated with oxygen and continuously fed into the column using a peristaltic pump at a flow rate of 0.15 ml/min. This flow rate was selected through initial optimization experiments based on determining an effective hydraulic retention time for decolorization of the dye. The bed was earlier saturated with the dye solution without mediator in order to rule out decolorization due to adsorption on the matrix. Percent decolorization was determined at different time intervals. The samples were also assayed for leached laccase activity.

Degradation of AR 27 (obtained from Sigma Aldrich) was performed using 100 μM AR 27 + 1 mM 1-hydroxybenzotriazole monohydrate (HOBT). The concentration of mediator was selected based on the previous reports [5]. The decolorization reaction was monitored at 515 nm.

Toxicity and mutagenicity assessment

Toxicity analysis was done by measuring decline in oxygen consumption rate of *Pseudomonas putida.* Mutagenicity was assessed by standard Ames test using *Salmonella typhimurium* TA 98 and 100 strains as described earlier [16].

Mass analysis

LC-MS analysis was performed for analysis of degradation products of AR 27 using Perkin Elmer HPLC coupled with Qstar Electrospray ionization (ESI) high resolution mass spectrometer (Applied Biosystems, USA) equipped with quadruple and Time of Flight mass analyzers. The decolorized dye samples were introduced into the reverse phase C-18 HPLC column (LiChrospher) using an autosampler at a flow rate of 0.2 ml/min. Elution was performed using a combination of solvent A(acetonitrile) and solvent B (water) in 1:1 ratio. The products were monitored at 210 nm using a UV-Visible detector. The flow coming from HPLC was introduced into the mass spectrometer (MS). The MS was operated in the negative ion mode and other parameters were set as follows: ion spray voltage −4500 V, Nebulizer gas: 35 lb/in^2, Curtain gas: 25 lb/in^2, Declustering potential: −80 V, Focussing potential: 265 V. Collision energy was varied from −20 to −80 V for analyzing product ion spectra.

Results and discussion

Laccase immobilization, leaching and bead stability

PVA-boric acid and PVA-nitrate gave 65 and 90% immobilization respectively. Leaching of laccase was monitored in both preparations and it was found that PVA-nitrate retained 75% laccase activity after 108 h of incubation at 30°C at 100 rpm (Table 1). Both the bead type were highly stable even up to 1000 rpm. While PVA-boric acid beads exhibited rapid loss in laccase activity, PVA-nitrate beads retained 20% laccase activity even after incubation at 1000 rpm (Table 1).

The major obstacles in implementing laccase-mediator systems for bioremediation at an industrial scale are the cost of the enzyme and the mediators. Immobilized enzymes allow for an efficient use of available catalyst. Multiple batches of substrates can be processed as the enzyme does not get washed off with the effluent [20].

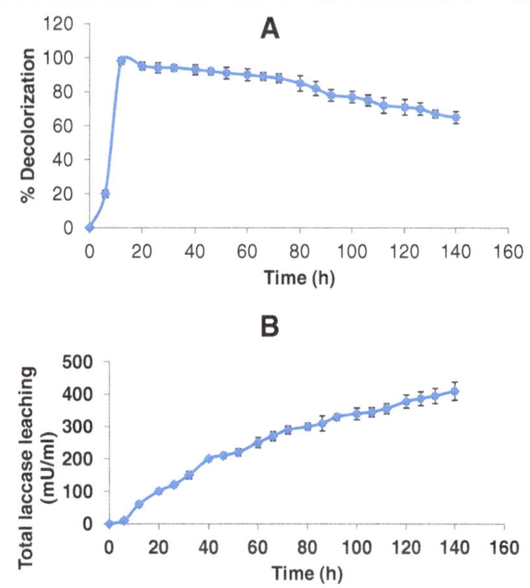

Figure 3 Continuous decolorization in packed bed column using PVA-nitrate beads. **(A)** Profile of dye decolorization with time and **(B)** Leaching of laccase from the column.

Table 3 Summary of the degradation products of Acid Red 27 after laccase-mediator action as observed by ESI-MS/MS

Product	Mass/charge (m/z)	Major product ion peaks as observed in MS/MS spectrum	
		Collision Energy (V)	m/z (% Relative intensities)
Dye alone	[M-H]- =536.97	−50	316.97 (100); 301.96 (50); 237.02 (40) 206.00 (30); 79.96 (30); 518.96 (20); 457.02 (10); 439.00 (10); 377.07 (5) 365.09 (5).
I	[M-H]- =535.9	−50	315.97 (100); 301.96 (50); 236.02 (40) 206.00 (30); 79.96 (30); 518.96 (20); 457.02 (10); 439.00 (10); 377.07 (5) 365.09 (5)
II	[M-H]- =235.01	−30	182.65 (100); 233.99 (45); 79.95 (30); 219.99 (20).
III	[M-H]- =207.01	−30	207.01 (100); 183.01 (50); 79.95 (40); 119.05 (20)
IV	[M-H]- =316.94	−40	316.9 (100); 289 (50); 245.06 (20); 79.95 (15); 255.20 (10); 145.0310 (10); 236.98 (5).

Also, immobilized enzymes generally exhibit higher resistance to inhibitors and high temperatures. The process of immobilization itself should also be less expensive, easy to perform and effective and this can generally be achieved using cheaper matrices and easy methods of immobilization. Any method which fulfills the above criteria qualifies to be used at large scale making the overall process economical.

The suitability of any immobilized enzyme preparation is largely dependent on the mechanical and chemical stability of the immobilization matrix. In earlier work, immobilization of laccase was carried out in alginate beads cross linked with calcium, copper or zinc. Copper alginate and zinc alginate showed high % immobilization and laccase retention but these preparations were unstable both mechanically and chemically. In the present study, laccase was immobilized in PVA-boric acid and PVA-nitrate beads, making it the first study involving PVA for this enzyme. The immobilization was performed by modifications in the protocol given by Chang and Tseng [21]. Also, this immobilization method fulfilled most of the criteria listed above The beads were found to be stable and of high mechanical strength, a property described earlier also [22]. Also, higher percentage of laccase was immobilized. The beads were stable enough to be used in continuous operations in a column reactor. Palmeiri et al. [13] and Brandi et al. [23] reported 65% immobilization of laccase in copper alginate beads. Lower values for immobilization also mean using higher quantities of initial enzyme making the process expensive [24]. Liu et al., [25] reported 91% retention of laccase activity during immobilization on carbon based mesoporous magnetic composites. The preparation of such composites, however, involved complicated procedures while entrapment by far is the easiest way to immobilize enzymes.

Characterization of immobilized laccase

It was found that immobilization of laccase did not confer any increased resistance to common inhibitors such as chlorides, EDTA and azide. The IC50 values were nearly equal to that of the free enzyme. The immobilized laccase, however, displayed lower affinity (K_m) for ABTS compared to the free enzyme. The K_{cat} values also decreased for the immobilized enzyme (Table 2). This observation has been made with several other immobilized enzyme preparations [4,23] and is attributed to mass transfer resistance displayed by enzymes in the entrapped

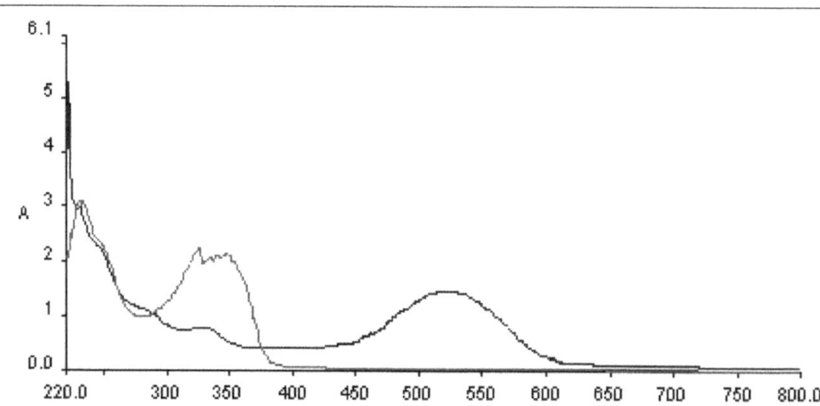

Figure 4 Absorption profile of AR 27 (—) and AR 27 treated with laccase and 1-HOBT (—).

state. This eventually leads to a decline in catalytic efficiency. In our study, this can also be attributed to the low permeability of PVA beads leading to low diffusivity of dyes as well as oxygen required for laccase action. The storage stability of the immobilized enzyme, however, was substantially increased as compared to the free enzyme and is the highest reported till now. The only other highly stable laccase was reported by immobilization in controlled porosity silica beads [26].

The enzyme stability increased for immobilized preparations with more than 80% activity remaining after 4 months of storage at 4°C. Free laccase, on the other hand, exhibited complete loss in activity in a period of less than a month. The optimum pH and temperature of the laccase remained the same with ABTS for immobilized preparations as that for the free enzyme.

Batch decolorization using the immobilized laccase

In order to determine the decolorization efficiency of the immobilized enzyme in batch mode, an acidic and a basic dye were selected. Acid Violet 17 decolorized at a much slower rate compared to Basic Green 4 [17]. Therefore, both the dyes offered different chemical structure and decolorization kinetics making them appropriate to determine the behavior of immobilized laccase. Decolorization of two different simulated effluents,

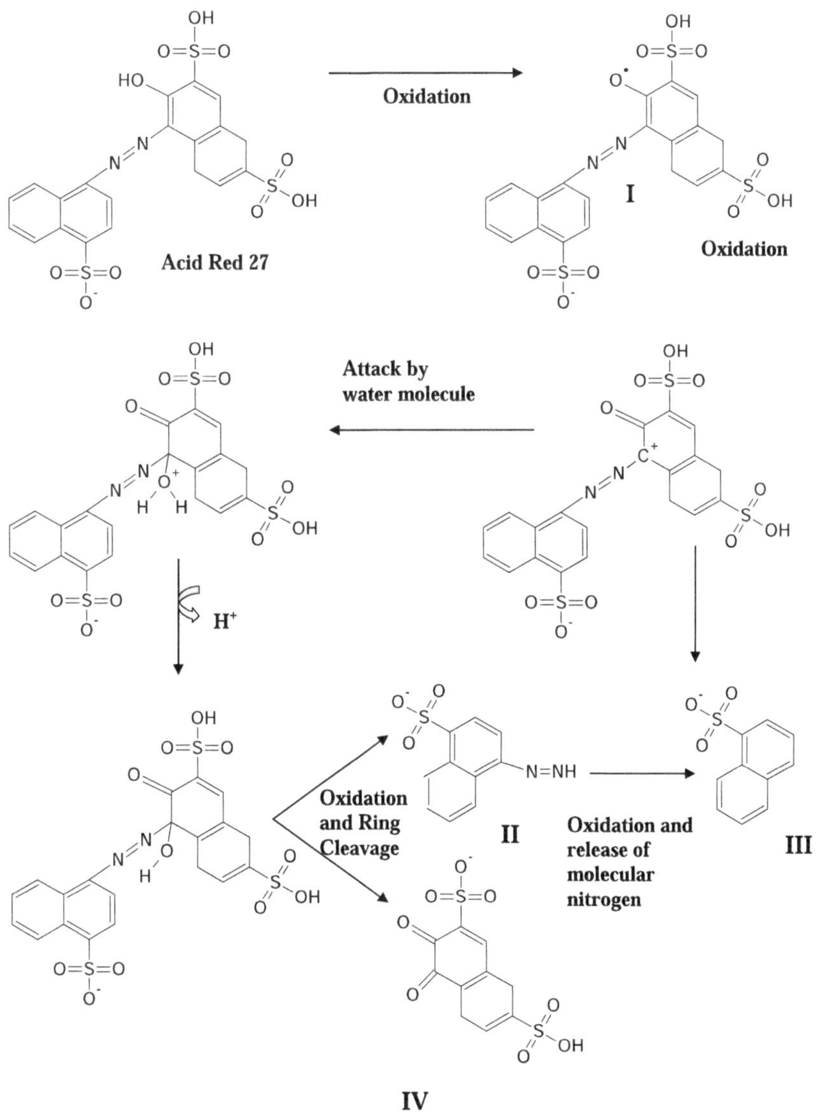

Figure 5 Proposed pathway of decolorization of AR 27 using laccase-mediator system.

one containing Acid Violet 17 and the other containing Basic Green 4 was performed. PVA-nitrate beads were chemically stable in both the dye solutions and led to 95% decolorization of Basic Green 4 even up to 20 cycles. In batch mode, the preparations also led to 90% decolorization of the effluent containing Acid Violet 17 up to 10 cycles (Figure 2A and B). The results were promising when compared to previous reports wherein the immobilized laccase was not highly reusable in batch studies. This was largely due to loss in laccase activity due to leaching or inactivation [13,23].

Continuous decolorization using immobilized laccase

The immobilized laccase preparations were used to perform decolorization of the simulated effluent. The decline in % decolorization achieved in packed bed column was directly correlated with remaining laccase activity in the beads. PVA-nitrate packed bed retained 60% of the initial laccase activity even after 120 h of use. The decolorization of the simulated effluent was also more than 70% up to 120 h (Figure 3A and B).

AR 27 decolorization and degradation using immobilized laccase

AR 27 was decolorized by laccase only in the presence of the mediator, as is characteristic of most high redox dyes. Mass spectra of the dye gave multiple peaks which were attributed to sodium and potassium adducts of the dye. Differential ionization of the dye molecule gives multiple peaks at different m/z. The analysis of the decolorized samples led to identification of four products, named from I to IV (Table 3). These persistent products were Compound I: 4-((2-oxo-3, 6-disulfo-2,3-dihydronaphthalen-1-yl)diazenyl)naphthalene-1-sulfonate at m/z 535.9, compound II: 4-diazenylnaphthalene-1-sulfonate at m/z 235.01, compound III: naphthalene-1-sulfonate at m/z 207.0, and compound IV: 3,4-dioxo-7-sulfo-2,3,4,4a-tetrahydronaphthalene-2-sulfonate at m/z 316.94. The UV-visible absorbance scan showed complete disappearance of peaks in the visible range (Figure 4).

Figure 5 shows the proposed decolorization pathway of AR 27 by laccase-mediator. The four degradation products which were identified are proposed to appear by the action of laccase and mediator in the following way. Compound I at m/z 535.9 resulted from the oxidation of the phenolic dye group to a quinine. This laccase-mediator assisted two electron oxidation was proposed to occur via phenoxy radical formation and is one of the important intermediates through which the dye undergoes further conversion. This is followed by a nucleophilic attack by a water molecule which creates a site for further action of laccase/mediator system. This explains the formation of other identified products. Compound I undergoes further oxidation resulting in splitting of the dye molecule and generation of compound II at m/z 235.01 and compound IV at m/z 316.94. Compound II gets further oxidized to compound III which was detected at m/z 207.0. The nitrogen in the azo bond was thus released as molecular nitrogen. This

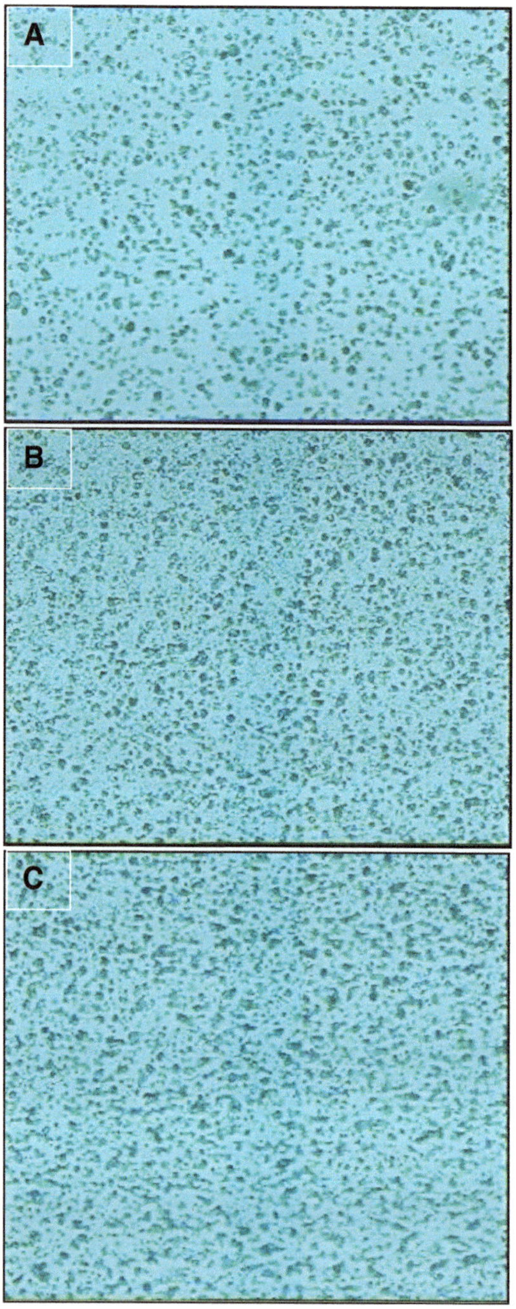

Figure 6 Background lawn of *S. typhimurium* TA 98 obtained during Ames test. **(A)** Untreated AR 27, **(B)** AR 27 treated with laccase and 1-HOBT, **(C)** Negative control-Distilled water.

form of azo bond disruption was earlier proposed by Chivukula and Renganathan [27]. The peak representing $[SO_3]^-$ at m/z 79.95 did appear in the decolorized dye spectrum. However, this was not attributed to laccase action as it also appeared in the spectrum taken with dye alone, although at slightly lower intensities. Thus, the appearance of these peaks is possibly due to ionization induced fragmentation of the dye molecule or its degradation products.

The high molecular weight polymerization products were absent which otherwise appear on treatment of dyes with free laccase on slightly long incubations. Zille et al. [28] observed high molecular weight molecules which were the polymerization products of intact dye and its degradation products obtained on longer incubations with free laccase.

Toxicity and mutagenicity assessment

AR 27 did not inhibit respiratory activity in *P. putida* but was found to be mutagenic as assessed by the Ames test. Similar observations have been made earlier [29]. Also, slight thinning of the background lawn of *S. typhimurium* occurred showing cytotoxic effect (Figure 6). Treatment with laccase-mediator did not add any toxicity but led to reduction in mutagenicity and also restored density of background lawns in the Ames test. This is possible as the mutagenic azo bond is disrupted by laccase-mediator action.

Conclusions

Laccase immobilized in PVA-nitrate was found to be stable for use in several decolorization studies using batch and immobilized systems. High mechanical strength of the beads allowed continuous operations in a column reactor. Immobilized laccase was also used for decolorization and degradation of the azo dye AR 27. The dye degradation studies showed that AR 27 underwent degradation to smaller moieties and a pathway of degradation was proposed. It was also concluded that polymerization products appearing during dye decolorization can be avoided through continuous decolorization process by providing just appropriate hydraulic retention time.

Competing interests

The authors declare that they have no competing interests.

Authors' contributions

MC has designed and performed the experiments and has written the manuscript. Authors SM and TRS were involved in the design of experiments and editing the manuscript. All authors read and approved the final manuscript.

Acknowledgements

Department of Biotechnology, Government of India is gratefully acknowledged for providing research grants to authors SM and TRS. Council of Scientific and Industrial Research, India is acknowledged for providing senior research fellowship to MC.

Author details

[1]Department of Biology, Indian Institute of Technology-Jodhpur, Jodhpur 342011, Rajasthan, India. [2]Department of Biochemical Engineering and Biotechnology, Indian Institute of Technology Delhi, Hauz-Khas, New-Delhi 110016, India.

References

1. Maciel MJM, Silva ACE, Ribeiro HCT. Industrial and biotechnological applications of ligninolytic enzymes of the basidiomycota: A review. Electron J Biotechn. 2010;13:6.
2. Gholami-Borujeni F, Mahvi AH, Naseri S, Faramarzi MA, Nabizadeh R, Alimohammadi M. Enzymatic Treatment and Detoxification of Acid Orange 7 from Textile Wastewater. Appl Biochem Biotechnol. 2011;165:1274–84.
3. Durán N, Rosa MA, D'Annibale A, Gianfreda L. Applications of laccases and tyrosinases (phenoloxidases) immobilized on different supports: a review. Enzyme Microb Tech. 2002;31:907–31.
4. Fernández-Fernández M, Sanromán MA, Moldes D. Recent developments and applications of immobilized laccase. Biotechnol Adv. 2013;31:1808–25.
5. Reyes P, Pickard MA, Vazquez-Duhalt R. Hydroxybenzotriazole increases the range of textile dyes decolorized by immobilized laccase. Biotechnol Lett. 1999;21:875–80.
6. Peralta-Zamora P, Pereira CM, Tiburtius ERL, Moraes SG, Rosa MA, Minussi RC, et al. Decolorization of reactive dyes by immobilized laccase. Appl Catal B Environ. 2003;42:131–44.
7. Kunamneni A, Ghazi I, Camarero S, Ballesteros A, Plou FJ, Alcalde M. Decolorization of synthetic dyes by laccase immobilized on epoxy-activated carriers. Process Biochem. 2008;43:169–78.
8. Ping W, Xuerong F, Li C, Qiang W, Aihui Z. Decolorization of reactive dyes by laccase immobilized in alginate/gelatin blent with PEG. J Environ Sci. 2008;20:1519–22.
9. Hu X, Wang P, Hwang HM. Oxidation of Anthracene by immobilized laccase from Tramates versicolor. Bioresour Technol. 2009;100:4963–8.
10. Champagne PP, Ramsay JA. Dye decolorization and detoxification by laccase immobilized on porous glass beads. Bioresour Technol. 2010;101:2230–5.
11. Osma JF, Toca-Herrera JL, Rodriquez-Couto S. Transformation pathway of Remazol Brilliant Blue R by immobilized laccase. Bioresour Technol. 2010;101:8509–14.
12. Asgher M, Kamal S, Muhammad H, Iqbal N. Improvement of Catalytic Efficiency, Thermo-stability and Dye Decolorization Capability of Pleurotus ostreatus IBL-02 laccase by Hydrophobic Sol Gel Entrapment. Chem Cent. 2012;6:110.
13. Palmieri G, Giardina P, Sannia G. Laccase-Mediated Remazol Brilliant Blue R Decolorization in a Fixed-Bed Bioreactor. Biotechnol Prog. 2005;21:1436–41.
14. Salony, Mishra S, Bisaria VS. Production and characterization of laccase from *Cyathus bulleri* and its use in decolorization of recalcitracnt textile dyes. Appl Microbiol Biotechnol. 2006;71:646–53.
15. Salony, Mishra S, Bisaria VS. Decolorization and detoxification of textile dyes and black liquor by laccase of *Cyathus bulleri*. J Sci Ind Res. 2007;66:684–8.
16. Chhabra M, Mishra S, Sreekrishnan TR. Mediator assisted decolorization and detoxification of textile dyes/ dye mixture using *Cyathus bulleri* laccase. Appl Biochem Biotechnol. 2008;151:587–98.
17. Chhabra M, Mishra S, Sreekrishnan TR. Laccase/mediator assisted degradation of triarylmethane dyes in a continuous membrane reactor. J Biotechnol. 2009;143(1):69–78.
18. Peijs T, Van Vught RJM, Govaert LE. Mechanical properties of poly(vinylalcohol) fibres and composites. Composites. 1995;26:83–90.
19. Yujian W, Xiaojuan Y, Wei T, Hongyu L. High-rate ferrous iron oxidation by immobilized *Acidithiobacillus ferrooxidans* with complex of PVA and sodium alginate. J Microbiol Methods. 2006;68:212–7.
20. Gholami-Borujeni F, Mahvi AH, Naseri S, Faramarzi MA, Nabizadeh R, Alimohammadi M. Application of immobilized horseradish peroxidase for removal and detoxification of azo dye from aqueous solution. Res J Chem Environ. 2011;15(2):217–22.
21. Chen KC, Lin YF. Immobilization of microorganisms or enzymes in polyvinyl alcohol beads. US patent. 1994;5:290–693.
22. Chang CC, Tseng SK. Immobilization of *Alcaligenes eutrophus* using PVA crosslinked with sodium nitrate. Biotechnol Tech. 1998;12:865–8.

23. Brandi P, D'Annibale A, Galli C, Gentili P, Pontes ASN. In search for practical advantages from the immobilisation of an enzyme: the case of laccase. J Mol Catal B: Enzym. 2006;41:61–9.
24. Ahn MY, Dec J, Kim JE, Bollag JM. Treatment of 2,4-Dichlorophenol Polluted Soil with Free and Immobilized Laccase. J Environ Qual. 2002;31:1509–15.
25. Liu Y, Zeng Z, Zeng G, Tang L, Pang Y, Li Z, et al. Immobilization of laccase on magnetic bimodal mesoporous carbon and the application in the removal of phenolic compounds. Bioresour Technol. 2012;115:21–6.
26. Champagne PP, Ramsay JA. Reactive Blue 19 decolourization by laccase on silica beads. Appl Microbiol Biotechnol. 2007;77:819–23.
27. Chivukula M, Renganathan V. Phenolic Azo Dye Oxidation by Laccase from *Pyricularia oryzae*. Appl Environ Microbiol. 1995;61:4374–7.
28. Zille A, Gornacka B, Rehorek A, Cavaco-Paulo A. Degradation of Azo Dyes by *Trametes villosa* Laccase over Long Periods of Oxidative Conditions. Appl Environ Microbiol. 2005;71:6711–8.
29. Hong Y, Guo J, Xu Z, Mo C, Xu M, Sun G. Reduction and partial degradation mechanisms of naphthylaminesulfonic azo dye amaranth by *Shewanella decolorationis* S12. Appl Microbiol Biotechnol. 2007;75:647–54.

Monitoring and assessment of water health quality in the Tajan River, Iran using physicochemical, fish and macroinvertebrates indices

Jaber Aazami[1], Abbas Esmaili-Sari[1*], Asghar Abdoli[2], Hormoz Sohrabi[3] and Paul J Van den Brink[4,5]

Abstract

Background: Nowadays, aquatic organisms are used as bio-indicators to assess ecological water quality in western regions, but have hardly been used in an Iranian context. We, therefore, evaluated the suitability of several indices to assess the water quality for an Iranian case study.

Methods: Measured data on biotic (fish and macroinvertebrates) and abiotic elements (28 physicochemical and habitat parameters), were used to calculate six indices for assessment of water quality and the impact of human activities in the Tajan river, Iran. GIS, uni- and multivariate statistics were used to assess the correlations between biological and environmental endpoints.

Results: The results showed that ecological condition and water quality were reduced from up- to downstream. The reduced water quality was revealed by the biotic indices better than the abiotic ones which were linked to a variety of ecological water quality scales.

Conclusion: The fish index showed a strong relationship with long-term database of physicochemical parameters (12 years (94%)), whereas macroinvertebrates index is more correlated with short-term data (76%). Meanwhile, the biotic and abiotic elements in this study were also classified well by PCA. Pulp and wood plants and sand mining are indicated to have the most negative effects on the river ecosystem.

Keywords: Water health quality, Bio-indicators, Tajan, Physicochemical parameters

Introduction

One of the greatest environmental challenges of this century is to sustain natural biological structural and functional attributes of aquatic ecosystems, rivers in particular. This goal requires that we know the condition of these dynamic systems and how they are being affected by specific factors and forces [1]. Nowadays, we can easily see that there are many pollutants in the environment due to anthropogenic activities. The destruction of natural habitats and the presence of environmental pollutants may affect the ecological balance of every ecosystem [2]. Among various ecosystems in the world, rivers which cross different areas such as agriculture and industry are the most threatened and affected by anthropogenic activities [3]. In developing countries such as Iran, water pollution is a common and widespread problem. Therefore, water resource management of rivers is of great importance and especially essential for semiarid and developing countries [4]. Assessment and classification of ecological water quality using indices-based approaches can help the conservation and management of rivers. The measurement of physicochemical parameters is usually time-consuming, cost-intensive and also dependent on special instruments. However, physicochemical parameters can only show water quality at the moment of measurement and these can change over time. Nowadays, indicators based on the presence or absence aquatic organisms have been developed to assess water quality and for the classification of ecological status. Norris and Thoms suggested that the effects on biota are usually the final point of

* Correspondence: esmaili@modares.ac.ir
[1]Department of Environment, Faculty of Natural Resources, Tarbiat Modares University, Tehran, Iran
Full list of author information is available at the end of the article

environmental degradation and pollution of rivers and thus are an important indication of ecosystem health [5].

Many living organisms (e.g. small mammals, fish, aquatic plants, algae, invertebrates) can be used to assess ecological water quality. Fish encompass different trophic levels, have a long life cycle, and high mobility, and can herewith be used to integrate the effects of habitat change and environmental pollution over a long period [6]. Macroinvertebrates are used for bioassessment because they are relatively easily sampled and are a very biodiverse group of species inhabiting waters that is contaminated to a different extent (from clean to highly polluted) [7-9]. They are important for the cycling of organic matter and provide food resources for higher trophic levels. The fluctuation of macroinvertebrate richness in the aquatic environment may result in the change of the ecosystem function. Moreover, the relatively low mobility and long life cycles of macroinvertebrates ensure that the presence of a given taxon reflects the past conditions. Many previous studies have shown the importance of biotic indices in the world, especially in Europe [10].

However, there are only a few studies using biotic indices in Asian countries such as Iran. The water quality monitoring programs in Iran are mainly based on the determination of some physicochemical parameters and water quality indices have generally not yet been use as a tool for the assessment and management of river ecosystems. Meanwhile, scientific efforts have often focused on improving freshwater resources that are of economic, cultural or spiritual importance. Unfortunately, most of these efforts have proceeded without documentation of the relative successes and failures of individual activities. Even though success is noted, there is often a lack of biotic data to identify specific results or endpoints for the river management activity [1].

In this study, we tried to employ the most widely used biotic and physicochemical indices to classify the ecological water quality in one of the most important Iranian rivers, the Tajan River. This river was chosen as a pilot river from the 115 rivers in the north of Iran because of having a good water-flow, discharge regime, catchment area, Accessibility and environmental condition [11]. Also, there are many similarities such as environmental landscape, climate and land uses between this river and European rivers. Besides, it is possible to select some stations as references because of being away of human activities.

The goals of this study were to determine and classify the ecological water quality of Tajan River based on different indices of water quality and to evaluate their performance, to zone the water quality based on these indices and GIS (Geographic Information System), to assess the effects of human land uses on the river and to compare the results between up- and down-stream parts of the river. This is the first study that compares biotic and physicochemical indices, as well as uses fish species as a bio indicator in an Iranian River.

Materials and methods

Study area

The Tajan River is one of the best freshwater ecosystems in the Mazandaran province located in the north of Iran. The river is 140 km long and originates from forested mountains and continues through different land uses, especially the agricultural areas of the coastal plain, where rice is extensively cultivated. The river eventually terminates in the Caspian Sea, the biggest land-locked aquatic ecosystem in the world (Figure 1). There are different land-uses including agriculture, aquaculture, dam construction, sand mining and industrial activities in the river [11,12]. It is chosen as a pilot river from the 115 rivers in north of Iran [13] because of having a good water-flow, discharge regime, catchment area and environmental condition. The Tajan River is divided to tow parts (Upstream and Downstream) by an old big dam (Shahid-Rajaie Dam, 1993). Water, macroinvertebrates and fish were sampled at 17 sites in September 2013. Eight sites are located in the upstream part and 9 sites in the downstream part. Also, five sites were selected as reference sites where there was no or slight pollution expected. Site selection was based on land use, accessibility and anthropogenic activities. Sites were subjected to non-point (i.e. agricultural runoff) and point influents (i.e. fish pond).

Physicochemical and habitat parameters

Width, length, depth, altitude (m), dissolved oxygen (mg/l), pH, water and air temperature (°C), conductivity (μS/cm), turbidity (NTU) and nutrients (NO3-N, NO2-N, NH4-N, and PO4-P, mg/l) were measured in situ by using Portable multi-parameter water analyser and UV–vis Spectrophotometry 8000. Biochemical Oxygen Demand (BOD5, mg/l) and Total Suspended Solids (TSS, mg/l) were determined according to APHA [14] with three replicate samples being measured in the laboratory. Some stations in this study overlapped with the stations of Iranian Water Resources Management (IWRM). Therefore, the accuracy of the collected data of physicochemical parameters was checked with the data of IWRM which had collected monthly data for twelve years. Average water flow (m/sec) was calculated by calculating the average of the recordings available from the IWRM for the same period as our sampling. According to the Field Manual for Water Quality Monitoring, the National Sanitation Foundation Water Quality Index (NSFWQI) surveyed 142 sites representing a wide range of locations at the local, state, and national level, using about 35 water quality tests which outcomes were combined in an index [15]. Nine parameters (dissolved oxygen,

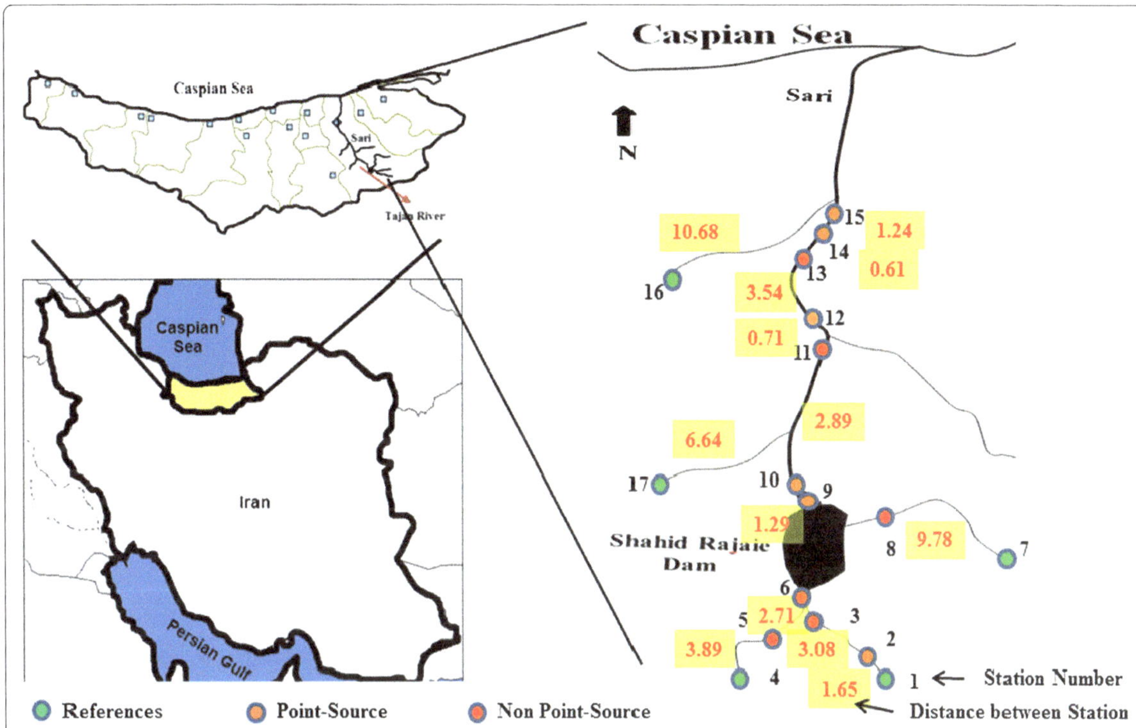

Figure 1 Sites sampled in this case study.

faecal coliform, biochemical oxygen demand, pH, nitrates, total phosphate, temperature, turbidity, and total solids) were chosen and some were judged to be more important than others, so the values were combined by calculating a weighted mean, based on the method described by Nikoo [16]. For this, field measurements were converted to index values using a questionnaire in which respondents were asked to estimate the level of water quality (0 through 100) corresponding to the field measurements (e.g., pH 2–12). The curves were then averaged and are assumed to represent the best professional judgment of the respondents. When the test results were not available for all 9 measurements, we preserved the relative weight for each factor and scale the total so that the range remains 0 to 100 [17]. The $IRWQI_{sc}$ index is a modification of NSFWQI based on the local condition in Iran. Habitat assessment was performed using 10 factors assessed by four experts and the Rapid Bioassessment Protocol (RBP) was used for river habitat assessment by visual observations at each site [18]. The range of each habitat parameters was from 0 (very perturbed) to 20 (pristine) and included Epifaunal Substrate/Available Cover, Embeddedness (Embed), Velocity/Depth Regime, Sediment Deposition, Channel Flow Status, Channel Alteration, Frequency of Riffles (Freq), Bank Stability, Vegetative Protection, Distance of References (DisRef), and Riparian Vegetative Zone Width. Finally, an average of the ranges of the parameters resulted in an overall RBP index.

Biological elements

A surber sampler (30*30 cm and 250 micron mesh) was used for collecting macroinvertebrates based on the RBP [18]. Three replicates were collected on each site and all the three replicates were composited as one sample. Benthic macroinvertebrates were preserved in 4% formaldehyde solution before being sorted, identified and counted to family level in laboratory. Electrofishing (200–300 V) with Blank net 6 mm based of FAME Protocol was used to sample fish [19]. Most of the sampled fish were identified, measured and released in situ. Some fish had to be identified further and were fixed in 10% formalin and transferred to the laboratory. Finally, the species and weight of each fish were recorded. RBP metrics were calculated for each sample and only mean site values are reported. In next step, mean values for each site were compared to references sites to generate a score for that site and classification. In Table 1, the classification of introduced indices and their colours are provided [20-25].

Physicochemical and biological indices of water quality

Physical habitat quality (type, stability, availability, etc.) and water quality are reflected by the diversity of stream

Table 1 Indices used to classify water and habitat quality

	Range	<15	15 – 29.9	30 – 44.9	45 - 55	55.1 - 70	70.1 - 85	85 ≤
$IRWQI_{sc}$	Quality	Very Polluted	Polluted	Enough Polluted	Moderate	Enough Good	Good	Excellent
	Colour	Violet	Red	Orange	Yellow	Olive-green	Green	Blue
NSFWQI	Range			0-25	25-50	50-70	70-90	90-100
	Quality			Very bad	Bad	Medium	Good	Excellent
MMIF	Range			0.00–0.29	0.30–0.49	0.50–0.69	0.70–0.89	0.90–1.00
	Quality			Bad	Poor	Moderate	Good	Excellent
Karr	% Index			<40	40-59	60-74	75-87	88-100
	Quality			Very poor	Poor	Fair	Good	Excellent
	Colour			Red	Orange	Yellow	Green	Blue
RBP	% Index				<60	74-60	89-75	≥90
	Quality				Poor	Marginal	Good	Excellent
BMWP/ASPT	Range				4 >	4-5	5-6	6 <
	Quality				Polluted	Moderate	Doubtful	Excellent
	Colour				Red	Yellow	Green	Blue

communities. Based on these relationships, environmental quality of aquatic systems can be described or categorized using integrated approaches that incorporate an evaluation of the physical, chemical and biological components. The guidelines associated with the RBP provide systematic approaches for identifying habitat quality and biotic integrity of river systems [26-28].

The NSFWQI is one of the first water quality indices that aggregate some water quality parameters through a weighted arithmetic mean function [15]. The Index is a usual water quality index method that was developed by paying great rigor in selecting parameters, developing a common scale and assigning weights. The water quality data are recorded and transferred to a weighting curve chart, where a numerical value of Qi is obtained. The mathematical expression for NSF WQI is given by [29]:

$$\text{NSFWQI} = \sum_{k=1}^{n} WiQi$$

Q_i = sub-index for each water quality parameter;
W_i = weight associated with each water quality parameter;
n = number of water quality parameters.

Also, the Ministry of Energy of Iran developed a local index, $IRWQI_{SC}$ (Iranian Water Quality Index for Surface Water Resource-Conventional Parameters), for the assessment of water quality [23,30]. This index is calculated by:

$$IRWQI = \sum_{1}^{n} (IiWi)^{1}/_{a}$$

$$a = \sum_{1}^{n} Wi$$

Ii = sub-index for each water quality parameter;
Wi = weight associated with each water quality parameter;
n = number of water quality parameters;
a = the sum of the weight.

Both indices are based on physicochemical parameters that are evaluated in this study. For assessment and classification of water quality by macroinvertebrate indices, we used the ratio of Biological Monitoring Working Party score to Average Score per Taxon (BMWP/ASPT) and Multimetric Macroinvertebrates Index Flanders (MMIF) [21]. These two indices are numerical expressions used to assess water quality based on the presence and, in many cases, the diversity of taxa with known environmental-pollution tolerances according to the following equation:

$$\text{BMWP/ASPT} = \left(\sum \text{B.n}\right)/\text{N}$$

Whereas:
B = the value for each species;
n = abundance of each species;
N = total number of species.

Subsequently, ecological water quality is assessed with Karr Biotic Index of Fish (KBI). KBI was firstly invented by Karr to study the River Trent basin in Champaign, then applied by other researchers, and is nowadays used all over the world and well documented by Karr and Ruaro [22,31]. The index is designed to assess the present the status of the community using twelve fish community parameters. These parameters can be roughly grouped into two sets: ecological factors (including number of individuals in samples, the proportion of omnivore individuals, proportion of insectivorous cyprinids, proportion of top carnivorous, proportion of individuals with disease, tumours, fin damage, and other anomalies) and species composition and richness (including number of species, presence of intolerant species, species richness and composition of darters, suckers and sunfish (except green sunfish) and proportion of green sunfish and hybrid individuals) [22]. Habitat condition is classified based on US EPA Rapid Bioassessment Protocol (RBP) [18].

Statistical analysis

Statistical analysis of the resulting data was performed using Excel and SPSS software version 19 (licensed by Tarbiat Modares University, Iran) and Canoco version 5 (licensed by Wageningen University, The Netherlands). Normality of data was checked by ShapiroWilk test. Because the data were not normally distributed, Mann–Whitney U test were used to assess the significance of the difference between up- and downstream values of physicochemical parameters. Cluster analysis was used for the classification of the stations on the basis of the indices and Casewise analysis was used for assessing the correlation between the classifications (with using SPSS). Principal component analysis (PCA) was used to analyse the correlations between abiotic (habitat and physicochemical) parameters and biotic (macroinvertebrates and fish) community. Before analysis, all the macroinvertebrates were divided into 5 groups based on the pollution tolerance owned by the dominant species in each family (Very sensitive, Sensitive, Neutral, Resistant and Very resistant) that was provided in Appendix B of RBP, EPA [18].

Results and discussion

For irrigating of the vast plain area of downstream rice farms, the dam valves are opened from spring to late summer and closed in early September, with a minimal flow at all time of 1 m^3/s. The rains start in early autumn and therefore the best time for sampling is September in accordance with local condition and some previous studies [32]. The main land-use in the upstream areas is not similar to that in the downstream areas, so no similar anthropogenic impact may be expected. Furthermore, there is a larger downstream flow, and herewith also a larger move of pollutants, most of the year because of the opening of the dam valves and irrigation. Habitat parameters show significant differences between up- and downstream sites (see Figure 1 for better understanding). Some land uses such as mining that affect the rivers' physical status are less development upstream compared to downstream. In the downstream part, there are many mining industries that have altered the physical status of the river. Humans have channelized, diverted, drained and filled streams because of the dredging of sand. In the upstream part of the Tajan River, the higher number of residents has no major impact on the physical structure of habitat. Also, the accessibility of the upstream part of the river is lower than downstreams. Therefore, there are more pristine habitats in upstream. Padmalal et al. showed that in the past 3–4 decades, rivers in the densely populated areas of the world are subjected to immense pressures due to various kinds of human interventions, among which indiscriminating mining for construction-grade sand from alluvial reaches is among the most disastrous one [32].

We identified 2426 fish individuals covering 17 species, 5 families and 7639 individuals of macroinvertebrates covering 45 geneses or families. Zoning of ecological river condition was performed with GIS based on the result of indices (Table 2 and Figure 2). All indices showed water quality reductions from upstream to downstream: the references stations (1,4,7,16,17) are indicated to have a good water quality (almost unpolluted) and in the most downstream stations, water quality was classified as moderate to polluted (14,15). We tried to get an even distribution of the stations along the river so we could analysis the effects of land uses on study's parameters, too. Human land use, development and urbanization can alter water flow and degrade stream habitat and biotic conditions through, for example, draining agricultural fields, channelizing streams and increasing sedimentation. The first station is located in an almost pristine area, with hardy any agriculture activities and covered with natural forest. All indices showed that the station was unpolluted. Upstream of the second station, there was a fish pond that probably affected the natural fish community, which is expressed by the low KBI value (green, Figure 2). Other indices, however, did not show a marked response. Because of the limitation to captured fisheries production, aquaculture has been developed worldwide especially in developing countries such as Iran. Aquaculture may have significant impacts on the environment and natural resources, and a number of concerns have been expressed by environmental activists and scientists [33,34]. Raczyńska et al. [35] reported that little attention is paid to the changes in water quality resulting from the inflow of effluent discharged from fish breeding ponds. They, however, report that the aquaculture effluent discharge from the carp breeding ponds had a significant impact on the physicochemical parameters of the river water. They observed an increase in concentrations of the organic and biogenic compounds immediately after the inflow of the polluted water from the ponds [35]. However, more studies are needed in order to understand the decrease in fish biodiversity observed at station 2. Downstream of the fish ponds, rice farms were located. In station 3, physical habitat parameters were altered probably because of the influence of traffic of agriculture. Therefore, the habitat index (RBP) descended from excellent to good levels (Figure 2). Also, the values of macroinvertebrates indices were reduced but no differences were observed for the KBI index. Agriculture may increase nutrient levels, erosion, and turbidity due to the use of fertilizers, planting and harvesting.

Table 2 The value of the calculated indices for each station

Station	Longitude	Latitude	KBI	BMWP/ASPT	MMIF	NSFWQI	IRWQIsc	RBP
1	712608	4004824	99	8.9	0.95	81.49	87.99	100
2	712059	4005260	75.03	7.5	0.95	70.91	74.06	95
3	709238	4006751	76.28	6.6	0.95	66.97	68.39	88.5
4	708189	4001145	81.48	8.2	0.95	81.74	81.26	100
5	707769	4006595	69.24	6.7	0.95	61.62	62.57	82
6	706166	4010260	75.94	5.7	0.85	59.6	54.84	79
7	726606	4007091	100	7.8	1	69.37	89.39	97
8	707556	4013824	70.95	5.8	0.85	63.59	72.06	73
9	700152	4014338	49.05	7.8	0.8	67.95	80.29	61
10	699857	4015151	52.03	5.2	0.8	59.67	62.2	58.5
11	695882	4025071	56.21	5.3	0.75	59.72	64.38	57.5
12	695429	4026319	51.96	5.5	0.7	52.02	52.36	44.5
13	689155	4037624	60.86	4.2	0.7	52.47	50.9	46.5
14	688694	4038040	52.08	4.2	0.65	53	46.82	38
15	686889	4039978	56.61	3.6	0.55	44.51	39.16	34.5
16	685112	4033640	90.53	7.6	0.95	67.84	86.57	96
17	689592	4023526	83.46	8.2	0.95	66.14	88.64	97

Especially rice farms may be important for the water quality because they require a lot of water for cultivation. Many scientists studied the effect of agriculture as a non-point pollutant on river basin and biota [36,37]. Traditional agriculture in the research area uses much water and the effluent of the farms often runs directly into the river. The indices did not show many changes for station 6 that is located after of accident tow stream and only the value of the MMIF index decreased. Between station 7 (reference station of the second stretch of the river) and 8, there are large farms and all indices may have decreased in station 8 because of the polluted discharge water from farming. This study is the first one reporting the abundance of fish population in two streams of the Tajan River, as there were 10 fish (*Barbus lacerta*) in station 7. *B. lacerta* is one of the most sensitive species to anthropogenic stress, similar to *Salmo trutta* [38]. *S. trutta* as an indicator species showing a high sensitivity to a variety of human pressures (e.g. water pollution, habitat degradation). Normally, it inhabits headwaters with high oxygen saturation, steep slope, fast flow, suitable temperature regimes and adequate food and due to anthropogenic influences, the brown trout was eliminated from many original habitats in Iran [39]. The comparison of indices in the 8 stations in the upstream catchment showed a comparable decrease in index values between the two streams of the river. A big dam is located between the eight upstream stations and station 9 which changed most of the parameters. Results show that values for pH, TSS, BOD and nutrients were significantly higher in the downstream area compared to the upstream area. The NSFWQI index showed a significant change due to the dam, although the local index ($IRWQI_{SC}$) did not change. This result is similar to other studies on the effect of a dam on physicochemical parameters [40]. The dam altered the diversity and density of macroinvertebrates, but had no effects on the macroinvertebrates indices, because after the dam the total density and total biomass increased, while taxa richness tended to decrease, which is in line with the findings of others [41]. Native fish assemblages have experienced profound changes as a consequence of the dam construction. We documented the decline of native fish and the rise of non-native fish which become the major component of the ichthyofauna in the downstream stations. RBP showed a reduction in the physical structure of the habitat in the downstream part. After the dam, there is a fish pond which may be responsible for the strong reduction in the physical habitat quality as shown by RBP (Table 2). The fish pond are situated very close to the river and also to the dam, and takes in clean water and releases the same amount of effluent water to the downstream part of the river. Downstream of the fish pond there are rice farms and small streams like the Zellem stream entering the main river. In the Zellem stream there were many sand mining and sandblast sites which polluted the water in the Tajan River. The stream is small and is located between station 11 and 12 and a number of indices show a marked decrease between these stations (Table 2). Station 13 and 14 were chosen to show the effect of sand mining which was expressed by difference in levels of $IRWQI_{SC}$. Sand mining increased TSS and negatively affected the physicochemical parameters, herewith affecting some sensitive species such as Salmonids. The most notable changes occurred in this

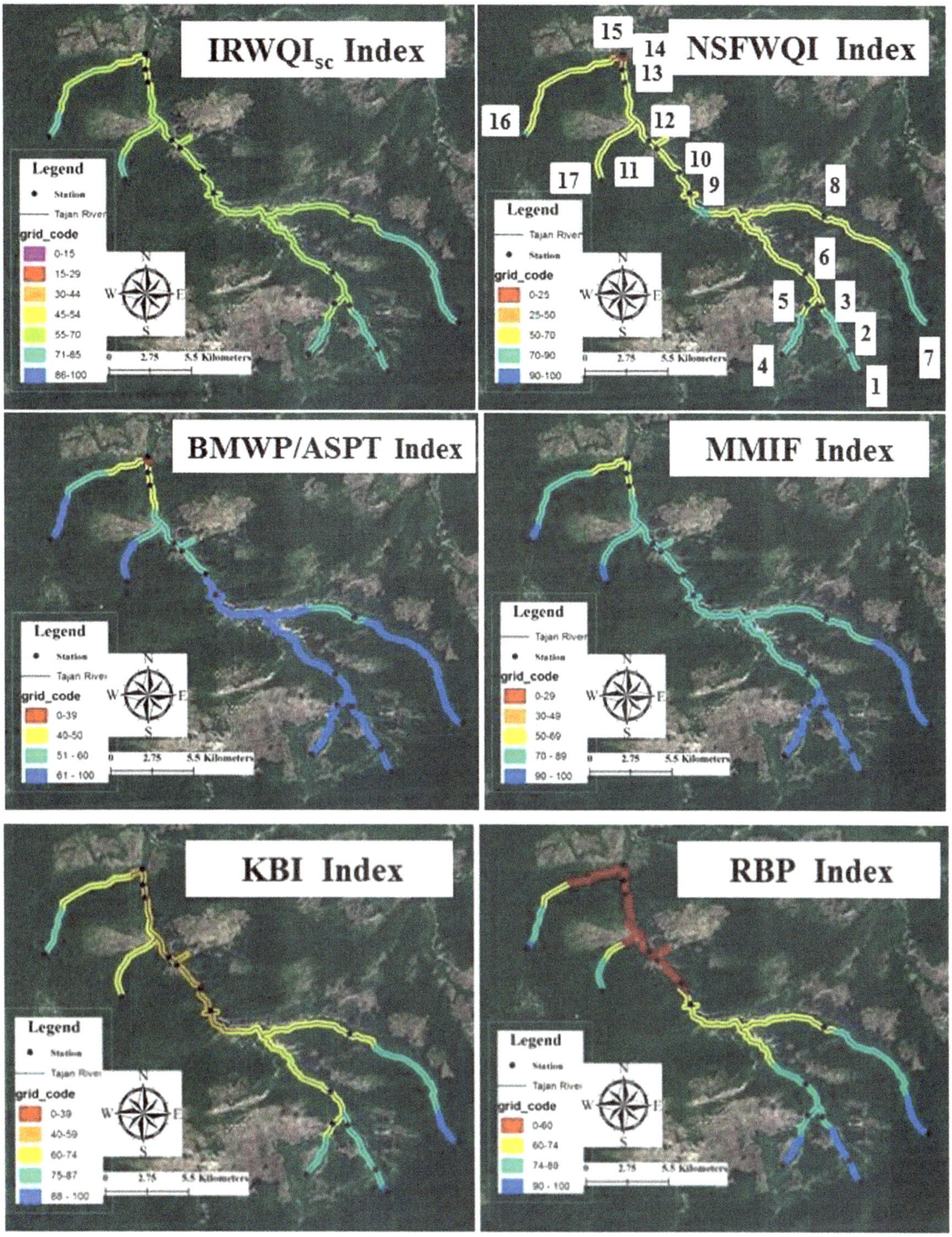

Figure 2 Water quality classification of the Tajan River by different indices and GIS.

study at station 15 which is located after a wood factory, because the wastewaters of this factory were directly discharged into river. Walker et al. [42] reported that the studies conducted in the early 1990s downstream of pulp and paper mills in Canada, found a number of changes in the physiological, biochemical and reproductive responses of wild fish and macroinvertebrates. The study assessed changes in the taste or odours of fish as a result of exposure to pulp and paper effluent as a measure of effects on the usability of fisheries resources by humans and their conclusion on the effects of the factory on fish and habitat parameters confirms our results [42]. Thompson et al. [43] also described the pulp and paper making industry as the third most water intensive industry in terms of freshwater withdrawal in the world, after the primary metals and the chemical industries. There is a large variation in the quality of the wastewaters from pulping and papermaking operations. This is due to the diversity of processes and the chemicals used [43]. Unfortunately, there has not been any exhaustive study on the effects of the plant's wastewater on the ecology of the Tajan River in Iran. Effluents of the plant were released without or with a preliminary treatment into the river, and hence the colour and odour of river's water were changed into brown and strong distinctive odour. There were large differences between station 15 and the others stations, especially the references stations with the same elevation (16, 17). pH and DO decreased and water temperature, BOD and nutrient increased in station 15. The diversity of animals decreased and some exotic species were present (i.e. *Rasborinae pseudorasbora parva*), while the sensitive species were absent. We suggest that more studies on the environmental effects of the paper plant on the ecology of the river are conducted. After station 15, there is a big city, Sari which is the capital of Mazandarn province. A major component of the human use of aquatic systems is the construction, maintenance, and use of roads that occur as part of human infrastructure and the road/stream interface is one of the main pathways for sediment to reach waterways. Stream crossings, often culverts, can alter in-stream sediment accumulation and the geomorphology of a stream. The effects of sedimentation on macroinvertebrates and fish have been well documented [1], as well as of traffic, excavating bottom sand from the river and changing of the riparian zone , which likely caused the obviously decrease in the habitat parameters at the station.

Figure 3 shows the classification of the stations of the Tajan River based on biotic indices (KBI, MMIF and BMWP/ASPT) and Figure 4 shows it based on abiotic indices (IRWQIsc, NSFWQI). The dendrogram of the biotic indices divides all stations into two major groups. All references stations are present in the first group and others stations are in the other group. The stations 12, 13, 14 and 15 are then subdivided in a separate group and these stations were located downstream with heavy human activities (Figure 3). The classification based on the abiotic indices shows the impact of land uses on river ecosystems very well. The classification groups the references stations together with station 9 which is located after dam because the dam improved the water quality by reducing sediment deposition, temperature go down and other physicochemical parameters partly improve. The results indicate also the percentage of group membership. On the basis of the results, KBI (fish index) could predict 61% of the changes in physicochemical parameters using our sampling data and 94% of these changes when using the 12 years of data. The prediction ability based on macroinvertebrates indices (MMIF and BMWP/ASPT) were 76 and 71% for our sampling data and 12 years of data, respectively. Fish are at the top of the food chain and could, therefore, show the effects of differences in water pollution at longer time scales than macroinvertebrates, explaining the higher predictive value of the fish index compared to the macroinvertebrate indices. Macroinvertebrate indices had a better correlation with physicochemical parameters in our sampling time, compared to the fish one, probably because of their shorter life cycle, greater dispersal ability, and lower position in the food chain than fish.

With the increase of the habitat parameters located at the left side of the diagram, the abundance of the very sensitive and sensitive fish species also increase (Figure 5). DO is the only physicochemical parameters that has a positive correlation with the abundance of the very sensitive and sensitive species. The other physicochemical parameters have negative correlations with both groups (Figure 5). Also the stations with a good status, such as 9, are located on the left side of the diagram. The arrow of neutral species indicates that this group was not strongly correlated with any of the parameters. In other words, the abundance of this group is not correlated with pollution and cannot be used as an indicator. On the right side there are the polluted stations and the groups of resistant and very resistant species. The arrows of the resistant species, pH and Ammonium point into the same direction, which means these two parameters correlate with the abundance of resistant species and less with the others parameters. The abundance of the very resistant species increases with most of the studied physicochemical parameters (except DO) and decrease of habitats parameters. In the upper right quadrant the highly polluted stations are located, with the placement of station 15 being extreme because it receives high pollution from a pulp and wood plant. The result of the PCA analysis complements was similar to the result of bio indicators and classification of stations.

Also, the result of the PCA analysis on the fish data and abiotic parameters shows that the placement of two

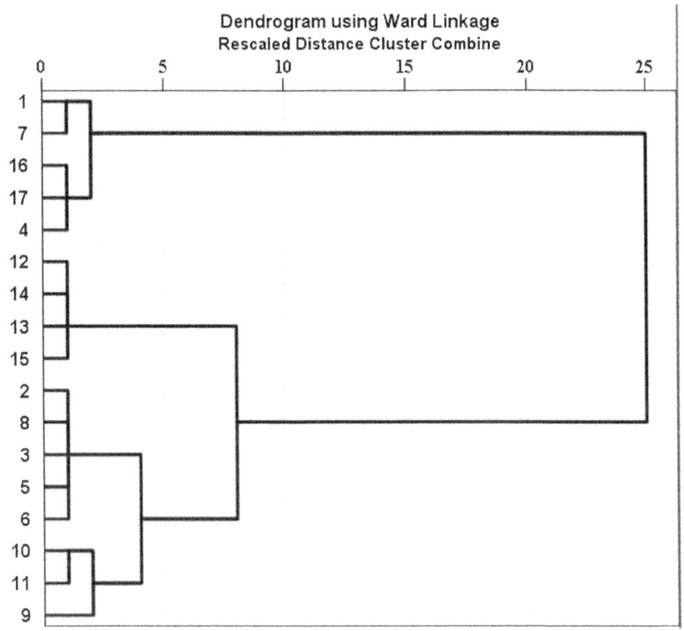

Figure 3 Classification of stations on based biotic indices.

fish species (i.e. *B. lacerta* and *S. trutta*) are different from the others (Figure 6). *B. lacerta* and *S. trutta* are correlate positively with high DO and habitat related parameters, while the opposite is true for most other species (Figure 6). On the other hand, the two species are indicated to have a negative correlation with most polluted related physicochemical parameters and therefore may be suitable to indicate clean water. As the physicochemical parameters except DO increase, the presence of the sensitive species decrease which is in accordance

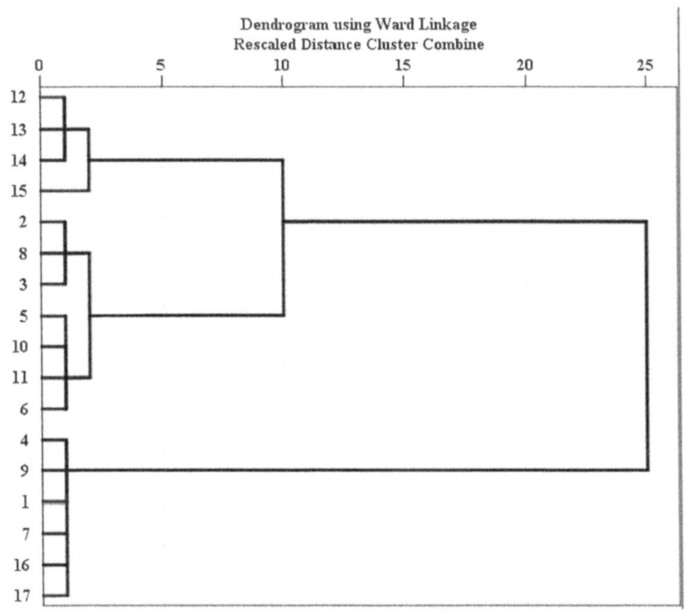

Figure 4 Classification of stations on based abiotic indices.

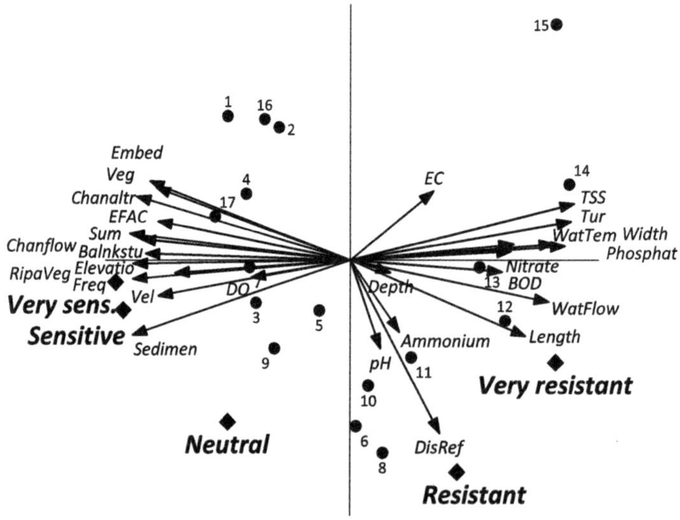

Figure 5 The correlations between abiotic parameters and macroinvertebrates sensitivity groups (PCA, Canoco 5). Of the variation in group abundances 75% is displayed on the horizontal axis and another 17% on the vertical one.

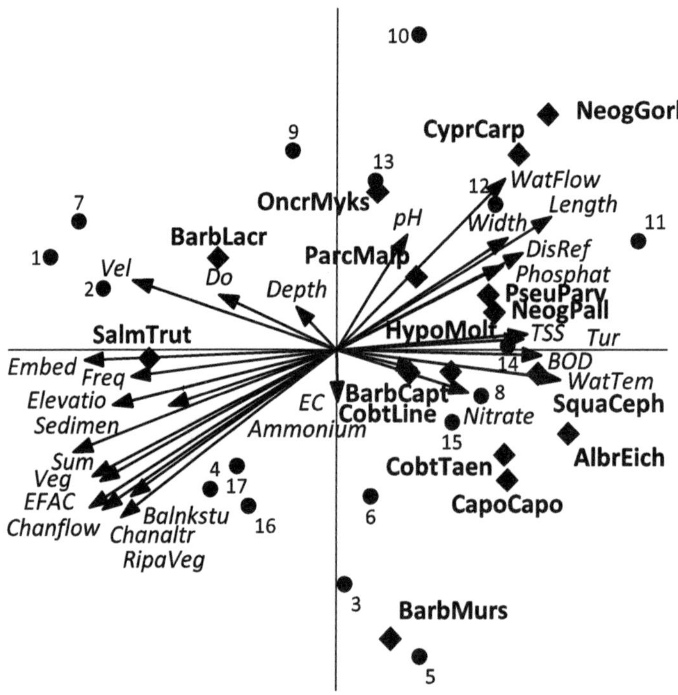

Figure 6 The relationship between abiotic parameters and fish species (PCA, Canoco 5). Of the variation in group abundances 31% is displayed on the horizontal axis and another 23% on the vertical one.

with Bengar et al. [44]. Habitat related parameters and DO had a negative correlation with most of the other fish species in the Tajan River and most fish species increase in abundance in the stations at the right side of the diagram (such as station 11, 12).

Conclusion

This study provides an assessment and comparison of biotic and abiotic indices-based approaches for the Tajan River. As well as Yazdian et al. concluded, we cannot claim that all of the indices would work in other regions as well [45], because of the different range in biodiversity, physicochemical parameters and land-uses. Actually, we have to use, modify and develop some native indices for Iranian ecosystems that they can be used for a rapid assessment of environmental health condition. A comparison between the indices shows that the classification based on biotic indices can show the long-term environmental condition better than those based on abiotic indices. We suggest that macroinvertebrates and fish can be used as indicators of water pollution with having the advantages of low cost, easy identification and it provides a better reflection of water quality than using physicochemical parameters alone. We also suggest to develop and modify the biotic indices for research and management of Iranian rivers, since Iran is located in the mid-dry area where water resources management is particularly urgent and important.

Competing interests

The authors declare that they have no competing interests.

Authors' contributions

All of the authors have the same contribution. All authors read and approved the final manuscript.

Acknowledgements

The manuscript is part of the PhD thesis of Jaber Aazami and is supported by Tarbiat Modares University, Iran. He would never have been able to do this investigation without the guidance of his committee members, help from friends, and support from his family.

Author details

[1]Department of Environment, Faculty of Natural Resources, Tarbiat Modares University, Tehran, Iran. [2]Department of Biodiversity and Ecosystem Management, Environmental Research Institute, Shahid Beheshti University, Tehran, Iran. [3]Department of Forestry, Faculty of Natural Resources, Tarbiat Modares University, Tehran, Iran. [4]Department of Aquatic Ecology and Water Quality Management, Wageningen University, Wageningen University and Research Centre, Wageningen, The Netherlands. [5]Alterra, Wageningen University and Research Centre, Wageningen, The Netherlands.

References

1. Ogren SA. Using indicators of biotic integrity for assessment of stream condition. Michigan Tech. 2014;7(5):10–9.
2. Begon M, Townsend CR, Harper JL. Ecology: from individuals to ecosystems. London, UK: Blackwell Publishing; 2009.
3. Leprieur F, Beauchard O, Blanchet S, Oberdorff T, Brosse S. Fish invasions in the world's river systems: when natural processes are blurred by human activities. PLoS biology. 2008;6(2), e28.
4. Kivaisi AK. The potential for constructed wetlands for wastewater treatment and reuse in developing countries: a review. Ecological Engineering. 2001;16(4):545–60.
5. Norris RH, Thoms MC. What is river health? Freshwater Biology. 1999;41(2):197–209.
6. Mathuriau C, Silva NM, Lyons J, Rivera LMM. Fish and Macroinvertebrates as Freshwater Ecosystem Bioindicators in Mexico: Current State and Perspectives. Water Resources in Mexico. USA: Springer; 2011. p. 251–61.
7. Cheimonopoulou MT, Bobori DC, Theocharopoulos I, Lazaridou M. Assessing Ecological Water Quality with Macroinvertebrates and Fish: a case study from a small Mediterranean river. Environmental management. 2011;47(2):279–90.
8. Li F, Cai Q, Fu X, Liu J. Construction of habitat suitability models (HSMs) for benthic macroinvertebrate and their applications to instream environmental flows: A case study in Xiangxi River of Three Gorges Reservior region. China Progress in Natural Science. 2009;19(3):359–67.
9. Schultz R, Dibble E. Effects of invasive macrophytes on freshwater fish and macroinvertebrate communities: the role of invasive plant traits. Hydrobiologia. 2012;84(1):1–14.
10. Ogleni N, Topal B. Water Quality Assessment of the Mudurnu River, Turkey. Using Biotic Indices Water Resour Manage. 2011;25(10):2487–508.
11. Ahmadi M, Khorasani N, Rafiee G. Investigation of pollution sources and water quality of Tajan River. J Nat Environ. 2011;63(4):327–17.
12. Shokri M, Rossaro B, Rahmani H. Response of macroinvertebrate communities to anthropogenic pressures in Tajan River (Iran). Biologia. 2014;69(10):1395–409.
13. Ahmadi-Mamaqani Y, Khorasani N, Talebi K, Hashemi SH, Rafiee G, Bahadori-Khosroshahi F. Diazinon Fate and Toxicity in the Tajan River (Iran) Ecosystem. Environmental Engineering Science. 2011;28(12):859–68.
14. Eaton AD, Franson MAH. Standard methods for the examination of water & waste water. USA: American Public Health Association (APHA); 2005.
15. Bharti NK. Water quality indices used for surface water vulnerability assessment. Int J Environ Sci. 2011;2(1).
16. Nikoo MR, Kerachian R, Malakpour-Estalaki S, Bashi-Azghadi SN, Azimi-Ghadikolaee MM. A probabilistic water quality index for river water quality assessment: a case study. Environmental monitoring and assessment. 2011;181(1–4):465–78.
17. Mitchell M, Stapp W. Field Manual for Water Quality Monitoring–An Environmental Education Program for Schools. Kendall: Hunt Publishing Company; 2000.
18. Water USEPAOo, Barbour MT, Gerritsen J, Synder BD, Stribling JB. Rapid bioassessment protocols for use in wadeable streams and rivers: periphyton, benthic macroinvertebrates, and fish. USA: EPA Publishing; 1999.
19. Consortium F. Manual for the application of the European Fish Index–EFI; A fish-based method to assess the ecological status of European rivers in support of the Water Framework Directive. Wat Fram Direc. 2005;10(1):12–9.
20. Extence C, Bates A, Forbes W, Barham P. Biologically based water quality management. Environmental pollution. 1987;45(3):221–36.
21. Gabriels W, Lock K, De Pauw N, Goethals PL. Multimetric Macroinvertebrate Index Flanders (MMIF) for biological assessment of rivers and lakes in Flanders (Belgium). Limnologica-Ecology and Management of Inland Waters. 2010;40(3):199–207.
22. Karr JR. Assessment of biotic integrity using fish communities. Fisheries. 1981;6(6):21–7.
23. Management IWR. Iranian Water Quality Index for Surface Water Resource-Conventional Parameters "IRWQI". Tehran: IWRM; 2013. p. 5–14.
24. Said A, Stevens DK, Sehlke G. An innovative index for evaluating water quality in streams. Environmental management. 2004;34(3):406–14.
25. Mali L, Boaf N. Environmental condition of rivers and streams in the ovens catchment. Freshwater Sci. 2003;909(1):10–65.
26. Barbour MT, Gerritsen J, Snyder BD, Stribling JB. Rapid bioassessment protocols for use in streams and river: Pryphyton, Benthic Macroinvertebrates and fish. 2nd edition. Washington D.C: USEPA; 1999:841-B-99-002. 408p.
27. Barbour MT, Stribling JB, Verdonschot PF. The multihabitat approach of USEPA's rapid bioassessment protocols: Benthic macroinvertebrates. Limnetica. 2006;25(3):839–50.
28. Winger PV, Lasier PJ, Bogenrieder KJ. Combined use of rapid bioassessment protocols and sediment quality triad to assess stream quality. Environmental monitoring and assessment. 2005;100(1–3):267–95.
29. Tyagi S, Sharma B, Singh P, Dobhal R. Water quality assessment in terms of water quality index. American Journal of Water Resources. 2013;1(3):34–8.

30. Nabizadeh R, Amin MV, Alimohammadi M, Naddafi K, Mahvi AH, Yousefzadeh S. Development of innovative computer software to facilitate the setup and computation of water quality index. Journal of Environmental Health Science and Engineering. 2013;10(1):32.
31. Ruaro R, Gubiani ÉA. A scientometric assessment of 30 years of the Index of Biotic Integrity in aquatic ecosystems: applications and main flaws. Ecological Indicators. 2013;29:105–10.
32. Padmalal D, Maya K, Sreebha S, Sreeja R. Environmental effects of river sand mining: a case from the river catchments of Vembanad lake, Southwest coast of India. Environmental geology. 2008;54(4):879–89.
33. Saremi A, Saremi K, Sadeghi M, Sedghi H. The effect of aquaculture effluents on water quality parameters of Haraz River. Iranian Journal of Fisheries Sciences. 2013;12(2):445–53.
34. Rempel LL, Gill G, Fisheries Do, Oceans W, MB. Freshwater Inst. Bioassessment of streams along the Mackenzie River Valley, Canada, using the reference condition approach: biological, habitat, landscape and climate data. Winnipeg, MB (Canada): DFO; 2011.
35. Raczyńska M, Machula S, Choiński A, Sobkowiak L. Influence of the fish pond aquaculture effluent discharge on abiotic environmental factors of selected rivers in Northwest Poland. Acta Ecologica Sinica. 2012;32(3):160–4.
36. McCarthy KA, Johnson HM. Effect of agricultural practices on hydrology and water chemistry in a small irrigated catchment, Yakima River basin. Washington: US Geological Survey; 2009.
37. Smakhtin V. Environmental water needs and impacts of irrigated agriculture in river basins. A Framework for a New Research Program IWMI Working Paper. 2002;42
38. Abdoli A, Naderi M. Biodiversity of fishes of the southern basin of the Caspian Sea. Tehran: Abzian Scientific Publication; 2009.
39. Mostafavi H, Pletterbauer F, Coad BW, Mahini AS, Schinegger R, Unfer G, et al. Predicting presence and absence of trout (Salmo trutta) in Iran. Limnologica - Ecology and Management of Inland Waters. 2014;46:1–8.
40. Nnaji J, Uzairu A, Harrison G, Balarabe M. Effect of pollution on the physico-chemical parameters of water and sediments of river Galma, Zaria, Nigeria. Research Journal of Environmental and Earth Sciences. 2011;3(4):314–20.
41. Benítez-Mora A, Camargo JA. Ecological responses of aquatic macrophytes and benthic macroinvertebrates to dams in the Henares River Basin (Central Spain). Hydrobiologia. 2014;728(1):167–78.
42. Walker SL, Hedley K, Porter E. Pulp and paper environmental effects monitoring in Canada: An overview. Water Quality Research Journal of Canada. 2002;37(1):7–19.
43. Thompson G, Swain J, Kay M, Forster C. The treatment of pulp and paper mill effluent: a review. Bioresource technology. 2001;77(3):275–86.
44. Banagar GR, Kiabi BH, Homayoonnezhad I, Piri I, Amirian P. Biodiversity of Fish Species in Haraz River (An Ecological Approach). World Applied Sciences Journal. 2008;5(1):05–11.
45. Yazdian H, Jaafarzadeh N, Zahraie B. Relationship between benthic macroinvertebrate bio-indices and physicochemical parameters of water: a tool for water resources managers. Journal of Environmental Health Science and Engineering. 2014;12(1):30.

11

Biosynthesis of planet friendly bioplastics using renewable carbon source

Roopesh Jain* and Archana Tiwari

Abstract

Plastics are uniquely flexible materials that offer considerable benefits as a simple packing to complex engineering material. Traditional synthetic polymers (often called plastics), such as polypropylene and polyethylene have been derived from non-renewable petrochemicals and known to cause environmental concerns due to their non-biodegradable nature. The enormous use of petroleum-based plastic compounds emphasized a need for sustainable alternatives derived from renewable resources. Bioplastics have attracted widespread attention, as eco-friendly and eco-tolerable alternative. But they have got certain limitations as well, such as high cost of production and unsatisfactory mechanical properties. In this study we have found agriculture waste (AW) as low-cost and renewable substrate for the production of bioplastics in bacterial fermentation. Improvement in tensile properties of produced bioplastic film has also been documented upon blending with Cellulose Acetate Butyrate (CAB).

Keywords: Polyhydroxyalkanoates, Biosynthesis, Biodegradation, Cellulose acetate butyrate, Agricultural waste

Introduction

Every activity in modern life has been influenced by plastics and many depend entirely on plastic products due to their useful material properties and low production cost. Traditional synthetic polymers (often called plastics), such as polypropylene and polyethylene have been derived from non-renewable petrochemicals and known to cause environmental concerns due to their non-biodegradable nature [1]. The use of non-renewable fossil fuels for the production of plastic products has not only diminishing fossil fuel stockpiles but believed 'eco-unfriendly' too. However the need for polymers and the products they constitute has been reported ever increasing. The enormous use of petroleum-based plastic compounds emphasized a need for sustainable alternatives derived from renewable resources [2].

A special class of optically active biopolymers called Polyhydroxyalkanoates (PHAs) has been one of the most talked bioplastics and appeared as a potential candidate in replacing some petroleum based plastics because of their biodegradable nature and physical properties [3]. PHAs have showed some of the extraordinary similarities to the well-known synthetic polymers i.e. low-density polyethylene and polypropylene; in addition their disposal as bio-waste made them increasingly attractive in the pursuit of sustainable development of biodegradable plastics [4]. These PHAs comprised a class of polyesters of natural origin which has been accumulated by a variety of microorganisms intracellularly in the form of intracellular granules and stored in response to an environmental stress or nutrient limitation as a reserve of carbon, energy, and reducing power [5]. Under the conditions of carbon source depletion, PHAs have been reported to degrade by intracellular depolymerases and subsequently metabolized as a carbon and energy sources [6]. Poly-3-hydroxybutirate (PHB) and its copolymers with 3-hydroxyvalerate (3HV) known as poly(3-hydroxybutyrate-co-3-hydroxyvalerate) (PHBV), have been documented as the best known representatives of PHA family [7].

PHAs produced from standard carbon sources have got certain limitations such as high cost of production. Plant residues have been known as eco-friendly in nature and did not cause any ill-effects to the both flora and fauna. The best reported use of these agricultural wastes (AW) was feeding the cattle. Due to sufficient availability and non-commercial importance these natural residues have created attention for production of cost effective bioplastics [8].

* Correspondence: science.roopesh@gmail.com
School of Biotechnology, Rajiv Gandhi Proudyogiki Vishwavidyalaya, Airport Bypass Road, Bhopal, Madhya Pradesh, India

In this study model strain *Cupriavidus necator* was used for the production of biopolymer. AW was studied as a cheap and renewable carbon source and compared with standard carbon source glucose in bacterial fermentation. Role of blending in produced bioplastics was also described.

Materials and methods

Growth media

Nutrient broth (NB) comprises distilled water, 5 g/L Peptic digest, 5 g/L Sodium chloride, 1 g/L Beef extract and 1/L Yeast extract.

Nutrient agar (NA) comprises distilled water, 5 g/L Peptic digest, 5 g/L Sodium chloride, 1 g/L Beef extract, 1 g/L Yeast extract and 15 g/L Agar.

Tryptone soya broth (TSB) comprises distilled water, 17 g/L Tryptone, 3 g/L Soy Peptone, 5 gl/L NaCl, 2.5 g/L K_2HPO_4 and 2.5 g/L D-glucose.

Tryptone soya agar (TSA) comprises distilled water, 15 g/L Tryptone, 5 g/L Soy Peptone, 5 g/L NaCl and 12 g/L Agar.

Basal salts medium (BSM) comprises distilled water, 1 g/L K_2HPO_4, 1 g/L KH_2PO_4, 1 g/L KNO_3, 1 g/L $(NH_4)_2SO_4$, 0.1 g/L $MgSO_4.7H_2O$, 0.1 g/L NaCl, 10 ml/L.

Trace element solution comprises 2 mg/L $CaCl_2$, 2 mg/L $CuSO_4.5H_2O$, 2 mg/L $MnSO_4.5H_2O$, 2 mg/L $ZnSO_4.5H_2O$, 2 mg/L $FeSO_4$, 2 mg/L $(NH_4)_6Mo_7O_{24}.4H_2O$ [9].

Microorganism

The gram-negative bacterium *Cupriavidus necator* (former *Ralstonia eutropha*) was used as production culture and revived in nutrient broth. Cultures were tested for purity by Gram stain and observation of cell morphology under a microscope at 1000 × [9].

Quick screening for PHB production

10 µl of 48 hr old culture of the isolate was transferred to an eppendorf tube containing 50 µl of acridine orange and incubated for 30 minutes at 30°C. After the incubation period, the culture was centrifuged at 4000 rpm, for 5 min. The pellet was collected and then resuspended in distilled water. A smear was prepared on a clean microscopic slide and observed in a fluorescent microscope at 460 nm [10].

Microbial fermentation

TSB was inoculated from a single colony of *Cupriavidus necator* and incubated for approximately 24 hours at 30°C. Cultures were checked for purity by Gram staining and observed under a microscope at 1000 ×. BSM was used for batch fermentation [9]. 25ml TSB inoculums was added to 250 ml of BSM resulting in an inoculation ratio of 10% (v/v). AW was added to the fermentation medium as test carbon source while glucose as control at three different concentrations i.e., 10, 20 and 30 g/l. Batch fermentation was carried out using rotary incubator at 30°C and the growth of bacteria was checked intermittently after 24 hrs, 48 hrs and 72 hrs. All experiments were performed in triplicate. The biomass was separated by centrifugation.

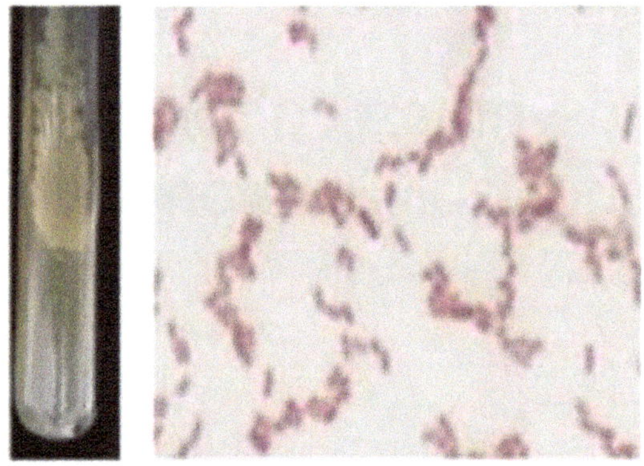

Figure 1 ***Cupriavidus necator*** **colonies on slant and Gram staining of the revived culture.**

Recovery of PHB

The recovery of PHB from dried biomass was carried out by digestion method using sodium hypochlorite as described by Hahn *et al.* [4].

Quantification of PHB

The Spectrophotometric method was used for quantification. Absorbance at 235 nm was measured against a sulphuric acid blank. Pure PHB was used to prepare the standard curve [11]. By referring to the standard curve, the quantity of produced PHB was determined.

Blending of polymers and film fabrication

The blend films were prepared by conventional solvent casting technique [12]. Cellulose Acetate Butyrate (CAB) was used as blending agent in different ratios (PHB:CAB - 50:50, 60:40, 70:30, 80:20, 90:10 and control was 100% PHB).

Table 1 Comparison of CDW and PHB concentrations

Carbon source	Cell dry weight (gL^{-1})	PHB concentration (gL^{-1})
Glucose	2.744 ± 0.005	1.518 ± 0.026
AW	2.251 ± 0.012	1.100 ± 0.125

Values are written as the mean (gL^{-1}) ± standard deviation (SD) of three replications.

Table 2 Tensile properties of PHB/CAB blends

Sample	Tensile strength (MPa)	Elongation at break (%)
Pure PHB	4.363 ± 0.002	3.298 ± 0.002
90:10	7.303 ± 0.014	4.773 ± 0.008
80:20	7.316 ± 0.002	4.671 ± 0.082
70:30	3.468 ± 0.002	2.149 ± 0.104
60:40	5.319 ± 0.010	3.839 ± 0.285
50:50	20.397 ± 0.368	5.158 ± 0.176

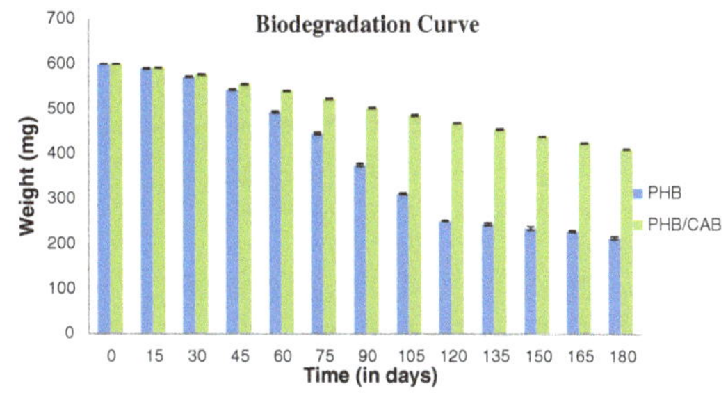

Figure 3 Biodegradation Curve.

Testing of tensile strength and biodegradation

Tensile strength determines the breaking point of various items. Testing of PHB film and blends was performed using an instrument called "INSTRON".

Laboratory testing is a common technique used to determine the susceptibility of compounds to biodegradation. In the lab test, solvent cast films of PHB and PHB blend (test samples), 100% biodegradable plastic (positive control) and synthetic plastic (negative control) of equal sizes were taken and weighed. All samples were added to separate container containing 100 g of garden soil and incubated for 6 months. Samples were periodically removed (once in 15 days), washed with distilled water and dried before analysis. The dried films were weighed for monitoring weight loss. All experiments were done in triplicate.

Toxicity test

Toxicity studies were performed by 7 day plant seedling test (Cress test) [13].

Results and discussion

In this study standard carbon source glucose was compared with AW for PHB production in *Cupriavidus necator* fermentation. Nutrient broth was used for the revival of bacterial culture. Sub-culturing was carried out using TSA after regular intervals of 15 days to maintain the stock & purity [9]. Microscopic analysis of heat fixed cells was performed after Gram staining. *Cupriavidus necator* was characterized as gram negative, short rods shaped aerobic bacteria as shown in Figure 1. Rapid screening of bacteria for PHB production was carried out using acridine orange staining. Microbial colonies were able to incorporate the dye and appeared yellow-orange indicating PHB production [1].

Different concentration of glucose and AW were checked in the fermentation to determine maximum PHB accumulation in *Cupriavidus necator* cells. Both glucose and AW showed best results at 20 g/l in BSM after an incubation of 72 hours with maximum production at 48 hours. Data shown in Table 1 indicated that the maximum cell dry weight (CDW) and PHB concentration were 2.744 ± 0.005 gL^{-1} and 1.518 ± 0.026 gL^{-1} (55% of CWD) respectively with glucose; while 2.251 ± 0.012 gL^{-1} and 1.100 ± 0.125 gL^{-1} (49% of CWD) respectively with AW.

Fermentation based commercial polymer production occurs in aerobic processes; therefore, only about 50% of the main carbon source has been used for production of biopolymers. A second cost factor in normal production processes for PHAs is the costs for complex nitrogen sources. Both cheap carbon sources and cheap nitrogen sources are available from agricultural waste and surplus materials [14]. Results encourages that the AW can act as a competent substrate to glucose for PHB production at 2% concentration.

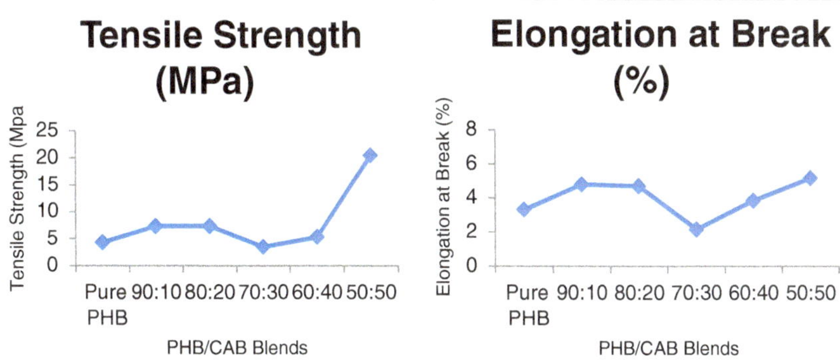

Figure 2 Tensile properties of PHB/CAB blends.

PHB was recovered through sodium hypochlorite digestion method and biopolymer film was prepared by dissolving the extracted PHB in chloroform. Film was found to be brittle and broke easily. To overcome these weaknesses blending with CAB was carried out. CAB was reported as a FDA approved biodegradable and transparent thermoplastic similar to cellulose acetate but tougher with lower moisture absorption and better weathering resistance [15]. A varying range of blending agent CAB was used and compared with 100% standard PHB film.

The main interest of plastic film preparation was to increase the amorphous content. The PHB was blended with CAB in ratios 50:50, 60:40, 70:30, 80:20, 90:10, while control was pure PHB. The tensile strength and elongation at break (Table 2) were tested to know the effect of blending against pure PHB film. Tensile strength value of Pure PHB was reported 4.363 ± 0.002 MPa, which was in the range of commercial PHB reported earlier [15-17]. Tensile strength of PHB/CAB (50:50 to 90:10) blends was varied from 7.303 to 20.397 MPa with best at 20.397 MPa for 50:50, whereas elongation at break varied from 2.149 to 5.158% with 5.158% as best for 50:50 blend (Figure 2). It was evident from results that blending had improved tensile propertied due to intermolecular bonding and increase in amorphous content [18,19].

The overall degradation mechanism involves abiotic factors followed by enzymatic degradation through PHB depolymerases and other degradation enzymes. These enzymes disrupt the inner molecular structure of biopolymer through microbial attack or microbial colonization. Therefore, the biodegradation efficiently begins from surface and gradually wrap the inner molecular conformation. Biodegradability analysis was carried out in lab tests for six months. PHB/CAB blend at ration 50:50 showed best physico-mechanical properties therefore was compared with pure PHB to find out biodegradation pattern. The readings for weight loss for blend and pure PHB were recorded after every 15 days and the same have been illustrated in biodegradation curve (Figure 3). PHB has showed 64.3% degradation in 6 months, whereas PHB/CAB blending exhibited only 31.5%. The miscibility of PHB and CAB was found very high that imparted highly amorphous nature to the blend, but at same times it made the polymer surface hydrophobic leading to a slow hydrolysis environment.

The toxicity study was performed by Cress Test, according to standard OECD guidelines. The cress, a dicotyledonous plant was chosen because of its fast germination rate and high sensitivity to the toxic chemicals [20]. At the end of the test, all emerged plants were found healthy with more than 90% germination rate for all treatment groups. Results of cress test demonstrated that the final degradation product of bioplastic was not toxic to plants.

Conclusion

As environmental concerns became more prominent, it encouraged bigger debate and greater research in the bioplastics sector expanding the range of cost effective biodegradable products. Bioplastics have now been taken seriously because of many documented benefits such as saving fossil resources, climate protection and creating jobs in future-oriented sectors. Since the bioplastics originated from a renewable source, they have been believed much more environmental friendly than 'non-renewable' conventional plastics.

The major drawbacks for PHAs production are their production cost, and lack of comparable properties to the extent of adequate commercialization. The development of cost effective bioplastics is required to be necessarily preceded by investigations on the production factors with particular reference to the nature of micro-organisms used in the fermentation and the raw materials such as carbon source. This study has emphasized use of AW as a cost-competitive carbon source. Blending with approved biodegradable materials such as CAB has also attracted as an approach to manufacture commercially compatible bioplastics.

Competing interests

The authors declare that they have no competing interests.

Authors' contributions

AT was the main investigator. She participated in the study design and data analysis. RJ was involved in experimental studies. Both authors read and approved the final manuscript.

Acknowledgement

The authors are thankful to School of Biotechnology, Rajiv Gandhi Proudyogiki Vishwavidyalaya, Airport Road, Bhopal, Madhya Pradesh, India for providing necessary assistance to carry out the research work.

References

1. Thompson RC, Moore CJ, vom Saal FS, Swan SH. Plastics, the environment and human health: current consensus and future trends. Philos Trans R Soc Lond B Biol Sci. 2009;364:2153–66.
2. Jain R, Kosta S, Tiwari A. Polyhydroxyalkanoates: a way to sustainable development of bioplastics. Chron Young Sci. 2010;1:10–5.
3. Tortajada M, da Silva LF, Prieto MA. Second-generation functionalized medium-chain-length polyhydroxyalkanoates: the gateway to high-value bioplastic applications. Int Microbiol. 2013;16:1–15.
4. Dua C, Sabirovab J, Soetaertb W, Lin SCK. Polyhydroxyalkanoates production from low-cost sustainable raw materials. Curr Chem Bio. 2012;6:14–25.
5. Lee SY, Choi JL, Han K, Song JY. Removal of endotoxin during purification of poly(3-hydroxybutyrate) from gram-negative bacteria. Appl Environ Microbiol. 1999;65:2762–4.
6. Jain R, Tiwari A. Homology modelling of PHA synthases in *Cupriavidus necator*. IJUPBS. 2014;3(3):214–23.
7. Khanna S, Srivastava AK. Recent advances in microbial polyhydroxyalkanoates. Process Biochem. 2005;40:607–19.

8. Rehm BH. Biogenesis of microbial polyhydroxyalkanoate granules: a platform technology for the production of tailor-made bioparticles. Curr Issues Mol Biol. 2007;9(1):41–62.
9. Verlinden RA, Hill DJ, Kenward MA, Williams CD, Piotrowska-Seget Z, Radecka IK. Production of polyhydroxyalkanoates from waste frying oil by Cupriavidus necator. AMB Express. 2011;1(1):11.
10. Kalaivani R, Sukumaran V. Isolation and identification of new strains to enhance the production of biopolymers from marine sample in Karankura, Tamil Nadu. Eur J Exp Biol. 2013;3(3):56–64.
11. Bhuwal AK, Singh G, Aggarwal NK, Goyal V, Yadav A. Isolation and screening of polyhydroxyalkanoates producing bacteria from pulp, paper, and cardboard industry wastes. Int J Biomater. 2013;2013:1–10.
12. Zakaria MR, Ariffin H, Abd-Aziz S, Hassan MA, Shirai Y. Improved properties of poly(3-hydroxybutyrate-co-3-hydroxyvalerate) produced by Comamonas sp. EB172 utilizing volatile fatty acids by regulating the nitrogen source. Biomed Res Int. 2013;2013:1–7.
13. Arcos-Hernandez MV, Laycock B, Pratt S, Donose BC, Nikolić MAL, Luckman P, et al. Biodegradation in a soil environment of activated sludge derived polyhydroxyalkanoate (PHBV). Polym Degrad Stab. 2012;97:2301–12.
14. Koller M, Bona R, Braunegg G, Hermann C, Horvat P, Kroutil M, et al. Production of polyhydroxyalkanoates from agricultural waste and surplus materials. Biomacromolecules. 2005;6(2):561–5.
15. Wang T, Cheng G, Ma S, Cai Z, Zhang L. Crystallization behavior, mechanical properties, and environmental biodegradability of poly(β-hydroxybutyrate)/cellulose acetate butyrate blends. J Appl Poly Sci. 2001;89:2116–22.
16. Aarthi N, Ramana KV. Identification and characterization of Polyhydroxybutyrate producing *Bacillus cereus* and *Bacillus mycoides* strains. Int J Environ Sci. 2011;1:744–56.
17. Ch'ng DHE, Sudesh K. Densitometry based microassay for the determination of lipase depolymerizing activity on polyhydroxyalkanoate. AMB Express. 2013;3:22.
18. Cai MM, Cui FY, Zhao QL, Chua H. Influence of substrate load on polyhydroxyalkanoates (Pha) accumulation by unenriched mixed cultures from excess sludge fermentation liquid. Adv Mater Res. 2012;456:856–61.
19. Kfoury G, Raquez JM, Hassouna F, Odent J, Toniazzo V, Ruch D, et al. Recent advances in high performance poly(lactide): from "green" plasticization to super-tough materials via (reactive) compounding. Front Chem. 2013;1:32.
20. Abe H, Doi Y. Structural effects on enzymatic degradabilities for poly[(R)-3-hydroxybutyric acid] and its copolymers. Int J Biol Macromol. 1999;25:185–92.

12

Sources of metal pollution in the urban atmosphere

Abdullah Aksu

Abstract

In this study, İstanbul-Tuzla Region atmosphere was selected as the working area for atmospheric pollution. Although the study area seems too local, this region contains shipyards and shipyard-related side product industries. It is also surrounded by aluminum (ASSAN) and copper (SARKUYSAN) facilities and tannery organized industrial district. For determined the atmospheric inputs, the sample collection was carried out as monthly in 2010. Particulate matter was filtered from aerosols via a high volume air sampler. The collected 46 ambient air samples were analyzed for Cr, Fe, Cu, Zn, Cd, Pb and Al using Atomic Absorption Spectrometry (AAS), Flame unit. Additionally, the volume of the air was drawn and meteorological data recorded.

Average individual heavy metal concentrations were found as Cd (0.06 ng/m^3) < Cr (0.09 ng/m^3) < Zn (0.21 ng/m^3) < Pb (0.23 ng/m^3) < Cu (0.48 ng/m^3). The concentrations of crustal elements Fe and Al were changed between 5.48 ng/m^3, 74 ng/m^3 and 14 ng/m^3, 284 ng/m^3 respectively during the sampling period. Except Cr and Fe anthropogenic contribution was seen on the concentrations of Zn, Cu, Pb and Cd in an increasing order. While the crustal element Fe was not show an appreciable change in concentration, but the Al concentration was display an important change in concentration depending on the wind transportation.

Keywords: Atmospheric pollution, Metal, İstanbul, Tuzla, Enrichment Factor (EF)

Introduction

Various chemicals are emitted into the air from both natural and man-made (anthropogenic) sources. The quantities may range from hundreds to millions of tonnes annually. Natural air pollution stems from various biotic and abiotic sources such as plants, radiological decomposition, forest fires, volcanoes and other geothermal sources and emissions from land and water. These result in a natural background concentration that varies according to local sources or specific weather conditions [1]. The emission of toxic substances in the environment has been spread from industrialized countries. Many industrial plants and also heavy traffic may produce heavy metal into atmosphere. Traffic pollutants include potentially toxic metals for health like lead, cadmium and zinc [2, 3]. Migon et al. [4] indicated after the implementation of antipollution policies on automotive Pb concentration levels decreased by 82 %. The changes in life style increases the levels at which trace metals are added in soil, water and air from anthropogenic sources [5, 6]. Heavy metal pollution is potentially a persistent problem all over the world in an increasing order [7].

Emissions from traffic contain many toxic heavy metals such as; Pb, Cd and Zn [2]. Meteorological conditions and local sources have an important role on trace element concentrations. Airborne particles are important carriers of metals [8].

In recent decades, major efforts have been made to reduce air pollution in the European Region [1]. So a considerable interest has been given to particulate matter (PM) of aerosols due to their impacts on visibility, human health, plants, aquatic life and materials [8–10].

While the industrialization and human activities intensify the emission of various pollutants into the environment and introduce various harmful substances into the atmosphere [3].

A few decades ago the study are Tuzla region in İstanbul was a coastal recreation area. The industrialization of the Tuzla region had been increased rapidly. Especially shipyards and shipping industry were growth in the region. Such areas were classified "transitional urban areas" had

Correspondence: aksu_abdullah@hotmail.com
Department of Chemical Oceanography, Institute of Marine Science and Management, Istanbul University, Vefa 34134-00, Istanbul, Turkey

rapid population growth between 1975 and 2005 but tapered projected growth between 2005 and 2015 such as İstanbul, Turkey [11]. The aim of this study is firstly to search the seasonal change of metals Cr, Fe, Cu, Zn, Cd, Pb and Al of concentrations as inorganic pollutants and secondly to find their sources and anthropogenic contributions for the first time for this region.

Material and methods

Sampling

The sampling area is located in Turkey/İstanbul, TUDEV-Piri Reis University Campus (40^0 49′ 05 66″N and 29^0 20′54 00″E) at a height of 22 m (Fig. 1). The sampling point is faraway 0.81 km from the Marmara Sea. A High Volume Air Sampler (HVAS) was fixed at the top of the University Building roof to carry out the aerosols sampling.

The sample collection was carried out at 2010. The filter papers divided into two parts; one halves were saved in plastic bag separately for metal analyses. Five samples were taken for February, March, April, May, June, July, August, September and October. Only one sample was obtained for December and this value assumed as the average of the month. The volume of the air was drawn and meteorological data recorded.

Analytical procedures

Particulate matter

Whatman glass fiber filter papers having a pore size of 0,1 μm are dried for 48 h in a desiccator and weighted before and after the sampling. So the daily concentrations of particulate matter with the size > 0.1 μm of collected over the filter paper were found.

Metal analyses

The halves of the filter papers put into the teflon (PTFE) tubes. A strong acid mixture of 5 ml HNO_3, 3 ml $HClO_4$ and 1 m l HF is added to the tubes to provide the decomposition metals at 180 °C in closed digestion unit (micro-wave) during one hour. Then diluted to a definite volume and then the analyses of Al, Fe, Pb, Zn, Cd and Cr are accomplished at Atomic Absorption Spectrometry (AAS), Flame unit. Duplicate metal analysis results are

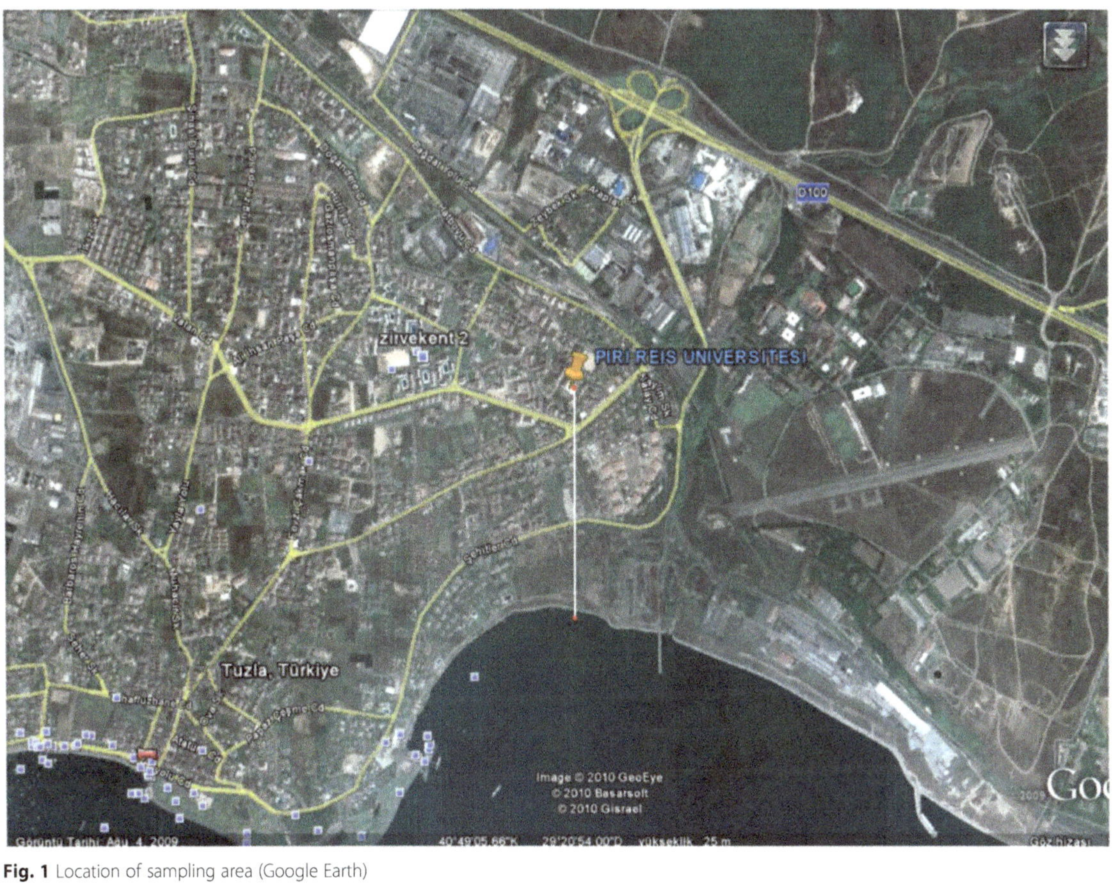

Fig. 1 Location of sampling area (Google Earth)

given in Table 1. Confidence interval = 1.96 × standard deviation/mean.

Results

Particulate matter

The air quality standard adapting to US EPA NAAQS (http://www.epa.gov/air/criteria.html) for PM_{10}particles in Turkey is 150 μg/m3 and required to decrease the concentration below 60 μg/m3until 2014 for human health [12]. As a result of the normal day-to-day variation in PM_{10} concentrations, 24-h averages of 100 μg/m^3 are regularly exceeded in many areas in Europe, especially during winter inversions. The effects have been observed at annual average concentration levels below 20 μg/m^3 (as $PM_{2.5}$) or 30 μg/m^3 (as PM_{10}). For this reason, no guideline value for long-term average concentrations is recommended [1]. In Turkey, it is required to reach these limit values at 2019 [12]. The range of PM (pore size >0.1 μm) concentrations measured at Tuzla region during the sampling period was 12.22 μg/m^3-209 μg/m^3 with the average value of 67.90 μg/m^3. The daily PM concentrations in each month were shown on Fig. 2. The tendencies in increase at PM concentrations were seen on from July to October in some measurements.

Cong et al. [13] studied with two filter paper simultaneously, one with polytetrafluoroethylene, PTFE, 0.5 μm, 47 mm) and the other polycarbonate (Nucleopor, Coaster,) filters with the pore size of 1.5 %. In the study the results were found in between 2.02-6.34 μg/m^3. Braga et al. [14] measured the concentration (μg/m^3) ranges between HV PM_{10} in 2.80-40.10 and Dichotomous (fine) in 12 0.46-16.02 and Dichotomous (coarse) in 1.51-24.42.

Table 2 shows the meteorological conditions and the measured PM concentrations (μg/m^3) ialong the sampling periods (Fig. 2). This may not be enough to explain all the measurement events, since the initial measurements generally depend on the previous days' meteorological conditions. It is necessary to develop a detailed study to explain the relationship between the meteorological parameters and the PM concentrations in the aerosol such as wind transportation. Although there are some deficiencies to explain the relationships between meteorological conditions and PM concentrations, this table will help to see some variations on PM concentrations. The weak correlations are found between PM concentrations with average temperature, relative humidity and average wind speed are 0.15, 0.08 and 0.18 respectively. To observe any relationship between meteorological conditions PM concentrations all parameters could be considered as a whole previous or sampling period. For instance, from Table 2, it is seen that the low concentrations were generally occurred at low temperatures and rainy days, e.g. December Sample No. 1 (1.22 μg/m^3) and March Sample No. 4 (12.87 μg/m^3). The high concentrations were seen at dry air and when humidity values are relatively low, e.g. July Sample No. 5 (209 μg/m^3) and September Sample No. 4 (154 μg/m^3). As an exceptional case, a dust concentration (0.25-0.50 mg/m^3) had been occurred at February 17 from Sahara Desert and continued a few days causing to decrease the visibility to 10–6 km (www.meteorgov.tr). This effect was seen on the PM concentrations on Sample No. 2 (93.11 ng/m^3) and Sample No. 3 (78.46 ng/m^3) even if the duration was rainy. After a dry period, the concentration increase was observed on the fifth sample depending on wind transportation. In Turkey, the wind systems generally cause to flow of air from west to 22 east [15, 16].

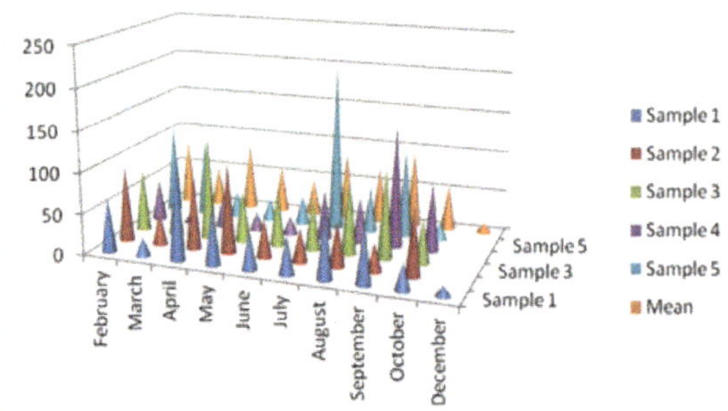

Fig. 2 Daily PM concentrations (μg/m^3)

Table 1 The sensitivity of duplicate metal analysis

Metal	Mean-st.deviation	Confidence interval
Cd	0.00735 ± 0.00021	0.0056
Pb	0.1153 ± 0.00113	0.019
Cu	0.34365 ± 0.01463	0.083
Cr	0.00825 ± 0.00007	0.0166
Zn	0.2044 ± 0.00806	0.077
Fe	13.3526 ± 0.32682	0.0479
Al	47.6121 ± 1.02968	0.0423

Metals

In this study metals are evaluated into two groups as crustal metals Al and Fe and non-crustal metals Cr, Cu, Zn, Cd and Pb. While the Al and Fe monthly average concentrations are shown in Fig. 2a, the others are shown Fig. 2b. Al concentrations begin to increase from July and reach a peak at September, then the concentration decreases at cold months (Fig. 3a). This situation shows some similarities with PM concentration changes (Fig. 2). Fe does not show any appreciable change in monthly average concentrations except December (Fig. 3a).

From Fig. 4b, it can be concluded that, while the monthly average concentrations of Cr, Zn, Cd, Pb show

Table 2 Meteorological parameters in the year 2010

Sampling date/duration	No	Average temp. (°C)	Wind-speed (Ave-Max) (km/h)	Wind direction	Precipitation (mm)	Relative humidity (%) (Ave-Max)
12.02.2010/(10:00)-13.02.2010/(09:30)	1	11	12-33	SW	0.2	72-83
16.02.2010/(16:00)-17.02.2010/(09:00)	2	7	7-15	NE	0.6	61-70
17.02.2010/(17:00)-18.02.2010(09:00)	3	12	10-23	SW	0.4	52-70
18.02.2010/(17:00)-19.02.2010/(09:30)	4	7	8-20	SW	-	80-90
22.02.2010/(17:00)-23.02.2010/(09:00)	5	8	6-18	S-SW	-	58-71
11.03.2010/(17:00)-12.03.2010/(09:00)	1	8	16-28	N-SW	showery	81-84
15.03.2010/(17:00)-16.03.2010/(09:00)	2	6	11-20	N-S	-	66-59
16.03.2010/(17:00)-17.03.2010/(09:00)	3	4	15-32	S	0.4	59-63
17.03.2010/(17:00)-18.03.2010/(12:00)	4	2	8-17	N-S	0.2	48-63
18.03.2010/(17:00)-19.03.2010/(09:00)	5	6	6-15	N-E	-	48-51
15.04.2010/(18:00)-16.04.2010/(10:30)	1	14	5-15	N-SW	-	86-90
16.04.2010/(18:00)-17.04.2010/(19:00)	2	14	20-32	N	-	71-57
19.04.2010/(09:30)-20.04.2010/(09:30)	3	14	10-20	N-S	0.6	71-100
20.04.2010/(17:30)-21.04.2010/(17:00)	4	13	15-35	N	0.8	80-100
21.04.2010/(17:00)-22.04.2010/(09:30)	5	13	11-22	W-SW	-	59-100
12.05.2010/(17:00)-13.05.2010/(09:00)	1	21	7-15	N-E	0.6	60-85
14.05.2010/(17:00)-15.05.2010/(17:00)	2	20	10-20	S-E	-	55-60
17.05.2010/(17:00)-18.05.2010/(17:00)	3	18	22-32	S-N	-	60-70
18.05.2010/(17:00)-19.05.2010/(16:00)	4	16	12-17	S-W	-	57-70
20.05.2010/(16:00)-21.05.2010/(13:00)	5	17	8-22	S-N	-	40-57
18.06.2010/(17:00)-19.06.2010/(18:00)	1	25	17-26	N	8	62-59
19.06.2010/(18:00)-20.06.2010/(16:00)	2	26	9-20	N-S	8	60-70
20.06.2010/(16:00)-21.06.2010/(09:00)	3	24	10-20	N-S	-	70-80
21.06.2010/(17:00)-22.06.2010/(17:00)	4	22	13-39	S	-	54-57
22.06.2010/(17:00)-23.06.2010/(17:00)	5	23	12-24	E-SW	-	54-65
19.07.2010/(18:00)-20.07.2010/(17:00)	1	26	15-37	N	-	62-89
20.07.2010/(17:00)-21.07.2010/(17:00)	2	25	11-24	N	-	69-83
21.07.2010/(17:00)-22.07.201/(18:00)	3	27	15-30	N	-	72-94
22.07.2010/(17:00)-23.07.2010/(17:00)	4	29	15-26	N	-	63-89
23.07.2010/(17:00)-24.07.2010/(09:00)	5	27	11-18	N-S	-	67-90
09.08.2010/(17:00)-10.08.2010/(11:00)	1	25	15-35	N	exist	75-89
10.08.2010/(17:00)-11.08.2010/(17:00)	2	30	14-39	N	-	70-89
11.08.2010/(17:00)-12.08.2010/(17:00)	3	28	14-28	N	-	67-83
12.08.2010/(17:00)-13.08.2010/(17:00)	4	39	9-16	N-S	-	69-94
13.08.2010/(17:00)-14.08.2010/(11:00)	5	30	9-22	Variable	-	69-50
20.09.2010/(17:00)-21.09.2010/(17:00)	1	22	14-39	N-NW	-	75-100
21.09.2010/(17:00)-22.09.2010/(17:00)	2	20	27-41	N-NE	-	70-88
22.09.2010/(17:00)-23.09.2010/(17:00)	3	21	21-33	N	-	58-63
26.09.2010/(17:00)-27.09.2010/(17:00)	4	26	8-20	S-SE	-	60-78
28.09.2010/(17:00)-29.09.2010/(17:00)	5	22	10-18	S-SW	-	61-78
13.10.2010/(17:00)-14.10.2010/(17:30)	1	17	7-15	N-S	5	87-94
14.10.2010/(17:00)-15.10.2010/(14:00)	2	17	13-18	NE	9	90-100

Table 2 Meteorological parameters in the year 2010 *(Continued)*

16.10.2010/(19:00)-17.10.2010/(19:00)	3	18	9-18	N-NE	2	90-100
17.10.2010/(18:00)-18.10.2010/(18:00)	4	18	5-13	S-SW	-	81-94
18.10.2010/(18:00)-19.10.2010/(19:00)	5	19	9-17	S	2	70-88
13.12.2010/(17:00)-14.12.2010/(17:30)	1	4	15-26	N	8	62-87

some extremums at different months, Cu shows an increase in average concentrations from beginning July reach to a maximum at September, then decreases. Copper concentration changes also show some similarities with PM and Al concentrations.

The daily metal concentrations of metals are given in Table 2 and Fig. 4a-g.

Crustal metals

In this study, aluminum concentration changes between 14 ng/m^3 and 248 ng/m^3 with an annual average of 71.13 ng/m^3 (Table 2, Fig. 4a). The highest concentration is found in the 5th sample of July. The PM concentration also has the highest concentration that sampling period. The correlation between Al and PM was found as r = + 0.64 (Table 3). The lowest Al concentration is found in March, Sample No. 1as14 ng/m^3. This sampling duration corresponds a rainy period. The second low Al concentration is found in a rainy period in April Sample No.4 as 21 ng/m^3. The increases in concentrations in September could be explained that there wasn't any precipitation. In October, the period was rainy and high relative humidity. So Al could be hold in the aerosol by high humidity.

In İstanbul (Tuzla region), iron concentration changes between 5.48 ng/m^3 and 74 ng/m^{3}with the average value of 37.22 ng/m^3 (Table 2 and Fig. 4b). Fe concentrations are generally low at rainy sampling periods e.g., August October Sample No.2 and Sample No. 3 are 7.66 ng/m^3 and 5.48 ng/m^3. The high concentrations were observed at dry periods, e.g. both April Sample No.1 and July Sample No.5 was measured as 74 ng/m^3. Fe concentrations were low in some measurements at August, October and December due to washing out of precipitation. The correlation between Fe and PM found as r = + 0.48.

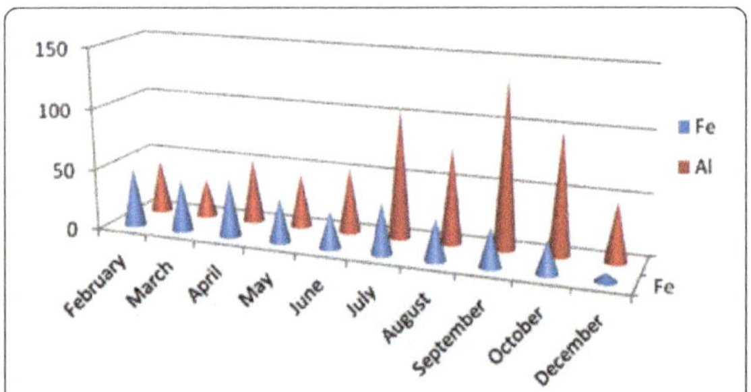

Fig. 3 Monthly average concentrations (ng/m^3) of Al and Fe

Non-crustal metals

In this study, chromium concentration range was in between 0.004 ng/m^3 and 0.28 ng/m^3, with an average value of 0.089 ng/m^3 (Table 2, Fig. 4c). The highest concentrations were observed at rainy and high relative humidity sampling days e.g., April Sample No. 1 and sample No. 3 were 0.28 ng/m^3 and 0.23 ng/m^3, July Sample No. 5 was 0.23 ng/m^3, September Sample No. 4 and Sample No.5 were 0.26 ng/m^3 and 0.22 ng/m^3. The lowest chromium concentration was found in July Sample No 1. (0.004 ng/m^3). There wasn't any precipitation and the relative high humidity value was low. Some low concentrations were observed at heavy rains and low relative humidity sampling periods e.g. June sample No 4 and 5 the concentration was measured as 0.1 ng/m^3. The correlation between Cr and PM was found as r = +.0.51 (Table 3).

The high Cr concentrations was observed when the prevailing wind directions are generally S and SW. Taşdemir et al. [8] was stated the leather industry caused in increase particularly on Cr levels depending on wind direction in Bursa In the study the correlation of Cr with average wind speed was found as r = + 0.38 that was the highest wind speed correlation (Table 3). In the study the correlation of high concentrations of chromium with average wind speed was found 6 as r = + 0.53.

In Tuzla region, the annual copper concentration was found as 0.48 ng/m^{3}within the range of 0.04 ng/m^3 and 2.68 ng/m^3 (Fig. 4d). The highest concentrations were observed at September. During the sampling period, there weren't any precipitation and the humidity values were high. The seasonal concentration fluctuations of Cu similar to Al. The correlation of Cu with Al was r = + 0.62 (Table 3).

In the study area, zinc concentration ranges were in between 0.01 ng/m^3 and 0.65 ng/m^3 with an average concentration of 0.18 ng/m^3. The highest concentrations were measured at April and October, when the relative humidity percentages were high (Fig. 4e). The highest correlation was found between Zn and relative humidity

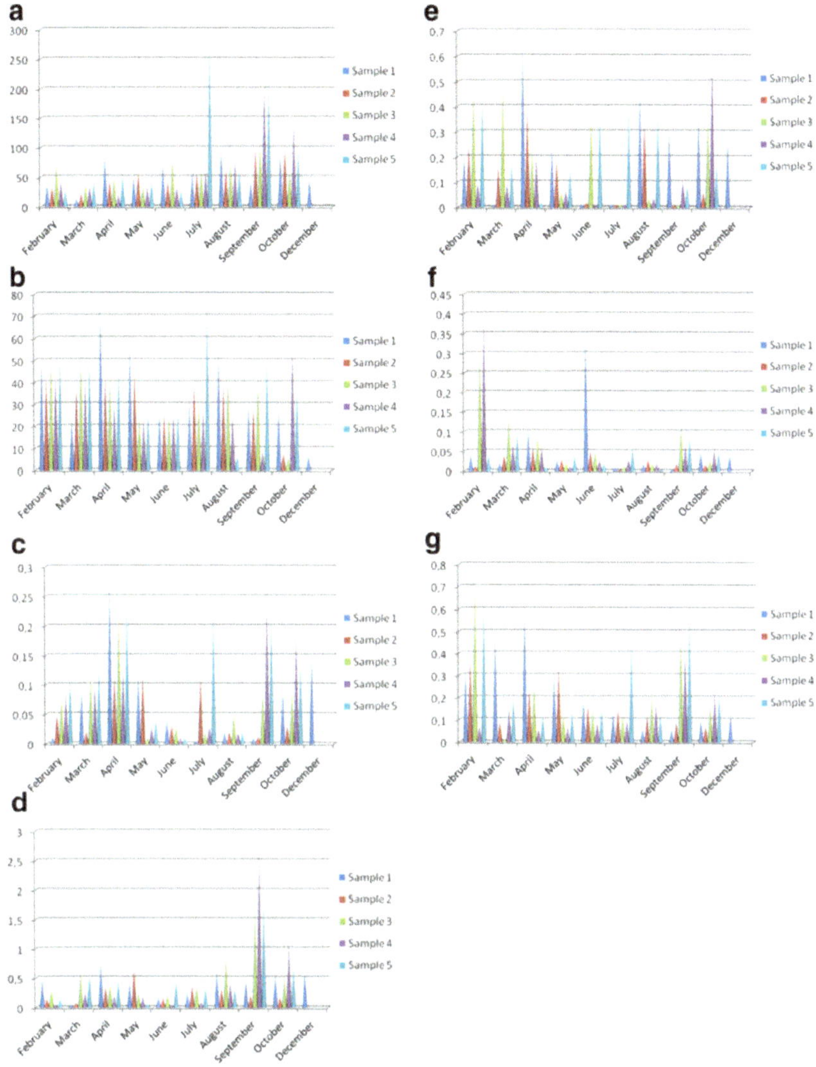

Fig. 4 a. Daily concentration changes of Al concentration (ng/m^3). **b**. Daily concentration changes of Fe concentration (ng/m^3). **c**. Daily concentration changes of Cr concentration (ng/m^3). **d**. Daily concentration changes of Cu concentration (ng/m^3). **e**. Daily concentration changes of Zn concentration (ng/m^3). **f**. Daily concentration changes of Cd concentration (ng/m^3). **g**. Daily concentration changes of Pb concentration (ng/m^3)

Table 3 Correlation coefficients among metals and meteorological data

	Fe	Cu	Zn	Cd	Pb	Al	PM	Temp.	R.Hum.	Wind speed
Cr	0.30	**0.54**	0.29	0.15	0.47	0.49	**0.51**	0.21	0.10	0.38
Fe	-	0.03	0.40	0.25	0.44	0.23	0.48	0.19	0.08	0.24
Cu		-	0.05	0.00	0.37	**0.62**	0.47	0.20	0.008	0.10
Zn			-	0.08	0.25	0.12	0.33	0.14	0.34	0.37
Cd				-	0.22	0.05	0.08	0.27	0.04	0.06
Pb					-	0.34	**0.56**	0.14	0.05	0.25
Al						-	**0.64**	0.43	0.17	0.08

percentage was 0.34 and the correlation between high concentrations of Zn with the precipitation was found as r = + 0.58 (Table 3).

In Tuzla aerosol, cadmium concentration range was 0.001 ng/m^3- 0.41 ng/m^3 with an average of 0.063 ng/m^3 (Fig. 4f). The high concentrations were observed two times at February and one times at June. These measurements correspond to rainy periods. The correlation between Cd and PM was r = + 0.48 (Table 3). Cizmecioğlu and Müezzinoğlu [17] found that the samples taken during rain incidents following elongated dry periods were high. And also parallel results were found by [18] who found high concentrations after a dry period and when rain continued for several days, concentrations were lower in rain in Amman, Jordan.

In Tuzla region lead concentration range was 0.01 ng/m^3 – 0.77 ng/m^3 with a mean of 0.23 ng/m^3 (Fig. 4g). Higher concentrations of Pb were seen on high PM concentration periods. The correlation between Pb and PM was found as r = + 0.56 (Table 3).

Discussion

Correlation coefficients among metals and meteorological data of this study are given in Table 3. The highest value (r = + 0.64) was found between Al and particulate matter contents in atmospheric aerosols. Additionally, high correlations were observed between Al and Cu contents, particulate matter and Pb, particulate matter and Cr amounts. The results are show the terrestrial origins on this metal distributions via atmospheric tranport. In contrast, the low correlation coeffecient values are indicate the different origins and factors on atmospheric metal transport.

Table 4 is prepared to compare this study results with other studies. Except Terra Nova Bay [19] the Cr and Fe concentrations of the region was very low from the other studies. The Al concentration of the region was higher than Mt. Qomolangma [13] and Terra Nova Bay [19] but the Al concentration of the Ankara having a continental climate [9] and West and East Black Sea aerosols were very high [16]. Zn and Cd measurements are similar to Terra Nova Bay [19]. Cu concentration was similar to Mt. Qomolangma [13] and Terra Nova Bay [19]. It could be concluded that out of Cr, Fe and Al, Tuzla region measurements show some approximate values with terra Nova Bay.

In general, most people take in very little aluminum from breathing. Levels of aluminum in the air generally range from 0.005 to 0.18 micrograms per cubic meter (μg/m^3), depending on location, weather conditions, and type and level of industrial activity in the area. Most of the aluminum in the air is in the form of small suspended particles of soil (dust). Aluminum levels in urban and industrial areas may be higher and can range from 0.4 to 8.0 μg/m^3 (http://www.atsdr.cdc.gov/toxprofiles/tp22-c1.pdf). The Occupational Health and Safety Administration (OSHA) has limited workers' exposure to aluminum in dusts to 15 mg per cubic meter (mg/m^3) (total dust) and 5 mg/m^3 (respirable fraction) of air for an 8-h workday, 40-h workweek (http://www.atsdr.cdc.gov/tfacts22.pdf). Aluminum is released to the environment by both natural processes and anthropogenic sources. Because of its prominence as a major constituent of the earth's crust, natural processes far exceed the contribution of anthropogenic releases to the environmental distribution of aluminum [20]. Sekhavatjou et al. [21] indicate that the Al level in industrial and heavily congested traffic sectors is higher than other areas. Coal burning, electroplating and paint are the source of Al [22]. Iron is not considered to cause harmful health effects in general; toxicological and epidemiological literature is limited. ATSDR [23] used a modification of the dose equation

Table 4 Mean and max-min values of metal concentrations (ng/m^3) measuredat this study and comparison with other studies

Metal	This study			Mt. Qomolangma[a]			Ankara[b]	Black Sea[c]		TerraNova Bay[d]
								West	East	
	Mean	Min	Max	Mean	Min	Max	Mean	Mean	Mean	Mean
Cr	0.089	0.004	0.28	3.0	2.3	12	3.2	9.0	8.3	0,043
Fe	37.22	5.48	74	61	28	89	100	420	420	2.31
Cu	0.48	0.04	2.68	0.31	0.21	0.69	-	-	-	0.422
Zn	0.18	0.01	0.65	1.4	0.81	3.3	16	46	26	0.081
Cd	0.063	0.001	0.41	-	-	-	0.11	-	-	0.016
Pb	0.23	0.01	0.77	0.43	0.30	0.59	71	-	-	-
Al	71.13	14	248	55	34	85	110	540	330	2.08

[a]Cong et al., [13]: Mt. Qomolangma,Everest, China – Atmospheric aerosols in May-June 2005
[b]Yatin et al., [9]: METU Campus, Ankara, Turkey – Atmospheric aerosols in February-June 1993
[c]Güllü et al., [16]: West and East Black Sea, Turkey – Over the atmosphere near sea level, June, August-September 1988
[d]Toscano et al., [19]: TerraNova Bay, Antartic – Coastal site aerosols, Summers of 2000–2001 and 2001–2002

(dose = concentration × intake rate) to calculate a daily consumption from exposure to iron in soil. Exposure to the maximum concentration of iron collected from a residential area (31,500 ppm). The National Academy of Sciences' (NAS's) dietary reference intake (DRI) for children 1 to 3 years old is 7 mg/day [24]. The median daily intake of dietary iron is roughly 11 to 13 mg/day for children 1 to 8 years old and 13 to 20 mg/day for adolescents 9 to 18 years old [23, 24]. (http://www.atsdr.cdc.gov/HAC/pha/oakRidge013107-TN/Final_Current_Screening_1-07_508.pdf. The vehicular activities cause to increase the level of Fe in the atmosphere [25].

The concentration level of Fe is highly influenced by the industrial activities and traffic of vehicle [21]. Chromium is ubiquitous in nature. Available data, generally expressed as total chromium, show a concentration range of 5–200 ng/m^3. There are few valid data on the valency and bioavailability of chromium in the ambient air [1]. Cr is emitted in the atmosphere due to coal and oil combustion, especially diesel-fed vehicles, refuses incineration [26]. The Occupational Safety and Health Administration (OSHA) requires that levels of copper in the air in work places not exceed 0.1 mg/m^3 copper fumes per cubic meter of air (0.1 mg/m^3) and 1.0 mg/m^3 for copper dusts [27]. The main source of Cu in the atmosphere is diesel engines, unleaded gasoline is the other source [28]. The other sources of the atmosphere are Cu-containing fungicides, metal working factories and electroplating materials [29]. The rural levels of airborne copper represented regional background levels in urban study sites with only episodic increases, depending on wind speed and direction and location relative to local point sources. In one urban study site (East St. Louis), smelters are the primary source of copper. Copper depositional fluxes follow an exponential decay as one transition from urban to rural settings [30]. Soil is not the major source of copper in cities or nearby rural soils, but is the predominant source for copper in the atmosphere over more remote areas [27, 31]. In general, levels of zinc in air are relatively low and fairly constant.

Average levels of zinc in the air throughout the United States are less than 1 microgram of zinc per cubic meter (μ g/m^3) of air, but range from 0.1 to 1.7 μg/m^3 in areas near cities [12, 32]. Zn is used in lubricating oil [30, 33] and Zn is also released from vehicular activities such as tire wear. A WHO air quality guideline for cadmium of 5 ng/m3 has been recommended in order to prevent any further increases in cadmium levels in agricultural soils [1]. By the end of the period studied, concentrations over western and central parts of Europe declined, ranging from 0.1 to 0.3 ng/m3 [34]. In Turkey, the limit value is 5 ng/m3. And it is planned to decrease the level 2 ng/m3 at 2020. Cd is originated from vehicular activities [33]. Average air lead levels are usually below 0.15 μg/m^3 at nonurban sites. Urban air lead levels are typically between 0.15 and 0.5 μg/m^3 in most European cities. Additional routes of exposure must not be neglected, such as lead in dust, a cause of special concern for children [1]. By 2003, the concentrations in air had decreased markedly and were within 2–15 ng/m^3 over most of Europe [34]. In Turkey, Pb limit value is 2 μg/m^3, and it is required to decrease the value to 1 μg/m^3 at 2019 [12]. Pb concentration levels in air are strictly depended on distance of the emission sources [35]. After the implementation of antipollution policies on automotive, Pb levels were decreased in the atmosphere by 82 % [4]. To improve the antiknock property, Pb is still being introduced in different forms to fuels in many countries [5].

Table 5 shows the mean and max-min values of EF_c values of heavy metals and their EF_c/Al correlations. Non-crustal contributions on heavy metal concentrations of the region aerosol are observed as Cd > Pb > Cu > Zn. Migon and Nicolas [36] had been found the importance of the recycling component decreases in the sequence Cd > Pb > Cu > Zn in the North-western Mediterrenean atmosphere. (These metals are also found in high levels in some studies [8, 37].

The enrichment factor (EF_{crust}) is used to define the origin of the metals in aerosol [38, 39]. EF_c (enrichment factor from crust to air) = $(C_x/C_{Al})_{air}/C_x/C_{Al})_{crust}$, here; $(C_x/C_{Al})_{air}$: the concentration ratio of the element X and Al in the measured air. $(C_x/C_{Al})_{crust}$: the concentration ratio of the element X and Al in the crust. The elements have EF_c values which range between 1 and 10, indicating that they are present in the aerosol in roughly crustal proportions. Except Cr, Fe the other heavy metal concentrations are independent of Al concentrations. So, it can be said that their source are non-crustal [9]. Anthropogenic contributions on heavy metal concentrations of the region aerosol are observed as Cd > Pb > Cu > Zn in a decreasing order. The EF_c values of Cd 8–9730 with an average of 804.11 indicates that the high anthropogenic contribution. EF_c values of Pb correspond to a range of 10–66, with an average value of 30.64 so the anthropogenic contribution could be considered besides the natural occurrence. The EF_c range of Cu

Table 5 Mean and max-min. EF values of heavy metals according to Al and EF/Al correlations

EF-Al correlations	Cr	Fe	Cu	Zn	Cd	Pb
Mean	0.13	0.87	11.06	4.67	804.11	30.64
Max.	0.54	2.2	25.35	18.27	9730	243
Min.	0.005	0.06	1.43	0.11	8	4
r (EF-Al)	0.30	0.59	0.05	0.34	0.17	0.30

1.43-25.35 with the average of 11.06 indicate that natural source was all, but also anthropogenic contribution was so high. The EF_c range of Zn was 0.11-18.27 with an average of 4.67 indicated that there were some anthropogenic contribution.

To observe the non-crustal (anthropogenic) contribution on measured metal concentrations, their EF_c-Al diagrams are prepared and shown in Fig. 5d. From figures it is seen that most of the elements are independent of Al concentrations when Al concentration > 67.50 ng/m^3. However, EF_c of these elements decreases with increasing Al concentration at lower Al concentrations which is a typical behavior of elements with non-crustal sources [9].

Conclusions

Based on the results of this research, the concentrations of Cr and Fe in the ambient air of İstanbul (Tuzla region) indicate that their presence in the aerosol comes from (depends on) crustal origin. The enrichment factor of Tuzla aerosol were found Cd > Pb > Cu > Zn in a decreasing order. It may not be possible to make any generalization about the episodic concentration changes of the metals due to their different origin. But, almost all the metal concentrations are low at June paralleling to PM concentrations. To determine the emission sources of these metals, it is recommended to do more detailed and comprehensive study. The planned study must include wind transportation from industrial smokestacks and organized industrial areas located near the residential location in Tuzla region. For this reason all the meteorological parameters especially wind direction data could be available locally. When metals are introduced to atmosphere, they are passing through water and soil. Then, they are taken by living organisms via ingestion, inhalation and skin absorption. In conclusion, this work results are a basic source for future studies and the study will be unique and a useful model for other industrial areas to determine pollution in a large range entirely on account of having examined inorganic pollutants.

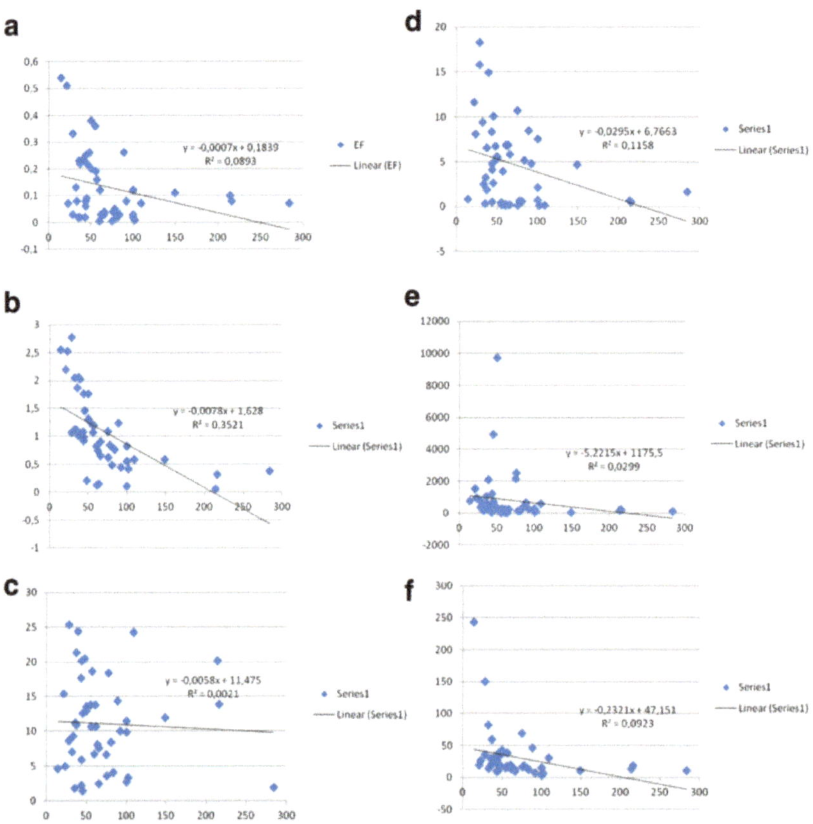

Fig. 5 a. Cr enrichment factor versus Al concentration (EF_c-Al)r = + 0.30. **b**. Fe enrichment factor versus Al concentration (EF_c-Al)r = + 0.59. **c**. Cu enrichment factor versus Al concentration (EF_c-Al)r = + 0.05. **d**. Zn enrichment factor versus Al concentration (EF_c-Al) r = + 0.34. **e**. Cd enrichment factor versus Al concentration (EF_c-Al)r = + 0.17. **f**. Pb enrichment factor versus Al concentration (EF_c-Al)r = + 0.30

Competing interests
The authors declare that he/she have no competing interests.

Acknowledgements
Author thank to Prof. Dr. M. C. BARLA and technicians of the Piri Reis University, for their help during the collection of aerosol samples. Author also thank Dr. Günay KURAL for her asistant during analysis and English edition.

References

1. WHO, Reginol Office for Europe. Air Quality Guidelines for Europe. Second ed. Copenhagen: WHO Reginol Publications, European Series; 2001. No. 91 ISBN 92 4 890 1358 3 (NLM Classification: WA 754) ISSN 0378–2255. http://www.euro.who.int/data/assets/pdf_file/0005/74732/E71922.pdf
2. Viard B, Pihan F, Promeyrat S, Pihan JC. Integrated assessment of heavy metal (Pb, Zn, Cd) highway pollution: Bioaccumulation in soil, graminaceae and land snails. Chemosphere. 2004;55:1349–59.
3. Onder S, Dursun Ş. Air borne heavy metal pollution of cedrus libani (A. Rich.) in the city centre of Konya (Turkey). Atmos Environ. 2006;4:1122–11 1133.
4. Migon C, Robin T, Dufour A, Gentili B. Decrease of Lead Concentrations in the Western Mediterrenean atmosphere during the Last 20 Years. Atmos Environ. 2008;42:815–21.
5. Olowoyo JO, Heerden EV, Fischer JL. Investigagting jacaranda mimosifolia tree as biomonitor of atmospheric trace metals. Environ Monit Assess. 2010;164:435–43.
6. Odukoya OO, Arawola TA, Bamgboe O. Pb, Zn and Cu Levels in tree bark as indicator of atmospheric pollution. Environ Int. 2000;26:11–6.
7. UNEP. Global Environment Outlook. London: Earthscan; 1999.
8. Taşdemir Y, Kural C, Cindoruk SS, Vardar N. Assesment of trace element concentrations and their estimated dry deposition fluxes in an urban atmosphere. Atmos Res. 2006;81:17–5.
9. Yatin M, Tuncel S, Aras NK, Olmez I, Aygün S, ve Tungel G. Atmospheric trace elements in Ankara, Turkey: 1. factors affecting chemical composition of fine particles. Atmos Environ. 2000;34:1305–18.
10. Güllü G, Ölmez I, Tuncel G. Temporal variability of atmospheric trace element concentrations over the eastern Mediterranean Sea. Spectrochim Acta B. 2000;29(55):1135–50.
11. Duh JD, Shandas V, Chang H, Gerge LA. Rates of urbanization and the resiliency of air and water quality. Sci Total Environ. 2008;400:238–56.
12. T. R. Ministry of Environment and Forestry, Air Quality Assessment and Management. 2010.
13. Cong Z, Kang S, Dong S, Liu X, Qin D. Elemental and individual particle analysis of atmospheric aerosols from High Himalayas. Environ Monit Assess. 2010;160:323–35.
14. Braga CF, Teixeia EC, Yomeama ML, Dias JF. Study of elemememtal composition of aerosols in the Candiota Region of Brazil Using the PIXE Tecnique. Nucl Instrum Methods Physi B. 2004;225:561–71.
15. Sen O. Variations of aerosols at İzmir, Turkey, Determined by neutron analysis. Atmos Environ. 1989;22:795–801.
16. Güllü G, Ulutaş F, Belli D, Erduran S, Keskin S, ve Tuncel G. Black Sea aerosol and the long-range atmospheric transport. J Eng Environ Sci. 1998;22:289–303.
17. Cizmecioğlu SC, Müezzinoğlu A. Solubility of deposited airborne heavy metals. Atmos Res. 2008;89:396–404.
18. Jaradat QM, Momani KA, Jiries AG, El-Alali A, Baterseh MI, Sabri TG, et al. Chemical composition of urban wet deposition in Amman Jordan. Water Air Soil Pollut. 1999;122:55–65.
19. Toscano G, Gambaro A, Moret I, Capodaglio G, Turetta C, Cesson P. Trace metals in aerosols at Terra Nova Bay, Antarctica. J Environ Monit. 2005;7(12):1275–80.
20. Lantzy RJ, MacKenzie FT. Atmospheric tracemetals: Global cycles and assessment of man's impact. Geochima Cosmochima Acta. 1979;43(4):511–25.
21. Sekhavatjou MS, Rostami A, Alhasemi ASH. Assesment of elemental concentrations in the urban air (case study: Tehran City). Environ Monit Assess. 2010;163:467–76.
22. Bajpai R, Upreti SK, Dwivedi SK. Passive monitoring of atmospheric heavy metals in a historical city of central India by Lepraria Lobificans Nly. Environ Monit Assess. 2010;166:477–84.
23. ATSDR. PublicHealthAssessment, Evaluation of Current (1990 to 2003) and Future Chemical Exposures in theVicinity of the Oak Ridge Reservation, U. S. Oak Ridge, Anderson County, Tennessee: Department of Energy Oak Ridge Reservation; 2007. EPA Facility ID: TN1890090003.
24. Nowak DJ. Institutionalizing urban forestry as a "biotechnology" to improve environmental quality. Urban Urban Greening. 2006;5:93–100.
25. Sadiqu M, Alam I, El-Mubarek A, Al-Mohtar HM. Preliminary evaluation of metal pollution from wear of auto tires. Bullatın Environ Contam Toxicol. 1989;42:743–8.
26. Monaci F, Bargagli R. Barium and other trace metals as indicators of vehicular emissions. Water Air Soil Pollut. 1997;100:89–98.
27. ATSDR. Agency for Toxic Substances and Disease Registry, Toxicological Profile for Copper. Atlanta, GA: U.S.: Department of Health and Human Services, Public Health Service; 2004.
28. Nriagu JO, Pacyana J. Quantitative assessment of worldwide contamination of air, water and soil by trace metals. Nature. 1998;333:134–9.
29. Baptista MS, Teresa M, Vasconcelos SD, Carbral JP, C M, Pacheo AMG. Copper, nickel, lead in lichens and tree bark transplants over different period of time. Environ Pollut. 2008;151:408–13.
30. Sloof JE. Lichens as quantitave biomonitors for atmosphere trace elelments deposition, using transplants. Atmos Environ. 1991;29:11–9.
31. Voutsa D, Samara C. Labile and bioaccessible fractions of heavy metals in the airborne particulate matter from urban and industrial areas. Atmos Environ. 2002;36:3583–90.
32. ATSDR. Agency for Toxic Substances and Disease Registary. Public Health Statement, Zinc. CAS#7440-66-6. 2005
33. Ward NI, Sampson KE. The use of bryophytes and lichens to monitor the rate of metal deposition along the London orbital (M25) motorway. In: Vernet JP, editor. Heavy metals in the environment. PProc. Of the Int. conference, Geneve 2. Edinburgh, UK: CEP Consultants; 1989. p. 444–7.
34. WHO, Regional Office for Europe. Joint WHO/Convention Task Force on the Health Aspects of Air Pollution, Health Risks of Heavy Metals from Long-Range Transboundary Air Pollution, Scherfigsvej.8, DK-2100 Copenhagen O Denmark, 33 ISBN 978 92 890 7179 6. http://www.euro.who.int/data/assets/pdf_file/0007/78649/E91044.pdf. 2007.
35. Dartey E, Adimado AA, Agyargo K. Evaluation of airborne lead levels in storage battery workshops and some environments in Kumasi Metropolis in Gana. Environ Monit Assess. 2010;164:1–8.
36. Migon C, Nicolas E. The Trace Metal Recycling Component in the north- western Mediterrenean. Mar Pollut Bull. 1998;36(4):273–7.
37. Muezzinoğlu A, Cizmecioğlu SC. Deposition of heavy metals in a Mediterenean climate area. Atmos Res. 2006;81:1–6.
38. Chester R. Marine Geochemstry. Hyman Ltd, London: Unwin; 1990.
39. Aksu A, Balkıs N, Erşan MS, Müftüoğlu AE, Apak R. Biogeochemical cycle of arsenic and calculating enrichment factor by using Li element. Environ Geochem Health. 2010;32:303–6.

13

CFD modeling of incinerator to increase PCBs removal from outlet gas

Kamyar Yaghmaeian[1,2], Nematallah Jaafarzadeh[3], Ramin Nabizadeh[1,2], Golbarg Dastforoushan[4] and Jalil Jaafari[5*]

Abstract

Incineration of persistent organic pollutants (POPs) is an important alternative way for disposal of this type of hazardous waste. PCBs are very stable compounds and do not decompose readily. Individuals can be exposed to PCBs through several ways and damaged by their effects. A well design of a waste incinerator will convert these components to unharmfull materials. In this paper we have studied the design parameters of an incinerator with numerical approaches. The CFD software Fluent 6.3 is used for modelling of an incinerator. The effects of several baffles inside the incinerator on flow distribution and heat is investigated. The results show that baffles can reduce eddy flows, increase retaining times, and efficiencies. The baffles reduced cool areas and increased efficiencies of heat as maximum temperature in two and three baffle embedded incinerator were 100 and 200 °C higher than the non-baffle case, respectively. Also the gas emission leaves the incinerator with a lower speed across a longer path and the turbulent flow in the incinerator is stronger.

Keywords: Incinerator, PCB, Fluent model, Baffle, Thermal degradation

Introduction

Waste incineration is a well-established treatment technology for municipal, industrial, hospital and hazardous wastes [1–3]. It is also one of the most frequently selected method of waste management for no-longer reusable or recyclable industrial products and materials [4–8]. In the some part of waste incinerators, persistent organic pollutants (POPs) are formed due to the presence of products of incomplete combustion, oxygen and chlorine at temperatures between 200 and 800 °C [9, 10]. The final solution for persistent organic pollutants (POPs) such as polychlorinated biphenyls (PCBs) that cannot be recycled or landfilled is to use an incinerator [11]. The minimum residence time suggested for removal of PCBs in an incinerator is about 2 s at 1200 °C or 1.5 s at 1600 °C [11]. This can be achieved only through increasing the residence time or improved heat distribution. Measurements indicate that most PCBs incinerators are not able to provide these conditions due to the presence of inefficient cold zone with low efficiency in terms of mixing and heat distribution [12, 13]. Furthermore, in the cold zone of waste incinerators polychlorinated biphenyls (PCBs) are formed [10, 14–16].

In recent years, many attempts have been made to model processes in incinerators. San José et al., investigated the effect of incinerator efficiency on the emissions in an industrial area with the help of MM5, CMAQ and EMIMO [17]. The results showed that the effect of emissions from incinerator is insignificant compared to the surrounding industries and highways. The effect was comparable just in the case of ozone. Stanmore et al., modeled the formation of PCDD/F in municipal and hospital incinerators and proposed a general empirical model to calculate the level of gaseous and solid PCDD/F [18].

In another study, Goh et al., modeled the combustion bed of a municipal incinerator and proposed a comprehensive flexible model [19]. The results of the model were used as boundary conditions for modeling upper gases in CFD models. The model results can also be used to optimize the incinerator and reduce the production of waste sludge and waste mixing [19]. In a similar study, the waste mixing was modeled before burning in an incinerator. In this study, a mathematical model was proposed for simulating waste mixing in the incinerator

* Correspondence: JalilJaafari@yahoo.com
[5]Department of Environmental Health, School of Public Health, Guilan University of Medical Sciences, Rasht, Iran
Full list of author information is available at the end of the article

Table 1 The technical specification of the incinerator

Value	Parameter
2270 (kg/h)	Maximum capacity
$C_{12}H_7Cl_3$	Input pollutant
20,277 (KJ/Kg)	Net thermal value
50 %	Excess air
2.76 m	Initial diameter
11 m	Initial height
0.5 m	Inlet and outlet diameter
2.7 s	Initial residence time

and the model results were compared with experimental results. Huang, used a kinetic model of reaction to model the formation PCDD/F in an industrial incinerator [20]. The model variables include the formation and removal rates of PCDD/F, carbon gasification, partial pressure of oxygen and equations for temperature and time. A good agreement was obtained between the experimental and model results. Khiari et al., proposed a mathematical model for dynamic simulation of an incinerator [21]. The lower part of the incinerator and waste pyrolysis were modeled. The model results were compared with the results of similar studies. Thomas offered an one-dimensional model to simulate the incineration of emissions in an incinerator [22]. Taking into account radiation, convection and conduction heat transfer processes, and the gas flow was simulated. The heat capacity of gases, thermal conductivity and viscosity effects were included considering the temperature dependence of the reaction.

The main objective of this study was to investigate the explores ways to optimize the efficiency of the PCBs removal in incinerators in the presence of baffles embedded in the combustion chamber. Modeling was performed first with 2 and then with 3 baffles.

Materials and methods

Incinerator specification

A rotational PCBs incinerator with one inlet and one outlet for pollutants was studied. The incinerator was designed for one of PCBs isomers called Arochlorine-1242 (a mixture of 1 % $C_{12}H_9Cl$, 13 % $C_{12}H_8Cl_2$, 45 % $C_{12}H_7C_{13}$, 31 % $C_{12}H_6C_{14}$ and 10 % $C_{12}H_5Cl_5$ with average Cl content of 42 %. Arochlorine-1242 is among the most commonly used types of PCBs, especially in power transformers in great plants. The specification of the incinerator was selected according to Theodore and Reynolds [23]. The Arochlorine-1242 incinerator was designed manually and with the help of software. The design principles in different parts of incinerator such as primary and secondary combustion chambers, furnaces, boilers, suppressor devices and air pollution control devices were similar to the incinerator modeled in [23]. Design calculations were performed according to Charles's law and Dulong's equation. Table 1 summarizes the specification calculated for the incinerator. The technical specifications presented in Table 1 were used as initial inputs to Fluent model for simulating the incinerator.

The equations governing the pollutant flow in the incinerator

Given the air flow velocity and the dimensions of the incinerator as well as the high temperatures, the flow regime is turbulent. Neglecting the net rotating flows, since all changes along the flow and in vertical direction are important, the k-ε turbulence model is a good model

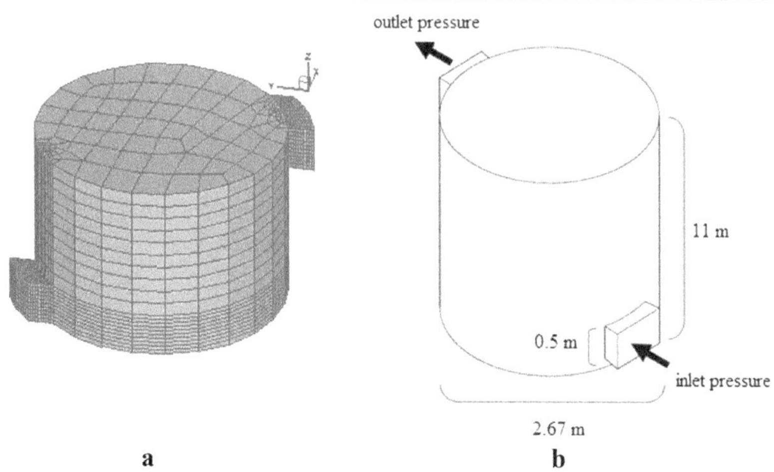

Fig. 1 **a** A view of the meshed model of the incinerator, (**b**) The geometrical charactersitics of the incinerator and boundary conditions

Table 2 The inputs to the Fluent model

Value	Parameter
3 %	Turbulent Intensity
Stationary	Wall Motion
No slip	Shear Condition
0.00004 (m)	Roughness Height
2.76 (m)	Hydraulic Diameter
0.5	Roughness Constant

for analyzing this problem. The equations required to solve the isothermal gas flow in the incinerator include time-averaged mass and momentum conservation equations [24]:

Mass conservation $\frac{\partial U_i}{\partial X_i} = 0$ (1)

Momentum conservation

$$\frac{\partial \rho U_i}{\partial t} + \frac{\partial \rho U_i U_j}{\partial X_i} = \frac{\partial p}{\partial X_i} + \rho \frac{\partial}{\partial X_j}\left[\nu\left(\frac{\partial U_i}{\partial X_i} + \frac{\partial U_i}{\partial X_i}\right)\right] - \frac{\partial U_i' U_i'}{\partial X_j} + S_{Mi} \quad (2)$$

Where U_i is velocity along i, i = 1, 2, 3, X_i is x, y, z coordinates along i, Y mass fraction of gas emissions, ρ air density, υ kinematic viscosity, U_i turbulent velocity component along i' and S_{Mi} is the momentum source along i'.As mentioned previously, since the Reynolds removal process and time-averaged equations will lead to unknown relationships for fluctuating velocity components, so a turbulent model is also needed. Thus, the k-ε model was used. This model requires the solution of two additional transport equations, one for turbulent kinetic energy, k and the other for its dissipation rate or ε [24]:

$$\frac{\partial}{\partial x_i}(\rho u_i k) = \frac{\partial}{\partial x_i}\left[\left(\frac{\mu + \mu_t}{\delta_k}\right)\left(\frac{\partial k}{\partial x_i}\right)\right] + P - \rho\varepsilon \quad (3)$$

$$\frac{\partial}{\partial x_i}(\rho u_i \varepsilon) = \frac{\partial}{\partial x_i}\left[\left(\frac{\mu + \mu_t}{\delta_\varepsilon}\right)\left(\frac{\partial \varepsilon}{\partial x_i}\right)\right] + C_1 \frac{\varepsilon}{k} P - C_2 \rho \frac{\varepsilon^2}{k} \quad (4)$$

Enthalpy conservation: $\frac{\partial}{\partial x_i}(\rho u_i h) = \frac{\partial}{\partial x_i}\left[\left(\frac{\mu+\mu_t}{\delta_h}\right)\left(\frac{\partial h}{\partial x_i}\right)\right] + S_h$ (5)

Chemical species conservation: $\frac{\partial}{\partial x_i}(\rho u_i m_s) = \frac{\partial}{\partial x_i}\left[\left(\frac{\mu+\mu_t}{\delta_h}\right)\left(\frac{\partial m_s}{\partial x_i}\right)\right] + S_s$ (6)

Equation of State: $\rho = \frac{P}{RT\sigma m_j / M_j}$ (7)

Incinerator simulation

To simulate the studied incinerator, a Cartesian coordinate system with the dimensions of 20 × 20 × 20 mm containing 10,424 control volumes was used. The x, y and z axes represent the length, width and height, respectively [24]. In the regions where sharp gradients in the variables are expected, fine meshing is considered as possible. Structured meshes were used for meshing the model, because the number and distribution of meshes in different solution regions can be controlled. In addition, the boundary conditions can be well defined [24]. The boundary layer mesh generator was used in the regions where thermal analysis was important. Fig. 1a and b show the model, the meshed model and boundary conditions.

Wall boundary condition was considered on the studied incinerator, because the solid surface is in direct contact with the fluid [24]. Given that the size and profile of the inlet pressure to the studied incinerator are known, the pressure inflow and pressure outflow were selected for inlet and outlet boundary conditions, respectively. The walls were made of steel with a thermal conductivity of 202 W/M.K. For the flows with a velocity less than the speed of sound, a turbulence of less than 5 % can be considered as assumed in the present model. Other inputs are described in Table 2.

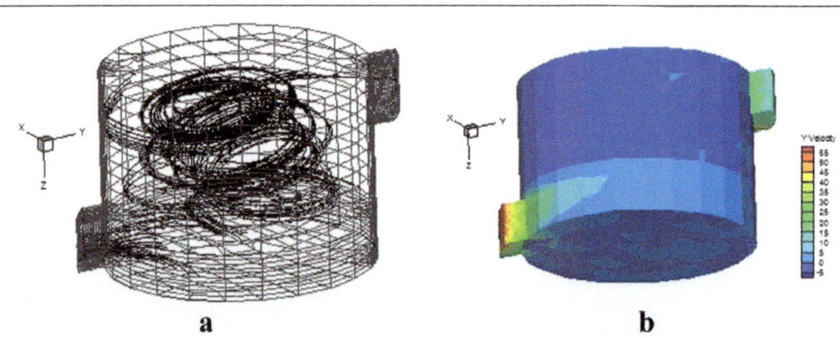

Fig. 2 **a** The motion path of gases in the incinerator in normal conditions, (**b**) Iso-velocity contours in the incinerator in normal conditions

Results and discussion

Figure 2a shows the motion path of emissions in the studied incinerator. As expected, due to the lack of baffles, incinerator geometry as well as the tangential flow of emissions, large rotating bubbles are formed in the incinerator. Rotating regions (eddies) are formed when the flow pressure reduces the kinetic energy of the fluid particles causing stagnant points in the flow. In addition, there is also a secondary flow that directs the flow downward [25]. Fig. 2b shows the iso-velocity contours in the initial state (without baffles). As seen, increasing the inlet gas temperature will increase the velocity (compared to the isothermal case). A maximum velocity of 55 m/s occurs in the incinerator outlet. In this case, the high speed gas particles quickly leave the incinerator causing a decrease in the residence time. In this case, the average residence time of gaseous emissions is 3 s.

Figure 3 shows the predicted temperatures. Temperature increases after the gaseous emissions entering the lower part of the studied incinerator. The temperature inside the studied incinerator reaches over 1300 °C. The predicted temperatures demonstrate cold zones in the incinerator where air is mixed with gas eddies and thus temperature decreases to 600 °C. The temperature drop in the incinerator may significantly affect the composition of the exhaust gases from the incinerator preventing the removal of PCB bonds.

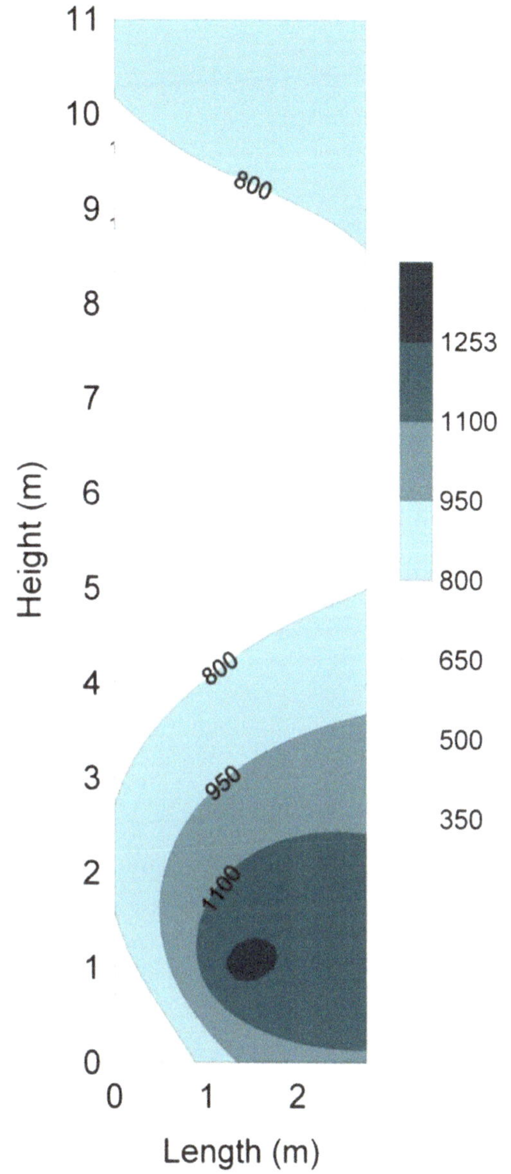

Fig. 3 The temperature profile (°C) inside the incinerator in normal condition

Efficiency Optimization

Baffles were embedded in the incinerator to remove large rotating currents (or eddy). In the first case, two baffles each with an angle of 90° were installed on the height of 3 and 5 m from the bottom of the studied incinerator. Fig. 4a, shows the two-dimensional motion of particles in this case. The baffles were made of steel as the studied incinerator body. Unlike the previous case, the motion path is divided into several parts. Removal of large eddies and the sudden displacements of gaseous emissions increase the residence time and thereby optimal air mixing. The residence time calculated for this case is 3.3 s. Fig. 4b, shows the temperature profile. The maximum temperature in this case was 1400 °C. Although the heat concentration can be observed in the middle of the studied incinerator, the temperature profile is not significantly different with the previous case.

In the second case, 3 baffles were installed on the height of 2, 4 and 8 m from the bottom of the studied incinerator with an angle of 90 °. Fig. 5a, shows the gas motion path in this case. As shown, the large volume of the incoming gas is located adjacent to the heat source. The gas emission leaves the incinerator with a lower speed across a longer path compared to the previous case. In this case, the turbulent flow in the incinerator is stronger than the previous case. The average residence time is 3.5 s. The temperature profile is shown in Fig. 5b, Little changes in the temperature profile are observed. Like the previous case, the heat concentration is observed in the middle of the incinerator. The maximum temperature in this case is 100 °C higher than the previous case. Other studies have shown that longer path incinerator or

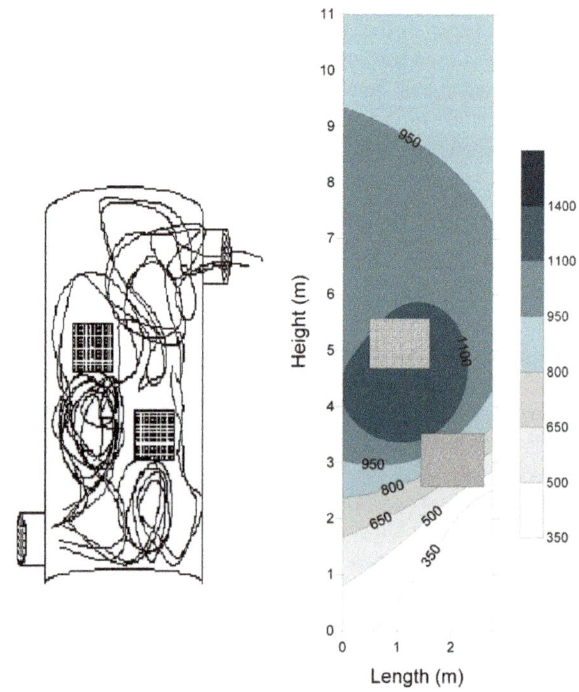

a b

Fig. 4 a The schematic motion path of gases in the incinerator for the Case 1, (**b**) The temperature profile (°C) inside the incinerator for the Case 1

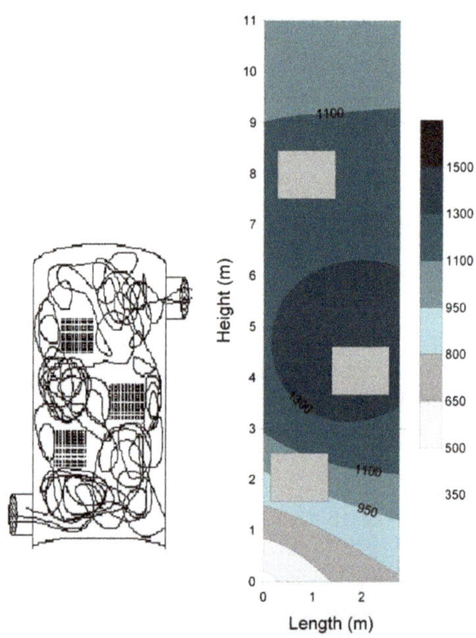

a b

Fig. 5 a Schematic motion path of gases in the incinerator for the Case 2, (**b**) The temperature profile (°C) inside the incinerator for the case 2 multiple chamber incinerator can be increased PCB removal rate [15, 26].

Conclusions

The present paper modeled a PCBs incinerator by computational fluid dynamics with the help of Fluent. First, the technical specifications of the incinerator were calculated considering the type of the pollutant. The calculated specifications were used as Fluent model inputs. Numerical modeling was performed in three different modes. In the normal mode, a cylindrical incinerator without any baffle was modeled. The results indicated the presence of vortices and cold zones in the incinerator which reduced the efficiency. Then, the impact of baffles on the heat distribution and mixing efficiency of gaseous emissions in the incinerator was studied. The baffles reduced eddies and improved the heat distribution in the incinerator. In the third case, the maximum residence time, temperature and turbulence of gaseous emissions were greater than the previous two cases. In this case, the reduced cold zones by adding baffles increased the thermal efficiency and pollutant removal by 10 % and 16 %, respectively.

Competing interests

The authors declare that they have no competing interests.

Authors' contributions

The overall implementation of this study were done by KY, RN, GD and JJ. Nj critically reviewed and revised the article. All authors read and approved the final manuscript.

Acknowledgements

The authors are most grateful to the Department of Environmental Health Engineering, School of Public Health, Tehran University of Medical Sciences, Iran, for their collaboration in this research.

Author details

[1]Department of Environmental Health, School of Public Health, Tehran University of Medical Sciences, Tehran, Iran. [2]Center for Solid Waste Research, Institute for Environmental Research, Tehran University of Medical Sciences, Tehran, Iran. [3]Environmental Technologies Research Center, Ahvaz Jundishapur University of Medical Sciences, Ahvaz, Iran. [4]Department of Environmental Engineering, Islamic Azad University, Tehran, Iran. [5]Department of Environmental Health, School of Public Health, Guilan University of Medical Sciences, Rasht, Iran.

References

1. Wu M-H, Lin C-L, Zeng W-Y. Effect of waste incineration and gasification processes on heavy metal distribution. Fuel Process Technol. 2014;125:67–72.
2. Sobiecka E, Obraniak A, Antizar-Ladislao B. Influence of mixture ratio and pH to solidification/stabilization process of hospital solid waste incineration ash in Portland cement. Chemosphere. 2014;111:18–23.
3. Consonni S, Giugliano M, Grosso M. Alternative strategies for energy recovery from municipal solid waste: Part A: Mass and energy balances. Waste Manag. 2005;25:123–35.
4. Wäger P, Hischier R, Eugster M. Environmental impacts of the Swiss collection and recovery systems for Waste Electrical and Electronic Equipment (WEEE): A follow-up. Sci Total Environ. 2011;409:1746–56.
5. Tsiliyannis C. End-of-life flows of multiple cycle consumer products. Waste Manag. 2011;31:2302–18.
6. Jaafari J, Mesdaghinia A, Nabizadeh R, Hoseini M, Mahvi AH. Influence of upflow velocity on performance and biofilm characteristics of Anaerobic Fluidized Bed Reactor (AFBR) in treating high-strength wastewater. J Environ Health Sci Eng. 2014;12:139.
7. Esfandyari Y, Mahdavi Y, Seyedsalehi M, Hoseini M, Safari GH, Ghozikali MG, et al. Degradation and biodegradability improvement of the olive mill wastewater by peroxi-electrocoagulation/electrooxidation-electroflotation process with bipolar aluminum electrodes. Environmental Science and Pollution Research. 2014;22(8):6288–97.
8. Jafari J, Mesdaghinia A, Nabizadeh R, Farrokhi M, Mahvi AH. Investigation of Anaerobic Fluidized Bed Reactor/Aerobic Mov-ing Bed Bio Reactor (AFBR/MMBR) System for Treatment of Currant Wastewater. Iran J Pub Health. 2013;42:860–7.
9. Altarawneh M, Dlugogorski BZ, Kennedy EM, Mackie JC. Mechanisms for formation, chlorination, dechlorination and destruction of polychlorinated dibenzo- p-dioxins and dibenzofurans (PCDD/Fs). Prog Energy Combust Sci. 2009;35:245–74.
10. Van Caneghem J, Block C, Van Brecht A, Wauters G, Vandecasteele C. Mass balance for POPs in hazardous and municipal solid waste incinerators. Chemosphere. 2010;78:701–8.
11. USEPA. Toxic Substances Control Act (TSCA). 2000.
12. Van Caneghem J, Block C, Vandecasteele C. Destruction and formation of dioxin-like PCBs in dedicated full scale waste incinerators. Chemosphere. 2014;94:42–7.
13. Nasserzadeh V, Swithenbank J, Scott D, Jones B. Design optimization of a large municipal solid waste incinerator. Waste Manag. 1991;11:249–61.
14. Liu P-Y, Zheng M-H, Zhang B, Xu X-B. Mechanism of PCBs formation from the pyrolysis of chlorobenzenes. Chemosphere. 2001;43:783–5.
15. Ishikawa Y, Noma Y, Yamamoto T, Mori Y, Sakai S-I. PCB decomposition and formation in thermal treatment plant equipment. Chemosphere. 2007;67:1383–93.
16. Van Caneghem J, Block C, Vermeulen I, Van Brecht A, Van Royen P, Jaspers M, et al. Mass balance for POPs in a real scale fluidized bed combustor co-incinerating automotive shredder residue. J Hazard Mater. 2010;181:827–35.
17. San José R, Pérez J, González R. The evaluation of the air quality impact of an incinerator by using MM5-CMAQ-EMIMO modeling system: North of Spain case study. Environ Int. 2008;34:714–9.
18. Stanmore B. Modeling the formation of PCDD/F in solid waste incinerators. Chemosphere. 2002;47:565–73.
19. Goh Y, Lim C, Zakaria R, Chan K, Reynolds G, Yang Y, et al. Mixing, modelling and measurements of incinerator bed combustion. Process Saf Environ Prot. 2000;78:21–32.
20. Huang H, Buekens A. Chemical kinetic modeling of de novo synthesis of PCDD/F in municipal waste incinerators. Chemosphere. 2001;44:1505–10.
21. Khiari B, Marias F, Zagrouba F, Vaxelaire J. Transient mathematical modelling of a fluidized bed incinerator for sewage sludge. J Clean Prod. 2008;16:178–91.
22. Tomaz E, Maciel Filho R. Steady state modeling and numerical simulation of the rotary kiln incinerator and afterburner system. Comput Chem Eng. 1999;23:S431–4.
23. Joseph J, Santoleri JR, Theodore L. Introduction to hazardous waste incineration. 2nd ed. John Wiley and Sons, New York: Wiley-Interscience; 2000.
24. Fluent M. Fluent Inc. Chapter. 2003;6:14–6.
25. Nasserzadeh V, Swithenbank J, Jones B. Three-dimensional modelling of a municipal solid-waste incinerator. J Inst Energy. 1991;64:166–75.
26. Wu J-L, Lin T-C, Wang L-C, Chang-Chien G-P. Memory effects of polychlorinated dibenzo-p-dioxin and furan emissions in a laboratory waste incinerator. Aerosol Air Qual Res. 2014;14:1168–78.

Use of mobile phone during pregnancy and the risk of spontaneous abortion

Fatemeh Shamsi Mahmoudabadi[1], Saeideh Ziaei[1,2*], Mohammad Firoozabadi[1] and Anoshirvan Kazemnejad[1]

Abstract

Background: Exposure to electromagnetic fields of cell phones increasingly occurs, but the potential influence on spontaneous abortion has not been thoroughly investigated.

Methods: In a case–control study, 292 women who had an unexplained spontaneous abortion at < 14 weeks gestation and 308 pregnant women > 14 weeks gestation were enrolled. Two data collection forms were completed; one was used to collect data about socioeconomic and obstetric characteristics, medical and reproductive history, and lifestyles. Another was used to collect data about the use of cell phones during pregnancy. For the consideration of cell phone effects, we measured the average calling time per day, the location of the cell phones when not in use, use of hands-free equipment, use of phones for other applications, the specific absorption rate (SAR) reported by the manufacturer and the average of the effective SAR (average duration of calling time per day × SAR). Analyses were carried out with statistical package state software(SPSS)v.16.

Results: All the data pertaining to mobile phones were different between the two groups except the use of hands free devices ($p < 0.001$).

Conclusion: Our result suggests that use of mobile phones can be related to the early spontaneous abortions.

Keywords: Abortion, Electromagnetic fields, Mobile phones, Pregnancy

Introduction

Widespread concerns have been raised about exposure to electromagnetic field (EMF) from sources used for mobile telecommunication [1-4]. Much of the concern arises because new technologies are introduced without adequate provision upon public information about their nature or discussion of the debate within the scientific community about possible health consequences. Indeed, mobile phone use has increased considerably along with reducing its costs, and developing countries are establishing mobile telecommunications rather than the more expensive fixed-line systems. Thus, if there is an impact on health from mobile telephones, it will affect everyone in the world. The major focus of research has been on radio-frequency (RF) radiation, which is mainly generated by phones, while some scientists are concerned about the possible impact of exposure to extremely low frequency electromagnetic field (ELF-EMF) fields generated by supply currents in the phone [5-7]. The device with the largest power consumption is the front-end amplifier. Consequently, the corresponding ELF-EMF has a spectrum similar to the pulse structure of RF signals.

The public discussion of the health hazards of EMF exposure has focused on the possible association with cancer, cardiovascular and immune systems. Less attention has been paid to the evaluation of evaluating the health hazards of EMF exposure on reproductive health such as spontaneous abortions [7-9].

The health effect of EMF on pregnancy has remained controversial despite efforts to reach consensus [10]. The main challenges in studying EMF are; 1) accurately assessing during the relevant time period and; 2) identifying susceptible population.

Therefore a case–control study has been designed o determine the effects of EMF exposure to mobile phones immediately after spontaneous abortions in selected patients.

* Correspondence: ziaei_sa@modares.ac.ir
[1]Faculty of Medical Science, Tarbiat Modares University, Tehran, Iran
[2]Reproductive Health Department, Tarbiat Modares University, P. O. Box: 14115-111, Tehran, Iran

Methods

For the evaluation of the effects of EMF exposure to mobile phones on spontaneous abortions, a case control study was chosen. Due to the lack of knowledge about these effects, a pilot study on 100 participants was carried out. Then, based on the primary data, with a significance level of $\alpha = 0.05$ and power of $1\text{-}\beta = 0.80$,a sample size of 220 participants for each group was calculated. In the current case–control study, 292 women who had an unexplained spontaneous abortion at < 14 weeks gestation (as early abortion) immediately after abortion(during first 48 hours) and 308 pregnant women at 14 weeks gestation were enrolled. Gestational age was calculated based on the last menstrual period that was confirmed by an ultrasonographic examination. The subjects were recruited from 10 hospitals in Tehran.

Pregnancy situations in the pregnant women in the control group were matched with maternal and paternal ages, gravidity, pre-pregnancy BMI, family relationship, duration from last delivery, educational level, occupation and history of previous abortions or preterm labor with the women in the case group.

The inclusion criteria were as follows:

1) singleton pregnancy; 2) 18–35 years of age; 3) spontaneous pregnancy without the use of assisted reproductive technology

The exclusion criteria were as follows:

1) chronic diseases (diabetes, hypertension,thyroid diseases, immune system diseases and cardiovascular diseases); 2) any genetic disorders (in the women, their husbands,and their first-degree relatives);3) vaginal bleeding in the first trimester of pregnancy in the control group; 4) history of birth defects in previous pregnancies; and 5) cigarette smoking, alcohol or any drug consumptions.

The women completed the data collection form, which was used to collect data about socioeconomic and obstetric characteristics, medical and reproductive histories, and lifestyles. The second data collection form which was completed by trained interviewer was designed to obtain information about the use of cell phones during pregnancy. For the consideration of cell phones effects, we measured the average calling time per day, the location of the cell phones when not in use (in handbags, clothing pockets, or ≥60-70 cm away from the body), use of hands-free equipment, use of phones for other applications (to send messages, to listen to music, and to play games), the specific absorption rate (SAR) reported by the manufacturer and the average of the effective SAR (average duration of calling time per day × SAR). SAR indicates how much electromagnetic energy is absorbed by body tissues. This parameter shows the estimation of total EMF energy. The effective SAR can show the exposure of ELF estimation from mobile phones indirectly. We investigated the data about the SAR of each mobile phone model from SAR.ir site.

For precise evaluation of the calling time, the receipts of mobile phone bills were considered. The information about the time spent for telecommunication in the specific section of new versions of cell phones were also taken into consideration.

The procedure was approved by the Ethics Committee of Tarbiat Modares University. All of the women participated voluntarily and provided written informed consent.

Statistical analysis

Demographic and obstetric characteristics, as well as the parameters which reflect the use of cell phones were compared between the two groups using t-tests and chi-square tests. A logistic regression model was used to assess the association between spontaneous abortions as a binary dependent factor and effective SAR, maternal age, paternal age, history of abortion and family relationship as independent factors. Analyses were carried out with statistical package state software(SPSS)v.16.A P-value <0.05 was considered statistically significant. All confidence intervals were calculated at the 95% level.

Results

The distribution of demographic and potential confounders for spontaneous abortions is presented in Table 1. As shown, there were no significant differences in maternal and paternal ages, pre-pregnancy BMI, occupation and educational status, socioeconomic level, family relationship, duration from last delivery and history of previous abortions or preterm labors between the two groups.

As shown in Table 2, all of the data pertaining to the use of cell phones such as average calling time per day, the location of the cell phones when not in use, use of phones for other applications, the specific absorption rate (SAR) and the average of the effective SAR), except the use of hands free devices were different between the two groups($p < 0.001$). Logistic regression analysis revealed a significant association between the effective SAR with the risk of spontaneous abortions after adjustment for maternal age, paternal age, history of abortions and family relationships (OR:1.11,$p < 0.001$) (Table 3).

Discussion

Exposure to EMF is increasingly common, but the potential influence on health has not been thoroughly determined, especially in pregnant women. Electromagnetic fields may produce biological stress and free radicals, which can make a susceptible population prone to congenital malformations

Table 1 The comparison of the maternal and paternal characteristics between the two groups

Parameters	Case group(n:226)	Control groups(n:246)	P
Maternal age(years)* Mean ± SD	27.81 ± 5.20	27.34 ± 4.3	0.2
Paternal age(years)* Mean ± SD	32.62 ± 5.46	31.90 ± 5.49	0.1
Family relationship** N(%)	62(21.24)	77(25.1)	0.3
Pre-pregnancy BMI*	25.14 ± 6.98	24.29 ± 5.24	0.09
History of abortion** N(%)	47(20.8)	36(14.6)	0.2
History of preterm labor**N(%)	3(1.3)	2(0.8)	0.6
Duration from last delivery(years)* Mean ± SD	6.61 ± 3.93	6.57 ± 3.46	0.9
Occupation N(%)**			0.2
Housewife	195(86.4)	225(91.5)	
Employee	31(13.6)	21(8.5)	
Educational status N(%)**			0.09
Primary school	88(38.9)	98(39.8)	
Secondary and high school	94(41.6)	100(40.6)	
University	44(19.5)	48(19.6)	

*T student test **Chi square test.

and tissue and cell damages [11]. Long-term exposure to electromagnetic fields may be linked to even higher levels of oxidative stress, with the aforementioned corresponding changes.

There is limited information on the association between EMF exposure during pregnancy and reproductive outcomes [10-18]. Some studies have reported increased risk of spontaneous abortions and congenital malformations, although these results were derived from poorly designed studies.

This study demonstrated an increased risk of spontaneous abortions associated with EMF exposure and confirmed the results of other researches [10,12]. Lee conducted a nested case–control study to evaluate the association of residential and personal magnetic fields with spontaneous abortions [12]. He reported personal magnetic field maximum exposure is associated with the risk of clinical spontaneous abortions. There was one main limitation to that study; specifically, the magnetic field measurements were obtained months after the first trimester and the occurrence of the spontaneous abortions. Perhaps the behavior and magnetic fields of women who had spontaneous abortions were different from the pregnant controls. But our study had this strength that, the data were obtained shortly after the spontaneous abortions occurred in the case group and in the early second trimester in the control group. This led to a better evaluation of the environmental conditions and the lifestyles of the participants

Table 2 The comparison of cell phone use during pregnancy between the two groups

Parameters	Case group(n:226)	Control group(n:246)	P(2-tailed)
Calling time per day* (minutes) Mean ± SD	9.31 ± 15.22	3.14 ± 3.59	<0.001
Use of hands free** n(%)	19(8.4)	11(4.5)	0.09
location of phones when not in use** n(%)			<0.001
on tables	86(38.1)	77(31.3)	
in handbags	101(44.7)	153(62.2)	
in pocket	39(17.3)	16(6.5)	
use of phone for other applications **n(%)	90(39.8)	44(17.9)	<0.001
Effective SAR* Mean ± SD	7.02 ± 13.92	2.28 ± 3.20	<0.001

*T student test, ** Chi square test or Fisher's exact test,SAR: Specific Absorption Rate.

Table 3 Association of spontaneous abortions with the effective SAR(Specific Absorption Rate) *

	OR(95%CI)	B	SE	P
Effective SAR	1.11(1.07-1.16)	0.106	0.21	<0.001

*Adjusted for maternal age, paternal age, history of abortion and the spouse relatives.

Though, the present findings are inconsistent with the results of some other works [14-18]. Some studies do not support an association between electric and magnetic fields and reproductive damages. However, many of them had investigated the effects of EMF either on animals or in vitro and not on human [14-16]. Also, lack of evidence for a strong association between EMF and adverse reproductive outcomes may be primarily the result of having little information data due to the lack of epidemiologic data in this area.

One potential limitation need to be kept in mind when one interprets the results of the current study. The study did not assess all the unknown risk factors for spontaneous abortions, such as balanced chromosomal abnormalities in one parent. Also, the data about unknown spontaneous abortion at very early stages were not collected. Other limitation of this study is its case control nature, and this can imply that caution should be taken in causal interpretations of the findings. The third limitation of this study is that cell phones may be not the only source of EMF. Since many parameters which can affect the EMF exposure and life styles like socioeconomic status were matched between the two groups, it is suspected that probably the effects that observed in this study are due to the use of cell phones.

The association between use of cell phones and the risk of spontaneous abortions against potential confounders was supported by evidence that despite adjustments for many known or suspected risk factors in logistic regression analyses, the estimation was not significantly altered.

Although the mechanisms underlying the effects of EMF on the risk of spontaneous abortions are not well understood, early embryos are known to be sensitive to environmental exposures. An adverse effect during early fetal development at the cellular level by EMF of cell phones could conceivably result in fetal death. EMF in cell phones is both ELF-EMF and RF-EMF. Based on the distance from outside the body to inside the uterus, the exposure reaching the fetus is likely to be extremely low frequency electromagnetic radiation.

Conclusion

Despite the lack of a clear understanding of the underlying mechanisms, the present result suggests that the use of cell phones may be related to early spontaneous abortions, thus further study is warranted.

Competing interests

The authors declare that they have no competing interests.

Authors' contributions

FSM carried out design and data collection of the study, SZ carried out design of the study and prepared the article, MF participated in design of the study, AK advised the statistical design of the study. All authors read and approved the final manuscript.

Acknowledgments

This study was carried out with the kind cooperation of the participating patients. Furthermore, this study was funded by Tarbiat Modares University, Tehran, Iran.

References

1. Repacholi MH. Health risks from the use of mobile phone. Occupational and Environmental Health. 1211 Geneva 27,Switzerland: World Health Organization; 2001.
2. Divan HA, Kheifets L, Obel C, Olsen J. Prenatal and postnatal exposure to cell phone use and behavioral problems in children. Epidemiology. 2008;19(4):1–7.
3. Toropainen A. Human exposure by mobile phones in enclosed area. Bioelectromagnetics. 2003;24(10):63–5.
4. Wilen J, Sandstrom M, Hansson MK. Subjective symptoms among mobile phone users-A consequence of absorption of radiofrequency field? Bioelectromagnetics. 2003;24(3):152–9.
5. Robert E. Intrauterian effects of electromagnetic fields(low frequency, mid frequency RF and microwave):review of epidemiologic studies. Teratology. 1999;59:292–8.
6. Bernabo N, Tellamanti E, Russo V, Martelli A, Turriani M, Mattoli M, et al. Extremely low frequency electromagnetic field exposure affects fertilization outcome in swine animal model. Theriogenology. 2010;73(9):1293–305.
7. Jauchem JR. Effects of low-level radio-frequency(3 KHz to 300GHz)energy on human cardiovascular, reproductive, immune, and other systems:A review of the recent literature. Int J Hyg Environ Health. 2008;211(1–2):1–29.
8. Erdogan Y. Electromagnetic pollution in the computer labs: The effects on the learning environment. Essays Educ. 2007;22:78–86.
9. Ahlbom A, Cardis E, Green A, Linet M, Savitz D, Swerdlow A. Review of the epidemiologic literature on EMF and health. Enviromental. Health Perspect. 2001;109(6):911–33.
10. Li D-K, Odouli R, Wi S, Janevic T, Golditch I, Bracken TD, et al. A population-based prospective cohort study of personal exposure to magnetic fields during pregnancy and the risk of miscarriage. Epidomiology. 2002;13:9–20.
11. Shaw GM. Adverse human reproductive outcomes and electromagnetic fields: a brief summary of the epidemiologic *literature*. Bioelectromagnetics. 2001;5:S5–18.
12. Lee GM, Neutra RR, Hristova L, Yost M, Hiatt RA. A nested case–control study of residential and personal magnetic field measures and miscarriages. Epidemiology. 2002;13(1):21–31.
13. Savits DA. Magnetic fields and miscarriage. Epidemiology. 2002;13(1):1–3.
14. Khaki AA, Khaki A, Garachurlou S, Khorshidi F, Tajadini N, Madinei N. Pre and post natal exposure of 50 HZ electromagnetic fields on prostate glands of rats:an electron microscopy study. IJRM. 2008;6(2):77–82.
15. Sun W, Tan Q, Pan Y, Fu Y, Sun H, Chiang H. Effects of 50-Hz magnetic field exposure on hormonal secretion and apoptosis-related gene expression in human first trimester villous trophoblasts in vitro. Bioelectromagnetics. 2010;31(7):566–72.
16. Huuskonen H, Saastamoinen V, Komulainen H, Laitinen J, Juutilainen J. Effects of low-frequency magnetic fields on implantation in rats. Reproductive Toxicology; 2000; 15(1):49–59
17. Marcus M, McChesney R, Golden A, Landrigan P. Video display terminals and miscarriage. J Am Med Womens Assoc. 2000;85(2):84–8.
18. Parazzini F, Chiaffarino F, Chatenoud L, Cipriani S, Ricci E, Chiantera V, et al. Exposure to video display terminals and risk of small-for- gestational-age birth. J Environ Health. 2010;72(6):24–7.

15

Surveillance of perchlorate in ground water, surface water and bottled water

Anupama Vijaya Nadaraja, Prajeesh Gangadharan Puthiyaveettil and Krishnakumar Bhaskaran*

Abstract

Background: Perchlorate is an emerging water contaminant that disrupts normal functioning of human thyroid gland and poses serious threat to health, especially for pregnant women, fetus and children.

Results: High level of perchlorate contamination in fresh water sources at places nearby ammonium perchlorate (rocket fuel) handled in bulk is reported in this study. Of 160 ground water samples analyzed from 27 locations in the State Kerala, 58 % had perchlorate above detection limit (2 μg/L) and the highest concentration observed was 7270 μg/L at Ernakulam district, this value is ~480 times higher than USEPA drinking water equivalent level (15 μg/L). Perchlorate was detected in all surface water samples analyzed ($n = 10$) and the highest value observed was 355 μg/L in Periyar river (a major river in the State). The bottled drinking water ($n = 5$) tested were free of perchlorate.

Conclusions: The present study underlines the need for frequent screening of water sources for perchlorate contamination around places the chemical is handled in bulk. It will help to avoid human exposure to high levels of perchlorate.

Keywords: Ground water, Ion chromatography, Kerala, Perchlorate, Thyroid disorders

Background

Perchlorate (rocket fuel) is an oxyanion (ClO_4^-), extensively used in arms and ammunition industries [1]. The chemical is reported as a potential thyroid disruptor by inhibiting iodide uptake causing hypothyroidism and associated health effects especially in infants, pregnant women and foetuses [2, 3]. A number of animal studies have reported ClO_4^- induced toxicities including delayed metamorphosis, haemolytic anaemia, thyroid tumor development etc. [4–6]. The current health advisory level for ClO_4^- is set at 15 μg/L based on the reference dose recommended by US EPA [7]. The World Health Organization (WHO) established provisional maximum tolerable daily intake (PMTDI) of 0.01 mg/kg body weight for ClO_4^- [8]. However, in many countries including India, drinking water/wastewater standard for ClO_4^- is yet to be defined. Detailed assessment and continuous monitoring of ClO_4^- in water sources have been reported from various countries such as USA [9], Canada [10], Europe and Middle East [11], Japan [12], Korea [13], India [14, 15] and China [16]. Perchlorate was detected in several food samples at concentrations above the Reference Dose (RfD) of 0.053 μg/kg bw/day proposed by National Academy of Sciences [17]. In USA, ClO_4^- was detected in 39 infant formulas at concentrations ranging from <0.4–13.5 μg/L [18].

An earlier study has reported ClO_4^- in 76 % of water samples collected from 13 locations in six States/Union territory (Tamilnadu, Karnataka, Bihar, Maharastra, West Bengal and Pondicherry) in India with concentrations ranged from <0.02–6.9 μg/L [14]. But, a more recent study has reported ClO_4^- in the range <0.005–7,690 μg/L in ground water samples from near cracker manufacturing industrial area (Sivakashi) in Tamil Nadu, India [15]. Unlike the states covered in previous two studies, Kerala has two known major inventories of ClO_4^-. One is the ammonium perchlorate experimental plant (APEP) at Aluva in Ernakulam district where this chemical is produced in bulk and other place is Vikram Sarabhai Space Research Centre (VSSC) at Thumba, Trivandrum district. A preliminary study conducted in our lab (in 2010) has revealed wide contamination of ClO_4^- in water samples from many districts in Kerala. In a different perspective, a study conducted in the coastal area of central Kerala revealed 10-15 % of iodine-sufficient population suffering from

* Correspondence: krishna@niist.res.in
Environmental Technology, CSIR-National Institute for Interdisciplinary Science & Technology, Thiruvananthapuram 695019, India

thyroid disorders [19]. Studies have also shown high incidence thyroid cancer in Kerala compared to major cities in India [20]. However, a proper reason for these serious health problems could not be identified yet. In view of this the present study focus on a detailed assessment of ClO_4^- in ground and surface water samples giving emphasis to places where this chemical is handled in bulk. The finding of this study underlines the need for regular screening of ground and surface water sources for perchlorate around places where this toxic chemical is handled in bulk.

Methods

Study area

Five sites in Kerala were selected in this study for detailed assessment of ClO_4^- contamination. One of the sites was near Ammonium Perchlorate Experimental Plant (APEP) at Aluwa in Ernakulam district. The second site was at Thumba, near Vikram Sarabhai Space Centre (VSSC) in Thiruvananthapuram district. The other two locations were in Kannur and in Palakkad districts. Our preliminary study has revealed high ClO_4^- level in water samples from these four districts. Location map of the study area is presented in Fig. 1. A fifth site, Koorg (latitude 12.33749° and longitude 75.80691°) in the nearby State Tamilnadu was selected as a control site. This is a hilly region (altitude ~ 5000 feet above sea level) and a far off place from any known inventories of perchlorate. Sampling was done from 27 locations in Kerala.

Sampling

Sampling was done during March – May 2012. Hundred ml of sample was collected from each point and filtered using 0.2 μm filters (Millipore) and brought to laboratory and stored at 4 °C till analysis. Repeated sampling was done from sites that showed high ClO_4^- values.

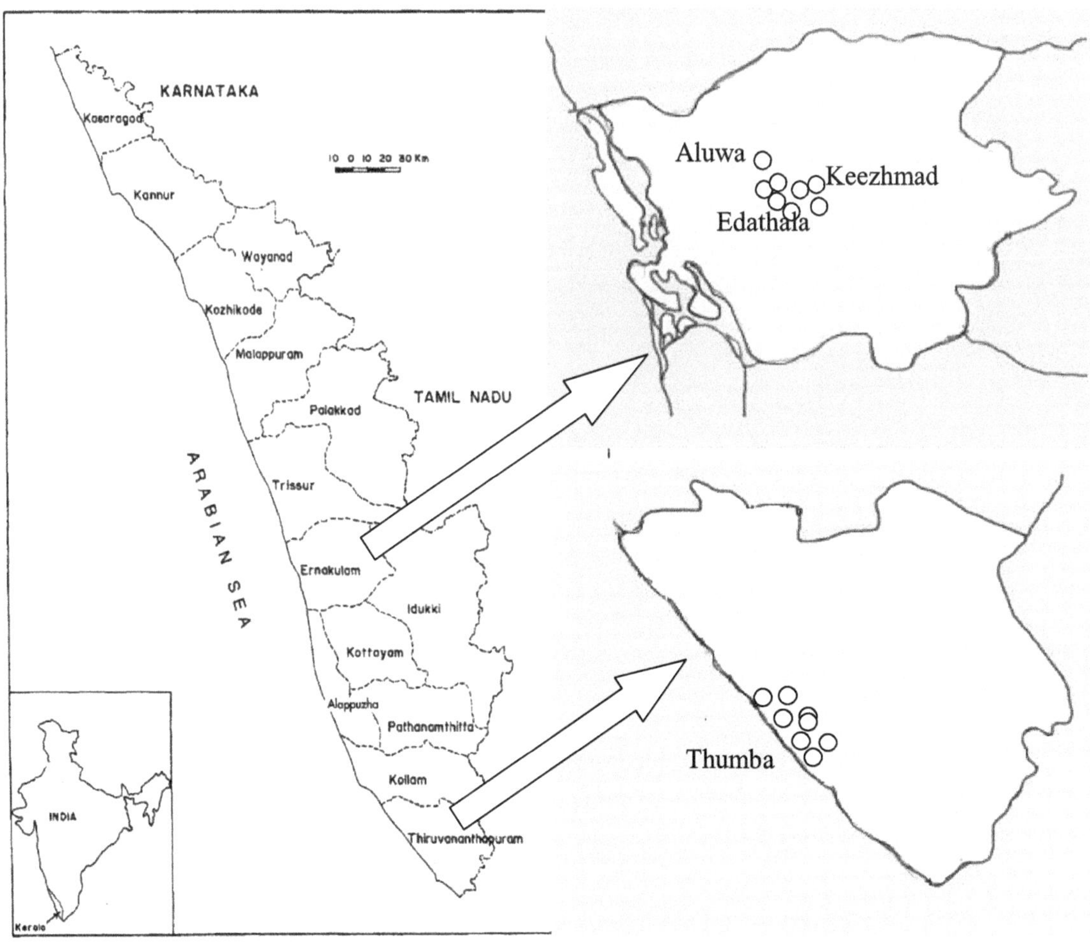

Fig. 1 Area map showing sample collection sites in Thiruvananthapuram and Ernakulam districts in Kerala, India

Perchlorate analysis in commercial drinking water

Perchlorate was also analyzed in 5 brands of bottled drinking water available in local market. These brands include Kinley, Aquafina, Neyyar aqua, Green valley and Surabhi.

Instrumental analysis

Water sample analysis was performed using an Ion Chromatography system (IC-1100, Dionex) with a separation column – Ion Pac AS 16 (2 × 250 mm and 4× 250 mm), guard column – Ion Pac AG 16 (2 × 50 mm and 4 × 50 mm) and an anion self-regenerating suppressor ASRS 300 (4 mm). The Ion Pac AS 16 column is specific for ClO_4^- ion with a lower detection limit of 2 ppb (μg/L). This method for ClO_4^- detection in drinking water is recommended according to USEPA methods 314.0 and 314.1. The eluent used was 50 mM Sodium hydroxide (NaOH, Fluka) at a flow rate 1.5 mL/min. The injection volume was 1000 μL. Calibration standards of ClO_4^- was prepared with high purity $KClO_4$ (Sigma Aldrich) by diluting 1000 mg/L primary standard. All solutions were prepared in ultra-pure milliQ water (Millipore).

Quality assurance and quality control (QA/QC) for IC

Three sets of calibration curves were generated ranging from 5–30, 50–100 μg/L and 100–500 μg/L. Laboratory reagent blank and fortified samples were also analyzed for QC. The mean recovery of ClO_4^- with the AS16 column and analytical condition was 100 ± 10 %.

Results and discussion

Detailed assessment of perchlorate in water samples

The present detailed assessment of ClO_4^- was based on our preliminary study in the past. We have observed ClO_4^- contamination of water samples from public drinking, open well and surface water sources at few districts in Kerala and a maximum of 91.4 ppb ClO_4^- was observed in open well water (for drinking) from Thumba (near VSSC) [21]. More detailed screening of samples was done during this study. Perchlorate concentration in ground and surface water samples collected from different locations in Kerala is presented in Table 1 and 2 respectively. Of the 160 samples analyzed from 27 locations, 58 % had ClO_4^- above detection limit (2 μg/L). The highest concentration

Table 1 Perchlorate (μg/L) in ground water samples from different districts in Kerala

District	Sampling site	No. of samples	Highest value	Mean value	Median value	Geometric mean
	Aluva town	4	BDL	BDL	BDL	BDL
	Keezhmad	5	2.76	0.52±1.2	BDL	1.22
	Marampalli	4	4.7	1.2±2.3	BDL	1.47
Ernakulam	Kulakkad site 1	7	666	243±218	185	170
	Kulakkad site 2	2	7270	7230±49.5	7230	7230
	Nalammile	15	667	243±278	45.4	28.3
	Edathala site 1	3	6420	4270±3700	6400	345.2
	Vazhakulam	5	6.74	1.67±2.9	BDL	1.6
	Edathala site 2	8	BDL	BDL	BDL	BDL
	Kazhakutom	10	7.07	4.05±1.2	3.8	3.9
	Manavila	7	10.9	6.46±5.2	4.83	3.9
Trivandrum	Kumapuram	12	9.19	3.66±4	2.8	2.66
	Thumba site 1	7	155	85±40	75	77.3
	Thumba site 2	2	300	300±80	294.7	294
	Veli	5	17.7	9.88±7.3	12.32	6.87
	Kannapuram	4	20.10	13.65±4.9	13.2	12.9
Kannur	Morazha	3	17	5.67±9.8	BDL	2.57
	Kannur town	4	BDL	BDL	BDL	BDL
	Thalassery	2	BDL	BDL	BDL	BDL
	Arayal	4	17	10.15±8.3	11.8	6.64
	Chittur	4	13.2	11.6±1.6	11.9	11.5
Palakkad	Palakkad town	5	15.0	14±1.02	13.6	13.9
	Olarakode	2	8.2	7.19±1.4	7.19	7.12
Koorg (control site)	Koorg	2	BDL	BDL	BLD	BLD

Table 2 Perchlorate (ug/L) in surface water samples from different districts in Kerala

District	Sampling site	No. of samples	Highest value	Average	Median	Geometric mean
Trivandrum	Veli lake	2	19.6	16.4	16.4	16
Ernakulam	Periyar river	4	355	122	35.4	66.8
Palakkad	Sokanashini river	2	21	17.5	15.75	17.14
Kannur	Vellikeel river	2	BDL	BDL	BDL	BDL

observed were 6420 μg/L and 7270 μg/L in samples from house hold open well at Edathala and from a public open well at Kulakkad respectively (both places are ~500 meters away from APEP). These values are ~480 times higher than drinking water equivalent level (DWEL), 15 μg/L established by USEPA. Perchlorate was detected in all the samples collected from Periyar river in the area with an average concentration 122 μg/L ($n = 4$) and the highest concentration observed was 355 μg/L. Periyar river (~3 km away from APEP) flows through the states of Tamil Nadu and Kerala, and it is one of the few perennial rivers in the region and provides drinking water to major town like Ernakulam in Kerala. Analysis of water samples from Trivandrum district also revealed high concentration of ClO_4^-. The highest value observed was 300 μg/L in a house hold open well water from Thumba (near VSSC). The average ClO_4^- level in this region was 85 ± 45 μg/L ($n = 7$). The mean concentration of ClO_4^- in ground water samples from all the study area is given in Fig. 2. Ground water perchlorate level was high in Ernakulam and low in Kannur district. Fig. 3 shows a comparison of perchlorate in all ground water and surface water sources studied. Compared to surface water (mean = 773), the ground water (mean = 79) perchlorate level is almost 10 times higher.

Perchlorate was detected in 60 % samples ($n = 18$) from various places in Kannur district with a highest value of 20.1 μg/L. Perchlorate was also detected in all the samples ($n = 5$) from Palakkad district with a peak value of 15 μg/L in sample collected from a rock mining site. As expected water samples from the public distribution system from Koorg had no detectable level of ClO_4^- due its geographical location.

Reports on perchlorate contamination in drinking water supplies from different countries in recent years indicate the growing concern about ClO_4^- contamination. Comparable to the high value (7270 μg/L) observed in this study, in a recent study 7,690 μg/ L ClO_4^- was reported in ground water samples collected from fire cracker manufacturing sites Tamil Nadu, India [15].

USEPA has reported ClO_4^- analysis data collected from 3,865 public water supplies between 2001 and 2005 from several states and territories. It was found that, ~4 % of samples had at least one analytical detection of ClO_4^- at ≥4 μg/L and the maximum detected level was 420 μg/L.

Probable sources of ClO_4^- contamination in water sources in Kerala

The two inventories of ClO_4^- could be considered as the major sources of its contamination in Ernakulam and Trivandrum districts. The NH_4ClO_4 manufacturing and using places in Ernakulam and Trivandrum districts respectively in this state could be one of the potential sources of ClO_4^- contamination. Our preliminary study revealed high level of ClO_4^- (91.4 μg/L) and chlorate (ClO_3^-) (177 μg/L) in water samples from Thumba, Trivandrum [21]. The present study also detected high level of ClO_4^- (300 μg/L) from household well water from Thumba. However, unlike the previous study, the

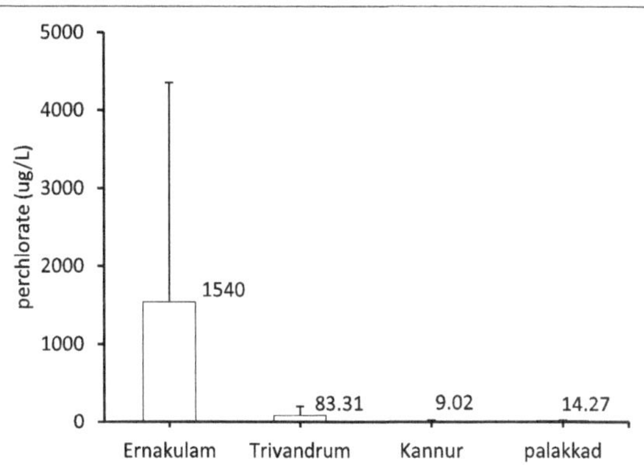

Fig. 2 Perchlorate concentration (mean/SD) at different districts in Kerala

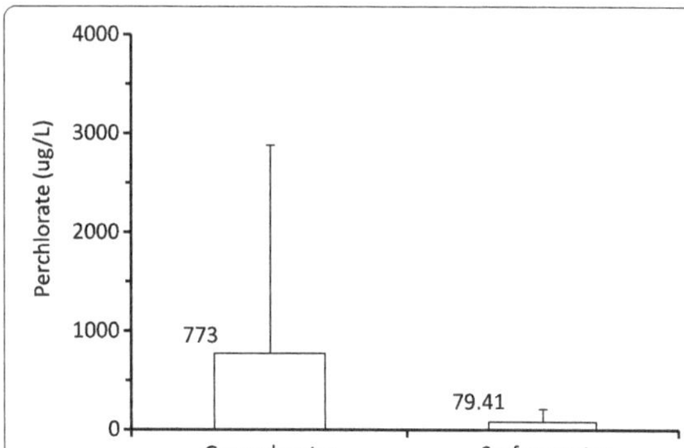

Fig. 3 Perchlorate concentration (mean/SD) in groundwater and surface water samples in Kerala

highest ClO_4^- concentration (7270 μg/L) was observed in public well water source located in close proximity to APEP in Ernakulam district. Ammonium perchlorate washout from these sources can be the potential source of high level contamination detected in ground water in this region. Isotope Ratio Mass Spectra (IRMS) analysis will provide more concrete information about the source of perchlorate in this region and other places in Kerala. Moreover, assessment of thyroid gland functioning of people around APEP will bring out health effects due to the exposure to high level of ClO_4^- in drinking water. Analyzing Thyroid gland functioning (TSH, T3, T4 etc. values) as well as Iodide and perchlorate level in urine samples will provide information about thyroid gland function and exposure level to perchlorate.

Perchlorate was also detected in water samples from the Northern districts, Kannur and Palakkad (20.11 and 15 μg/L) which are ~220 and ~150 km away from the APEP facility. The presence of ClO_4^- in these regions points to other possible sources like usage of ClO_4^- in fireworks, explosives etc. There are a large numbers of fire work display occurs in Kerala especially during April-May which is the festival season in the state. Several small scale fire work manufacturing units are operated during this period, however only few data about these industries are available as most of these are unauthorized. An increase in deposition of ClO_4^- by 18 fold was observed due to fallout from fireworks in USA [22]. In another study, surface water ClO_4^- level increase in the range of 24–1028 times was observed following fireworks display in Oklahoma Lake (USA) [23]. Perchlorate contamination due to firework production and display was also reported from China and Japan [12, 16]. As already mentioned, high levels of ClO_4^- was detected recently from ground water samples from a cracker manufacturing industrial area in Tamil Nadu [15]. A number of rocks mine operating in most of the districts in Kerala. The explosives used in these mines may include ClO_4^- salt also. Traces of ClO_4^- from these sites may also contribute to ClO_4^- contamination of waters in Kerala. This could be a potential source of ClO_4^- in the water sample from near rocks mining areas of Palakkad. No information is available on the consumption of ClO_4^- in rock mines, primarily due their unauthorized operational status similar to small scale fire-cracker industries. Perchlorate can also form naturally under rare environmental conditions like ozone oxidation of aqueous chloride or through electric discharging of chloride aerosol [24]. Natural ClO_4^- deposits were found at relatively high concentration in Atacama Desert in Chile [25]. High concentration of ClO_4^- (1000 mg/Kg) was detected in natural mineral ores like potash ore from places like New Mexico, Canada, Bolivia and California [26]. Stable isotope analysis of chlorine and oxygen in ClO_4^- ($^{17}O/^{16}O$, $^{18}O/^{16}O$ and $^{37}Cl/^{35}Cl$) can distinguish the nature of ClO_4^- (synthetic or natural) present in environmental samples [27].

Screening of commercial drinking water for perchlorate contamination

Perchlorate was not detected in any of the bottled drinking water samples tested in this study. Previous report from India also had a similar observation where ClO_4^- was not detected in 5 of the branded of bottled drinking water analysed [14]. Perchlorate concentration was also very low (<0.002–0.22 μg/L) in bottled water from China [16]. However ClO_4^- was found in 10 of the 21 bottled water samples in USA with a mean concentration of 0.16 μg/L [28].

Conclusions

The present study reveals high level ClO_4^- contamination in ground and surface water around places where ClO_4^- is handled in bulk. The contamination was more severe in ground water (max. value 7270 μg/L) compared to surface water (max. value 355 μg/L), both from Aluva in Ernakulam district, Kerala. Findings of this study points to the need for frequent monitoring of ground water samples around places where ClO_4^- is handled in bulk and necessitate epidemiological study in the contaminated area to assess the status of thyroid gland functioning. This study also underlines the need for defining water quality standards for perchlorate in India and also for controlling the environmental release of perchlorate especially from point sources like the manufacturing and using sites.

Competing interests

The authors declare that they have no competing interests.

Authors' contributions

AVN & PGP: Sample collection from field, chemical analysis and manuscript preparation. KKB Conceived the study, data analysis, interpretation of results and manuscript preparation. All authors read and approved the final manuscript

Acknowledgements

The present study was conducted with the financial assistance from the CSIR 12FYP project INDEPTH (BSC 0111). The infrastructural support from CSIR-NIIST is also acknowledged in this study.

References

1. ITRC (Interstate Technology and Regulatory Council). Perchlorate: overview of issues, status, and remedial options, PERC-1. Washington: Interstate Technology and Regulatory Counci; 2005.
2. Wolff J. Perchlorate and the thyroid gland. Pharmacol Rev. 1998;50:89–105.
3. Henrichs J, Bongers-Schokking JJ, Schenk JJ, Ghassabian A, et al. Maternal thyroid function during early pregnancy and cognitive functioning in early childhood: the generation study. J Clin Endocrinol Metab. 2010;95:4227–34.
4. Stettler R. Utilisation del'ozone dans le traitment de eaux de boisson. Gas Wass Abwass. 1977;57:81–95.

5. Greer MA, Goodman G, Pleus RC, Greer SE. Health effects assessment of environmental perchlorate contamination: The dose response for inhibition of thyroid radioiodine uptake in humans. Environ Health Perspect. 2002;110(9):927–37.
6. Hu F, Sharma B, Mukhi S, Patino R, Carr JA. The colloidal thyroxine (T_4) ring as a novel biomarker of perchlorate exposure in the African clawed frog *Xenopus laevis*. Toxicol Sci. 2006;93:268–77.
7. USEPA (United States Environmental Protection Agency). Interim drinking water health advisory for perchlorate. Washington D.C: United States Environmental Protection Agency; 2008.
8. WHO (World Health Organization). Joint FAO/WHO expert committee on food additives seventy-second meeting. Rome: WHO; 2010.
9. Hogue C. Rocket-fueled river. Chem Eng News. 2003;81:37.
10. Backus SM, Klawuun P, Brown S, D'sa I, Sharp S, Surette C, et al. Determination of perchlorate in selected surface waters in the Great Lakes Basin by HPLC/MS/MS. Chemosphere. 2005;61:834–43.
11. El Aribi H, Le Blanc YJC, Antonsen S, Sakuma T. Analysis of perchlorate in foods and beverages by ion chromatography coupled with tandem mass spectrometry (IC-ESI-MS/MS). Anal Chim Acta. 2006;56:39–47.
12. Kosaka K, Asami M, Matsuoka Y, Kamoshita M, Kunikane S. Occurrence of perchlorate in drinking water sources of metropolitan area in Japan. Water Res. 2007;41:3474–82.
13. Kim HK, Kim JH, Lee BC, Yu SJ, Kim HJ. Occurrence of perchlorate in drinking water sources in Korea. Water Sci Technol. 2009;9:133–9.
14. Kannan K, Praamsma ML, Oldi JF, Kunisue T, Sinha RK. Occurrence of perchlorate in drinking water, ground water, surface water and human saliva from India. Chemosphere. 2009;76:22–6.
15. Isobe T, Ogawa SP, Sugimoto R, Ramu K, Sudaryanto A, Malarvannan G, et al. Perchlorate contamination of groundwater from fireworks manufacturing area in South India. Environ Monit Assess. 2013;185:5627–37.
16. Wu Q, Zhang T, Sun H, Kannan K. Perchlorate in tap water, ground water, surface waters and bottled water from China and its association with other inorganic anions and with disinfection byproducts. Arch Environ Contam Toxicol. 2010;58:543–50.
17. NAS (National Academy of Science). Health implications of perchlorate ingestion. Washington, D C: National Academies Press; 2005.
18. Wang Z, Lau BPY, Tague B, Sparling M, Forsyth D. Determination of perchlorate in infant formula by isotope dilution ion chromatography/tandem mass spectrometry. Food Addit Contam. 2011;28:799–806.
19. Usha MV, Sundaram KR, Unnikrishnan AG, Jayakumar RV, Nair V, Kumar H. High prevalence of undetected thyroid disorders in an iodine sufficient adult south Indian population. J Indian Med Assoc. 2009;107:72–7.
20. Jayalekshmi P, Gangadharan P, Mani KS. Cancer in women in Kerala - a transition from a less-developed state. Asian Pac J Cancer Prev. 2006;7:186–90.
21. Anupama VN, Kannan K, Prajeesh PVG, Rugmini S, Krishnakumar B. Perchlorate, Chlorate and Bromate in water samples from the South-West coast of India. Water sci Technol-Water supply. 2012;12:595–603.
22. Munster J, Hanson NG, Jackson AW, Rajagopalan S. The fall out of fireworks: Perchlorate in total deposition. Water Air Soil Poll. 2009;198:149.
23. Wilkin RT, Fine DD, Burnett NG. Perchlorate behavior in municipal lake following fireworks display. Environ Sci Technol. 2007;41:3966–71.
24. Dasgupta PK, Martinelango PK, Jackson WA, Anderson TA, Tian K, Tock RW, et al. The origin of naturally occurring perchlorate: Role of atmospheric processes. Environ Sci Technol. 2005;39:1569–75.
25. Ericksen GE. The Chilean nitrate deposit. Sci Am. 1983;71:366–74.
26. Orris GJ, Harvey GJ, Tsui DT, Eldrige JE. Preliminary analyses for perchlorate in selected natural materials and their derivative products, Open-file report. USA: US Geological Survey; 2003. p. 03–314.
27. Sturchio NC, Hoaglund JR, Marroquin RJ, Beloso AD, Heraty LJ, Bortz SE, et al. Isotopic mapping of groundwater perchlorate plumes. Ground Water. 2012;50:94–102.
28. Shi Y, Zhang P, Shi J, Cai Y, Mou S, Jiang G. Perchlorate in sewage sludge, rice, bottled water and milk collected from different areas in China. Environ Int. 2007;33:955–62.

16

The relationship between bone health and plasma zinc, copper lead and cadmium concentration in osteoporotic women

Naficeh Sadeghi[1], Mohammad Reza Oveisi[1], Behrooz Jannat[2*], Mannan Hajimahmoodi[1], Masoomeh Behzad[1], Abdolazim Behfar[3], Fatemeh Sadeghi[2] and Sahereh Saadatmand[1]

Abstract

Osteoporosis is a multi factorial disease with dimension of genetic and nutritional considerations. The aim of this study was to present data from the association of plasma zinc, copper and toxic elements of lead and cadmium levels with bone mineral density in Iranian women. 135 women gave their information and enrolled. Fasting plasma was used for measurement of trace elements and heavy metals by Differential Pulse Anodic Stripping Voltammetry. Control group (n = 51) were normal in both lumbar spine (L_1-L_4) and femoral neck density (T-score ≥ −1), but just femoral neck T-score was considered as criterion in selection of patient group (n = 49, Tscore < −1). No differences were found in the nutritional status, number of diseases, drugs and functional activities between these groups. Plasma Zn, Cu, Pb, Cd levels were analyzed by, a method of voltammetry. Mean ± SD levels of copper and zinc was 1.168 ± 0.115, 1.097 ± 0.091 μg/ml in control group, 1.394 ± 0.133, 1.266 ± 0.11 μg/ml in total patient (TP) and 1.237 ± 0.182, 1.127 ± 0.176 μg/ml in Mild patients(−1 > T-score > −1.7), 1.463 ± 0.174, 1.327 ± 0.147 μg/ml in Severe patient group (T-score < −1.7); respectively. Mean ± SD plasma level of lead and cadmium was 168.42 ± 9.61 ng/l, 2.91 ± 0.18 ng/ml in control group, 176.13 ± 8.64 ng/l, 2.97 ± 0.21 ng/ml in TP, 176.43 ± 13.2 ng/l, 2.99 ± 0.1 ng/ml in mild patients, 221.44 ± 20 ng/l and 3.80 ± 0.70 ng/ml in severe patient group, respectively. In this study plasma zinc, copper, lead & cadmium concentrations were higher in the patients than in the control, though differences were not significant. However, differences were higher between the controls and patients with severe disease (T-score < −1.7). In addition adjusted T-score of femur with age and BMI showed negative significant correlation with plasma levels of zinc and lead in total participants ($p < 0.05$, r = −0.201, $p = 0.044$, r = −0.201). It seems that more extensive study with larger ample size might supply definite results about this association for copper and cadmium.

Keywords: Zinc, Copper, Lead, Cadmium, Osteoporosis, Voltammetry

Background

Osteoporosis is a skeletal disorder characterized by compromised bone strength predisposing to an increased risk of fracture [1,2]. The introduction of bone density measurement methods makes it possible to diagnose osteoporosis before fractures occur. The World Health Organization (WHO) has defined osteoporosis as a bone mineral density (BMD) at the hip or spine that is ≤ −2.5 SDs (standard deviation) below the mean value for young women, and osteopenia was defined as a BMD that is between −1 and −2.5 SDs below this value [3].

* Correspondence: janatbhr@sina.tums.ac.ir
[2]Halal research center, Ministry of Health and Medical Education, Tehran, Iran
Full list of author information is available at the end of the article

Zinc is essential element for growth of human. Role of zinc is demonstrated in the growth and mineralization and preservation of bone tissue. Copper has an essential role in the normal maturation of collagen, particularly in the important steps to the formation of lysine- derived cross-link [4,5]. Lead is a highly toxic metal that may damage directly bone-forming cells (increasing chondrogenesis, delaying cartilage mineralization) and alter the hormonal regulation of Calcium and vitamin D3. Lead has a negative effect on the processes regulating bone turnover mechanisms and bone maturation or skeletal growth. These changes may affect bone fragility and the risk of fractures [6]. Lead will be retained in bone by the replacement of calcium. Cadmium is a widespread environmental pollutant, present in food and tobacco. Cadmium may have both direct and indirect effects on bone turnover [7,8]. Stripping voltammetry (SV) analysis is powerful and simple tool to determination of trace elements. In trace analysis of metal ions, Anodic stripping voltammetry (ASV) is the most popular SV technique [9]. The aim of this study is determination of plasma levels of lead, cadmium, zinc and copper in osteoporotic women comparison to control by voltammetric method and how environmental toxins, lead and cadmium leads to the pathogenesis of osteoporosis.

Methods

Apparatus

The polarograph set was Trace analyzer model 746 instruments (Metrohm Ltd. Switzerland). Hanging Mercury Drop Electrode (HMDE), Reference silver chloride electrode (Ag/Agcl/kcl 3 M) and platinum (Pt) auxiliary electrode were used.

Reagents

The chemical, $CH_3COONa.3H_2O$, $Pb(NO_3)_2$, $Cd(NO_3)_2$, $Cu(NO_3)_2$, $Zn(NO_3)_2$ and HNO_3 used in analytical grades and were purchased from Merck Company Germany. Stock solution (1 g/L) of 0.1 Molar nitric acid was prepared. Hydrochloric acid (HCL) used to acidity the plasma samples was Suprapur grade of Merck Company. The stock solution (1 g/L) was prepared in 2.5 normal HCL. The pH acetate buffer was adjusted between 4.6 and 4.8.

Sample preparation

In the present study, people were screened through a total of approximately 1000 women that referred to bone mineral densitometry division of Jami Clinic in Tehran, Iran. 135 women gave their information and enrolled. Participants were separated into three groups according to T-score values suggested by WHO. The control group T-score is ≥ −1 and the total patients (mild osteopenia + severe osteopenia + osteoporosis) T-score < −1. The participants underwent a fasting blood withdrawal in 10-ml heparinized tubes on the day of the bone scan and then plasma was seprated. A number of 200 μL of plasma was added to 800 μL of 2.5 N hydrochloric acid to percipitate plasma proteins and centrifuged for 15 min at 6000 rpm. Then the clear supernatant passed from 0.2 micron filter before transferring to the polarographic cell. The complex composition of the plasma may cause changes in the calibration slopes; that is why, the zinc, copper, lead & cadmium concentrations were determined using the standard addition technique with three additions of standard stock solution. First 10 ml of 0.1 M acetate buffer was used to set the blank polarogram. Next 500 μL of sample solution was added to polarographic cell and the polarogram was plotted. After that 100 μL of the standard stock solution was added to the cell for three times and each time the polarogram was done. The cell solution was deoxygenated by nitrogen for 300 seconds. Determination was made on a Hanging mercury drop electrode (HMDE). Determination of zinc and copper was made after a collection period at a potential of −1200 and −500 mV respectively. Method was able to give estimation of cadmium and lead simultaneously when a sweep potential was applied between −800 and −100 mV. The potential was swept, using a differential pulse anodic stripping voltammetry (DPASV) [9,10]. Three determinations were done for each sample, and the standard deviations were calculated [10].

Table 1 Comparison of age and BMI among control and patient groups

Group	Control T-score > −1 (Mean ± SD)	Total patients T-score ≤ −1
	n = 51	n = 49
Age (year)	48.17 ± 11.16	55.8 ± 12.38[a]
BMI (Kg/m2)	28.25 ± 50.02	25.8 ± 12.38[a]

[a]$p < 0.01$.

Ethical approval

The project was approved by Ethics Committee of Tehran University of Medical Sciences (reference number of approval 89-01-33-10034).

Consent

We obtained in formed consent from all participants but no patient was reported as a case.

Results and discussion

In the current study the level of zinc, copper, lead and cadmium were determined in plasma of women with osteoporosis. The total number of subjects was 135. The T- score of both lumbar spine (L1-L4) and femoral neck were measured in all participants. The control group

Table 2 Femoral neck T-score and plasma concentrations of Zn, Cu, Pb, Cd in smokers and non-smokers

	Smokers (Mean ± SD)	Non-smokers
Participants (%)	12	88
Average T-score femur	−1.49 ± 0.99	0.62 ± 1.3
Zinc (ng/ml)	1512 ± 258	1108 ± 59
Copper (ng/ml)	1572 ± 282	1290 ± 79
Lead (ng/l)	196.71 ± 20.93	169.47 ± 5.81
Cadmium (ng/ml)	2.98 ± 0.12	2.69 ± 0.42

Table 3 Concentration of Zn, Cu, Pb, Cd in postmenopausal, Pre-menopausal, athlete and non-athlete participants

	Athlete	Non-athlete (Mean ± SD)	Pre-menopausal	Post-menopausal
Zinc (μg /ml)	1.079 ± 0.138	1.171 ± 0.067	1.138 ± 0.080	1.163 ± 0.090
Copper (μg /ml)	1.107 ± 0.155	1.369 ± 0.088	1.233 ± 0.105	1.428 ± 0.110
Lead (ng/l)	166.63 ± 11.82	173.54 ± 6.51	174.66 ± 8.53	169.11 ± 6.91
Cadmium (ng/ml)	2.74 ± 0.16	3.09 ± 0.14	2.87 ± 0.16	3.05 ± 0.16

was normal in both femur and spine with T-score ≥ −1 including 51 women (37.8%). The total patient (mild osteopenia + severe osteopenia + osteoporosis) had femoral T-score < −1 including 49 women (36.3%) and women with −1.7 ≤ T-score < −1 as mild osteopenia, patients with −2.5 < T-score < −1.7 as severe osteopenia including 41 women (30.3%). The total number of samples from 35 (25.9%) were not including in the classification of cases and controls; because of femoral T-score > −1 but lumbar T-score < −1, so they were analyzed in total participants data. No difference in menopause, drugs, nutritional habit, functional activities, stress and exposure time in the sun was found between groups, but the difference was significant for smoking ($P < 0.05$), age and BMI ($P < 0.01$) (Table 1). The concentrations of zinc, copper, lead and cadmium in each patient division were compared with control. The regression relationship between the amount of zinc, copper, lead and cadmium with femoral T-score in control and patient groups were compared. No significant relationship was detected between age and BMI with the concentration of zinc, copper, lead and cadmium. Zinc, copper, lead and cadmium levels are higher in smokers than non-smokers, but the difference is not significant ($p = 0.266$, $p = 0.150$, $p = 0.146$, $p = 0.462$ respectively). Results are shown in Table 2.

Also exercise and inactivity, menopause and pre-menopausal status showed no significant effect on zinc, copper, lead and cadmium levels. An inverse relationship between age and femoral T-score is illustrated. Results are shown in Table 3.

Morefurther, a direct and significant relationship between BMI and T-score of femur was detected. In this study, plasma mean levels of zinc, copper, lead and cadmium were obtained for 135 samples. Results are shown in Table 4.

The relationship between the concentrations of zinc, copper, lead and cadmium were studied with the T-score of femur Table 5.

For total participant's of135 after matching with age and BMI, plasma concentrations of Zn and Pb have a statistically significant negative relationship with T-score of femur. But the relationship between plasma concentrations of Cu and Cd with femoral T-score can be observed probably after sample size increasing. The results are shown in Table 6.

In this study, plasma zinc, copper, lead & cadmium concentration among Iranian osteoporotic women showed that: plasma zinc, copper, lead & cadmium concentration were higher in the patients than in the control, though differences were not significant. However, differences were higher between the controls and patients with severe disease (T-score < −1.7). In addition, T-score of femur adjusted with interfere factors of age and BMI, showed negative significant correlation with plasma levels of Zn and Pb in the entire study population ($p < 0.05$, $r = -0.201$, $p = 0.044$, $r = -0.201$ respectively). It seems that more extensive study with larger ample size might supply definite results about this reverse linear association for copper and cadmium. Bone zinc content is decreased by development of aging, bone loss, and post menopausal conditions. The metal directly activates Aminoacyl-tRNA synthetase in osteoblastic cells and it stimulates cellular protein synthesis. Zinc may act on the process of bone-resorbing factors induced protein kinase C activation, which is involved in Ca^{+2} signaling in osteoclastic cells [11]. Zinc is an essential trace element that is a cofactor of more than 200 enzymes [12]. A decrease in the formation of cross-linking amino acids is thought account in part for the increased fragility of bone from copper-deficient [13]. Cadmium can interfere with vitamin D metabolism [7].

Table 4 Plasma Mean ± SD levels of Zn, Cu, Pb, Cd in different groups

	T-score > −1	T-score ≤ −1	−1.7 < T-score < −1	−2.5 < T-score < −1.7	T-score < −2.5
Zn (μg/ml)	1.1 ± 1.09	1.26 ± 0.11	1.13 ± 0.17	1.33 ± 0.15	1.27 ± 0.42
Cu (μg/ml)	1.17 ± 0.11	1.39 ± 0.13	1.24 ± 0.18	1.46 ± 0.17	1.71 ± 0.46
Pb (ng/l)	168.42 ± 0.01	176.13 ± 0.26	176.43 ± 13.2	183.06 ± 10.91	221.4 ± 20
Cd (ng/ml)	2.91 ± 0.18	2.97 ± 0.21	2.99 ± 0.19	3.13 ± 0.29	3.80 ± 0.70

Table 5 Correlation between plasma concentrations of Zn, Cu, Pb, Cd and femoral T-score in different groups without age and BMI matching

	T-score < 2.5	T-score > −1	T-score ≤ −1	−1.7 < T-score < −1	−2.5 < T-score < −1.7	T-score < −2.5
Zn	p: 0.041[a]	0.125	0.356	0.736	0.669	0.799
	r: −0.181	−0.218	−0.135	−0.095	−0.076	−0.135
Cu	p: 0.205	0.980	0.278	0.202	0.610	0.899
	r: −0.113	−0.004	−0.188	−0.349	−0.091	−0.067
Pb	p: 0.072	0.179	0.093	0.230	0.363	0.933
	r: −0.160	−0.191	−0.243	−0.330	−0.161	−0.045
Cd	p: 0.596	0.781	0.160	0.652	0.440	0.994
	r: −0.047	−0.040	−0.204	−0.127	−0.137	−0.004

[a]$p < 0.05$.

For many years, it has been thought that Cd affects bone only at high-level long-term exposure. Recent epidemiological data indicate that Cd can damage the skeleton at considerably lower exposure than previously anticipated; however, the critical level of the exposure is still unknown. Decreased bone density with increased risk of fractures has been reported in conditions of low to moderate environmental exposure to Cd taking place in industrialized countries. Until now, various mechanisms for the Cd-induced bone damage involving both direct and indirect action of this metal have been suggested, they might explainthe indirect mechanism involves the Cd-induced disorders in the metabolism of vitamin D and minerals due to kidney and gastrointestinal tract damage [14].

In 2009 Smith *et al.* [15] reported that cadmium causes apoptosis in human osteoblast-like Saos-2 cells. They showed that cadmium exposure induces oxidative stress which leads to decreased RUNX2 mRNA expression and increased apoptotic death, antioxidant NAC alleviates the damaging effects of cadmium. Furthermore the osteoblast transcriptional factor RUNX2 is reported to play a protective role against osteoporosis in postmenopausal women. Cadmium can indirectly induce oxidative stress through depletion of antioxidant molecules or inhibition of antioxidant enzymes. Mounting worldwide epidemiological research indicates that chronic, low level exposure to lead and cadmium leads to increased risk of bone fractures. The importance of the age factor in bone density is obvious and known to increase the risk of osteoporosis and bone density decreases [16]. The analysis of the results of this study indicated that the plasma concentrations of zinc, copper, lead and cadmium are not affected by age. Plasma zinc and copper concentrations were higher in those who did not exercise. Physical activity has a negative association with plasma concentrations of lead, though it is not significant. However, the plasma concentration of cadmium was lower in athletes, and this difference was significant ($p < 0.01$). In 2011 Margaret E. Sears [17] studied the amount of lead, cadmium, mercury and arsenic in sweat, so it was seen as much more sweat as more excretion of these elements. In other words more exercise means more perspiration and excretion of these elements, so plasma levels of these harmful elements reduce. Smoking is known as a cause of bone mineral loss and may increase the risk of hip fracture [18]. Polidori reports that α, γ tocopherol concentrations, α, β cryptoxanthin, retinol and ascorbate to increases after 4 weeks smoking, significantly. In the present study, we also observed an increase in plasma antioxidant capacity in smokers, though is not significant [19]. Kido *et al.* [20] study has shown an association between renal tubular dysfunction and decreased BMD in examination of women environmentally exposed to cadmium. A study in 1988 by Wittmers *et al.* [21] was performed for survey the effects of lead injection in bone and skeletal tissue. It was seen that Lead makes some spaces in bone. Lead cause snaping of Hydroxy apatit crystals in calcification process. This toxic substance remains in bone until remodeling. In a clinical study, Saltman [22] has demonstrated the efficacy of Ca, Cu, Mn and Zn supplementation in spinal bone mineral density in postmenopausal women and the necessity of trace elements for optimal bone matrix development and bone density sustenance. Yamaguchi *et al.* [23] investigated the effects of zinc on bone cells. Those cells that were exposed to more zinc concentration, significantly showed increased consumption of zinc, calcium, alkaline phosphatase and ATPase activity, also collagen was increased significantly. These findings suggest the direct

Table 6 Correlation between concentrations of Zn, Cu, Pb and Cd with femoral T-score in total population after age and BMI matching

	Zn	Cu	Pb	Cd
Total participants	p: 0.030[a]	0.063	0.044[a]	0.306
(T-score <2.5)	r: −0.216	−0.186	−0.201	−0.103

[a]$p < 0.05$.

effect of zinc on the mineralization of bone *in vitro* and stimulation of bone protein synthesis. Insufficient copper intake in the diet can cause osteoporosis, bone calcium and copper deficiency, bone abnormalities and curvature of the spine. Suttle [24] showed that bone formation activity is one of the first activities which are disrupted in copper-deficient lambs. Our results showed more zinc and copper levels in osteoporotic patients than controls, although not significantly. Some studies indicated that free radicals increase osteoporosis as well as the role of antioxidants in their lack of support, so the disease can progress. But it seems that the in oxidative stress disease, antioxidants have always decreased, although this may still be inadequate to meet the condition. Getting more antioxidants can help prevent the progression of osteoporosis.

Conclusions

The comprehensive database shows that plasma levels of zinc, copper, lead and cadmium in the osteoporotic women is higher than control, though the differences were not significant. It seems that more extensive study with larger size might supply definite result about this association.

Competing interests

The authors declare that they have no competing interests.

Authors' contributions

NS designed of the study, also coordinated research activities, and revised the manuscript. MRO participated in data gathering. BJ carried out technical analysis of data, participated in final revision of the manuscript. MB gathered data and drafted the manuscript. MH and AB participated in the design of the study. FS and SS participated in data analysis. All authors read and approved the final manuscript.

Acknowledgements

Authors are grateful from the research deputy of Tehran University of Medical Sciences for financial support in a project and thank the Pharmaceutical Sciences Research Center for their help.

Author details

[1]Department of Drug and Food Control, School of Pharmacy, Tehran university of Medical Sciences, Tehran, Iran. [2]Halal research center, Ministry of Health and Medical Education, Tehran, Iran. [3]Department of Drug and Food Control, Faculty of Pharmacy, Jondishapour Ahvaz University of Medical Sciences, Ahvaz, Iran.

References

1. Handa R, Kalla AA, Maalouf GH: **Osteoporosis in developing countries.** *Best pract Res cl Rh* 2008, **22**(4):693–708.
2. Oveisi MR, Sadeghi N, Jannat B, Hajimahmoodi M, Hadjibabaie M, Behfar A: **Evaluation of antioxidants in bone mineral density of Iranian osteoporotic women.** *IJBMS* 2011, **14**(2):158–166.
3. Jannat B, Oveisi MR, Sadeghi N, Hajimahmoodi M, Jamshidi AR, Izadi Dolagh G, Behfar A, Gholami K, Behzad M: **The erythrocyte catalase enzyme activity in Iranian osteoporotic women.** *IJPS* 2010, **6**(1):19–24.
4. Hyun TH, Barrett-Connor E, Milne DB: **Zinc intakes and plasma concentration in men with osteoporosis : the rancho Bernardo study.** *Am J Clin Nutr* 2004, **80**:715–721.
5. Rucker R, Murray J: **Cross-linking amino acids in collagen and elastin.** *Am J Clin Nutr* 1978, **31**:1221–1236.
6. Rodríguez-Estival J, Álvarez-Lloret P, Rodríguez-Navarro AB, Mateo R: **Chronic effects of lead (Pb) on bone properties in red deer and wild boar: relationship with vitamins a and D3.** *Environ Pollut* 2013, **174**:142–149.
7. Wang H, Zhu G, Shi Y, Weng SH, Jin T, Kong Q, Nordberg G: **Influence of environmental cadmium exposure on forearm bone density.** *J Bone Miner Res* 2003, **18**(3):553.
8. Akesson A, Bjellerup P, Lundh T, Lidfeldt J, Nerbrand C, Samsioe G, Skerfving S, Vahter M: **Cadmium-induced effects on bone in a population-based study of women.** *Environ Health Perspect* 2006, **114**(6):830–834.
9. Chanlon S, Joly-Pottus L, Chatelut M, Vittori O, Cretier JL: **Determination of carmoisine, allura red ponceau 4R in sweets and soft drinks by differential pulse polarography.** *J Food Comp Anal* 2005, **18**:503–515.
10. Thunus L, Lejeune R: **Determination of Zn, Cu, Ni and Co by voltammetry in plasma, without mineralisation.** *Anal Lett* 1990, **23**(5):901–921.
11. Yamaguchi M: **Role of zinc in bone formation and bone resorption.** *J Trace Elem Exp Med* 1998, **11**:119–135.
12. Otsuka M, Oshinbe A, Legeros RZ, Tokudome Y, Ito A, Otsuka K, Higuchi W: **Efficacy of the injectable calcium phosphate ceramics suspensions containing magnesium, zinc and fluoride on the bone mineral deficiency in ovariectomizad rats.** *J Pharm Sci* 2008, **97**:421–432.
13. Opsahl W, Zeronian H, Ellison M, Lewis D, Rucker RB, Riggins RS: **Role of copper in collagen cross-linking and its influence on selected mechanical properties of chick bone and tendon.** *J Nutr* 1982, **112**:708–716.
14. Brzo'ska MM, Moniuszko-Jakoniuk J: **Low-level lifetime exposure to cadmium decreases skeletal mineralization and enhances bone loss in aged rats.** *Bone* 2004, **35**(5):1180–1191.
15. Smith SS, Rodriguez Reyes J, Arbon KS, Harvey WA, Hunt LM, Heggland SJ: **Cadmium-induced decrease in RUNX2 mRNA expression and recovery by the antioxidant N-acetylcysteine (NAC) in the human osteoblast-like cell line, saos-2.** *Toxicol in Vitro* 2009, **23**:60–66.
16. Gallagher JC, Riggs BL, Jerpbak CM, Arnaud CD: **The effect of age on serum immunoreactive parathyroid hormone in normal and osteoporotic women.** *J Lab Clin Med* 1980, **95**(3):373–385.
17. Margaret ES, Kathleen JK, Riina IB: **Arsenic, cadmium, lead, and mercury in sweat.** *J Environ Public Health* 2012, **2012**:1–10.
18. Law MR, Hackshow AK: **A meta-analysis of cigarette smoking, bone mineral density and risk of hip fracture: recognition of a major effect.** *Brit Med J* 1997, **315**:841–860.
19. Polidori MC, Mecocci P, Stahl W, Sies H: **Cigarette smoking cessation increases plasma levels of several antioxidant micronutrients and improves resistance towards oxidative challenge.** *Brit J Nut* 2000, **90**:147–150.
20. Kido T, Nogawa K, Honda R, Tsuritani I, Ishizaki M, Yamada Y, Nakagawa H: **The association between renal dysfunction and osteopenia in environmental cadmium-exposed subjects.** *Environ Res* 1990, **51**:71–82.
21. Wittmers LE, Aufderheide AC, Wallgren J, Rapp G, Alich A: **Lead in bone. IV. Distribution of lead in the human skeleton.** *Arch Environ Health* 1988, **43**(6):381–391.
22. Saltman PD, Strause LG: **The role of trace minerals in osteoporosis.** *J Am Coll Nutr* 1993, **12**(4):384–389.
23. Yamaguchi M, Oishi H, Suketa Y: **Stimulatory effect of zinc on bone formation in tissue culture.** *Biochem Pharm* 1987, **36**(22):4007–4012.
24. Suttle NF: **Osteoporosis in copper-depleted lambs.** *J Comp Pathol* 1972, **82**(1):93–96.

17

Effectiveness of biostimulation through nutrient content on the bioremediation of phenanthrene contaminated soil

Roshanak Rezaei Kalantary[1], Anoushiravan Mohseni-Bandpi[2], Ali Esrafili[1], Simin Nasseri[3], Fatemeh Rashid Ashmagh[1], Sahand Jorfi[4,5*] and Mahsa Ja'fari[6]

Abstract

Bioremediation has shown its applicability for removal of polycyclic aromatic hydrocarbons (PAHs) from soil and sediments. In the present study, the effect of biostimulation on phenanthrene removal from contaminated soil via adding macro and/or micronutrients and trace elements was investigated. For these purposes three macro nutrients (as N, P and K), eight micronutrients (as Mg, S, Fe, Cl, Zn, Mn, Cu and Na) and four trace elements (as B, Mo, Co and Ni) in 11 mineral salts (MS) as variables were used. Placket-Burman statistical design was used to evaluate significance of variables (MS) in two levels of high and low. A consortium of adapted microorganisms with PAHs was used for inoculation to the soil slurry which was spiked with phenanthrene in concentration of 500 mg/kg soil. The optimal reduction resulted when a high level of macro nutrient in the range of 67-87% and low level of micro nutrient in the range of 12-32% were used with the nitrogen as the dominant macronutrient. The Pareto chart showed that NH_4NO_3 was the most effective variable in this experiment. The effect of elements on phenanthrene biodegradation showed following sequence as N > K > P > Cl > Na > Mg. Effectiveness of the other elements in all runs was less than 1%. The type and concentration of nutrient can play an important role in biodegradation of phenanthrene. Biostimulation with suitable combination of nutrient can enhance bioremediation of PAHs contaminated soils.

Keywords: Polycyclic aromatic hydrocarbons, Phenanthrene, Bioremediation, Biostimulation, Macro/Micro nutrient

Background

Polycyclic aromatic hydrocarbons (PAHs) are chemical compounds made up of more than two fused aromatic rings in a linear or clustered arrangement, usually containing only carbon (C) and hydrogen (H) atoms, although nitrogen (N), sulphur (S) and oxygen (O) atoms may readily substitute in the benzene ring to form heterocyclic aromatic compounds. They are produced due to incomplete combustion of hydrocarbons and fossil fuels. Furthermore, natural occurrences also contribute in PAHs production. PAHs are highly hydrophobic which make them persistent and toxic to the environment and human [1,2]. Soil contamination to PAHs causes great health concerns because their persistency, toxicity, mutagenicity and carcinogenicity effects have been proved [3]. Different approaches like solvent extraction [4], phytoremediation [5], chemical remediation with various oxidants [6], photocatalytic remediation [7], electrochemical remediation [8], thermal destruction [9] and microbial degradation (bioremediation) [10,11] have been experimented for removal of PAHs from contaminated soils which bioremediation has been considered the most suitable for remediation of soils contaminated to PAHs [3,12]. Being environmental friendly and less ecologically damaging, less physical, chemical and biological changes in environmental conditions, less addition of chemicals, lower operational costs and proven efficiency are the main advantageous of bioremediation [13]. Both physical and chemical factors of reaction medium are effective on process efficiency. These factors include temperature, pH and accessibility of substrate to microorganisms, oxygen, nutrients, presence of electron acceptors and

* Correspondence: jorfi-s@ajums.ac.ir
[4]Environmental Technology Research Center, Ahvaz Jondishapour University of MedicalSciences, Ahvaz, Iran
[5]School of Public Health, Ahvaz Jondishapour University of MedicalSciences, Ahvaz, Iran
Full list of author information is available at the end of the article

addition of macro and micro elements. Addition of nutrients, macro and micro elements or oxygen to the polluted site to enhance the microbial degradation ability is called biostimulation. According to literature, two processes including biostimulation and bioaugmentation are usually considered to enhance the bioremediation of soils contaminated by hydrocarbons. Biostimulation increases the bacterial activity of various strains present in the contaminated soil through the addition of nutrients [14], humic compounds [15] or other chemicals which could affect on the bacterial condition.

The needs of bacteria and other microorganisms to nutrients are approximately similar to their cell composition. Three main categories of nutrients for microbial metabolism (macro and micro nutrient, and trace elements) were studied to determine the best nutritional composition in bioremediation of PAHs contaminated soils. However, carbon is usually needed in higher amounts and can be provided by target pollutants. Bailey and Ollis (1986) said that nitrogen and phosphorus as macronutrients are 14% and 3% of dry weight of a typical microbial cell, respectively [16] but Liebeg and Cutright (1999) reported in their research that phosphorus was the dominant macronutrient in bioremediation of PAH. However micronutrients such as sulfur, calcium and magnesium in microbial cell are only 1, 0.5, and 0.5%, respectively [17], but the concentration of these micronutrient and the others in mineral salt medium for bioremediation were very different. For investigating the effect of several macro, micro and trace nutrients the Plackett–Burman experimental design was used in optimization of liquid culture medium because of its potential in considering many variables.

Selection of the most efficient nutritional mixture can be investigated by the experimental or statistical approaches. Statistical experimental designs have some advantages which directed researchers to consider those in their bio studies such as their reliability, time saving (being rapid), cost saving and their reduction in the total number of experiments [18]. Different experimental design approaches are developed for such optimization of process conditions. Approaches like multi factorial designs are difficult because high number of variables should be screened. Also the orthogonal nature of Plackett–Burman gives pure effect of each variable [18]. Many studies are implemented according to Plackett–Burman experimental design. Chauhan et al. (2007) used Plackett–Burman statistical design for lactic acid production by *Lactobacillus sp. KCP01* using date juice [18]. Zhou et al. (2011) studied phenol degradation according to Plackett–Burman experimental design [19]. In the current study, the addition of different macro and/or micronutrients and trace elements in mineral salts medium (MSM) for phenanthrene removal from contaminated soil sample was experimented according to Plackett–Burman experimental.

Materials and methods

Chemicals

Acetone, methanol and acetonitril in HPLC grade were purchased from ROMIL Company. Phenanthrene (Purity > 98%), the salts for nutrient solutions were purchased from Fluka, Sigma Aldrich and Merck Company. Nutrient Broth and R_2A Agar were supplied by Difco and BIOMARK Company respectively.

Phenanthrene biodegradation investigation

Soil was collected from a depth of 5–20 cm of ground's surface and was passed through a 2-mm sieve. To get free of any organic matter it was washed with acetone several times and then distilled water was used for removing residual acetone. The soil was consisted of 83.1% sand, 11.9% silt and 5% clay. Total nitrogen and phosphorus were 0.025% and 0.0012%, respectively. The pH and electrical conductivity (EC) of the soil were 7.4 and 3.2 ds/m, respectively.

Two grams of dry soil was placed into a 50 mL Erlenmeyer flask as bioreactor. The bioreactors containing clean soil were autoclaved. A measured weight of phenanthrene was dissolved in acetone then it was used to spike the soil to have 500 mg phenanthrene/kg dry soil. For evaporation of the residual acetone the bioreactors were placed in a shaker (Heidolph, ProMax 2020 model) at the velocity of 180 rpm in room temperature and dark condition to have a uniform dispersion of phenanthrene.

The soil was inoculated with a culture of bacteria with an optical density of 1 at 630 nm [20] using CECIL UV/vis spectrophotometer (model 7100) in different concentrations of nutrients according to Table 1. The culture consisted of five types of bacteria; *Bacillus sporogenes, Bacillus licheniformis, Capnocytophaga ochracea* (presumably), *Acinetobacter sporogenes* and *Staphylococcus xylosus* which enriched with Phenanthrene in our previous study [21]. At the end, the soil liquid ratio was 10% w/v and all the samples and their similar blanks were put in the shaker at the velocity of 180 rpm in the room temperature (22 ± 3°C) with pH adjusted at 6.8 ± 0.2 for 8 weeks.

Experimental design

The liquid medium was processed for assessing the biodegradation of phenanthrene. Parametric optimization for biodegradation was studied with respect to three macro nutrients (as N, P and K), eight micronutrients (as Mg, S, Fe, Cl, Zn, Mn, Cu and Na) and four trace elements (as B, Mo, Co and Ni) in mineral salt mediums. For this purpose eleven mineral salts were used according to the literature [23].

Plackett-Burman design is an efficient method to identify the important factors among a large number of variables. In this study, a 12 runs Plackett-Burman design

Table 1 Twelve-trial Plackett–Burman design to study eleven factors in phenanthrene removal from soil: a comparison of experimented and predicted removal [22]

Run	A	B	C	D	E	F	G	H	I	J	K
	KH_2PO_4	K_2HPO_4	NH_4NO_3	$MgSO_4$	$FeCL_3$	NaCl	$ZnSO_4.H_2O$	$MnSO_4.H_2O$	$CuSO_4.5H_2O$	$FeSO_4.7H_2O$	Trace elements
1	+	-	+	-	-	-	+	+	+	-	+
2	+	+	-	+	-	-	-	+	+	+	-
3	-	+	+	-	+	-	-	-	+	+	+
4	+	-	+	+	-	+	-	-	-	+	+
5	+	+	-	+	+	-	+	-	-	-	+
6	+	+	+	-	+	+	-	+	-	-	-
7	-	+	+	+	-	+	+	-	+	-	-
8	-	-	+	+	+	-	+	+	-	+	-
9	-	-	-	+	+	+	-	+	+	-	+
10	+	-	-	-	+	+	+	-	+	+	-
11	-	+	-	-	-	+	+	+	-	+	+
12	-	-	-	-	-	-	-	-	-	-	-

was used to screen the important variables that significantly influenced phenanthrene degradation.

Each variable was applied at two levels of high (+) and low (−). The corresponding amount of variables and the level of them in 12 trials were shown in Tables 1, 2 and 3.

Determination of microbial population

The population of the inoculated culture was determined by the most probable number (MPN) method. The bacterial suspension was diluted tenfold serially in a sterile ringer solution (8.5 g NaCl per 1 L DW) and added to the sterile Nutrient Broth in the ratio of 10% of volume in triplicates in five series then they were incubated in 30°C. After 48 hours the turbidity of positive growth was seen in direct observation. The population of bacterial consortium was estimated according to statistical table of MPN [21].

Table 2 Variables showing medium components used in Plackett–Burman design

Factor	Variables	Maximum level g/L	Minimum level g/L
A	KH_2PO_4	3	0.5
B	K_2HPO_4	3	0.5
C	NH_4NO_3	6.1	0.4
D	$MgSO_4$	0.5	0.1
E	$FeCL_3$	0.2	0.01
F	NaCl	0.8	0.01
G	$ZnSO_4.H_2O$	0.00005	0.02
H	$MnSO_4.H_2O$	0.004	0.0002
I	$CuSO_4.5H_2O$	0.0004	0.00002
J	$FeSO_4.7H_2O$	0.001	0.1
K	Trace elements	1 mL	1 mL

The bioremediation efficiency in naturally contaminated soil

In order to investigate the optimized process efficiency, a soil sample naturally contaminated to PAHs was transferred to lab and the bioremediation efficiency was tried for, according to the most efficient expremental results and optimized conditions exactly like section 2–2. The preliminary investigations by GC-MS revealed that the soil was contaminated to phenenthrene, pyrene, anthracene, flourene and different aliphatic hydrocarbons. After 8 weeks the removal efficiency of PAHs was determined.

Extraction and analysis

The residual phenanthrene in the soil was extracted with methanol by ultrasonic homogenizer (Bandelin Sonoplus HD 2070) according to EPA 3550B (EPA) [24]. The extracted sample was then centrifuged (Hettich D7200) for 15 minutes at 6000 rpm , filtered through 2–3 cm of glass wool and then A portion of the filtered solution was used for analysis.

The extract was quantified by a HPLC from CECIL Company equipped with an Adept CE 4100 dual piston high pressure solvent delivery pump, a sample injection

Table 3 The trace elements of nutrient solutions

Salts for trace elements	Maximum level (+) g/L	Minimum level (−) g/L
H_3BO_3	13×10^{-3}	5×10^{-3}
Na_2MoO_4	1×10^{-5}	1.4×10^{-6}
$CoCl_2$	1×10^{-4}	1×10^{-4}
$NiCl_2$	2×10^{-4}	2×10^{-4}

valve having a 20 μL sample loop and an Adept CE 4201 UV-Visible variable wavelength detector with 8 μL × 10 mm flow cell set at the wavelength of 220 nm. Separations were carried out on a C18 column, and the mobile phase was a mixture of methanol/deionized (90:10, v/v). The flow rate was 1.0 mL/min and the retention time of phenanthrene was 10.0 min. The concentration of phenanthrene was determined after the calibration of the method with standard phenanthrene samples.

Results

The phenanthrene removal efficiency in different nutrient solutions

The results of average phenanthrene removal for different nutrient solutions are presented in Figure 1. The most removal efficiency of 85.7% was observed for run 1 with higher amount of nitrogen, phosphorus, Zn, Mn and trace elements in liquid medium folloewd by number 4, 3, 7 and 8 with removal values of 78.9%, 68.1%, 66.5% and 65.3% respectively.

Individual effect of factors on phenanthrene removal

The importance and effectiveness of each macro or micro nutrient and its positive or negative effect on removal efficiency is shown in Figure 2a (Pareto chart) and b (the main effect). For factors affecting the process positively, the nitrogen source showed the most importance followed by phosphorus sources; trace elements solution, Zn, $FeSO_4$, Mn and mg. Factors affecting the phenanthrene removal negatively included $FeCl_2$, NaCl and $CuSO_4$.

Analysis of ANOVA

The ANOVA table partitions the variability in removal into separate pieces for each of the effects (Table 4). It then tests the statistical significance of each effect by comparing the mean square against an estimate of the experimental error. In this case, 11 effects have P-values less than 0.05, indicating that they are significantly different from zero at the 95.0% confidence level.

The R-Squared statistic indicates that the model as fitted explains 99.9218% of the variability in removal. The adjusted R-squared statistic, which is more suitable for comparing models with different numbers of independent variables, is 99.8979%. The standard error of the estimate shows that the standard deviation of the residuals is 0.707696.

The phenathrene removal in optimized conditions for naturally contaminated soil

A soil sample naturally contaminated to different hydrocarbons was used to investigate to optimized process efficiency for PAHs removal. The GC-MS analysis on the soil sample is presented in Figure 3. According to analysis the picks of GC-MS, different hydrocarbons and four PAHs (to phenenthrene, pyrene, anthracene, flourene) were detected in the soil. The same inoculums like section 2–2 and optimized culture conditions were applied on the samples during 8 weeks. Results are presented in Table 5. The initial phenenthrene, pyrene, anthracene, flourene concentrations were 72, 61, 92 and 46 mg/kg, respectively. After 8 weeks, the phenenthrene, pyrene, anthracene, flourene concentrations were decreased to 31, 29, 26 and 27 mg/kg, respectively. The most removal efficiency of 71.7% was observed for antheracene, followed by 56.9% for phenanthrene, 52.4% for pyrene and 41.3 for flourene. The lower removal efficiency can be referred to the presence of other hydrocarbons and interfering factors of natural soil.

Discussion

Bioremediation is often limited by environmental, physical and chemical factors. One of the most important problems in the bioremediation of PAHs is nutrients

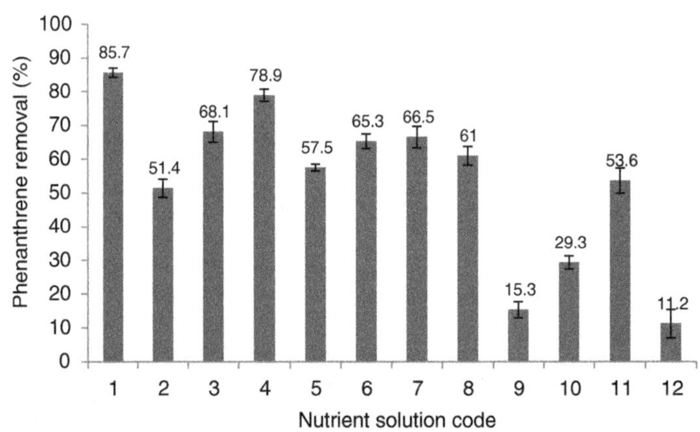

Figure 1 Removal efficiency of phenanthrene for various nutrient solutions.

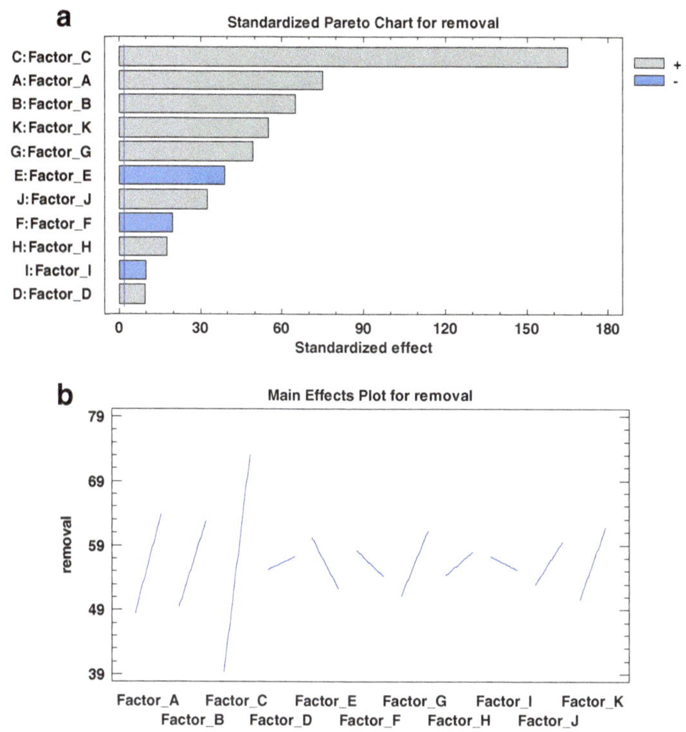

Figure 2 The individual effect of each factor on phenanthrene removal efficiency, a) Pareto chart and b) the main effect plot.

limitation in the soil and sediment. Biostimulation via addition of inorganic nutrients has been used as a strategy to enhance the biodegradation rate of PAH contaminated soils. Since the optimum values of macro and micro nutrient is highly dependent on the type of contaminant, microbial consortium and soil conditions, there isn't a unique and clear explanation in the literature about the type and concentrations of nutrients for bioremediation of soils contaminated by hydrocarbons. Therefore, the most effective macro and micro nutrients and their concentrations for each specific application should be determined separately to enhance the removal efficiency of hydrocarbons present in the soil. In the present study, the addition of three macro nutrients, eight micronutrients and four trace elements in two high and low levels (according to literature) was investigated to optimize the combination of these three categories of nutrients. Comparison with similar blanks showed that nutrient biostimulation through nutrients enhanced the phenanthrene removal from soil slurry and this was proved by the previous studies [25-27]. The highest of phenanthrene removal efficiency was observed in run 1 with the presence of high level of KH_2PO_4 and NH_4NO_3 in mineral salt medium. The second run with the best removal efficiency was run 4 with high level of the same macronutrient as run 1. In these two runs the micro nutrients and trace elements conditions were not in the same form showing the importance of factors C and A as the macronutrients in phenanthrene biodegradation. A relation between mineralization rates of phenanthrene and the initial concentrations of nitrogen and phosphorus

Table 4 Analysis of variance for phenanthrene removal

Source	Sum of squares	Df	Mean square	F-ratio	P-value
A:Factor_A	2836.69	1	2836.69	5663.94	0.0000
B:Factor_B	2120.02	1	2120.02	4232.99	0.0000
C:Factor_C	13608.1	1	13608.1	27170.85	0.0000
D:Factor_D	46.0208	1	46.0208	91.89	0.0000
E:Factor_E	756.841	1	756.841	1511.16	0.0000
F:Factor_F	196.021	1	196.021	391.39	0.0000
G:Factor_G	1230.19	1	1230.19	2456.28	0.0000
H:Factor_H	159.141	1	159.141	317.75	0.0000
I:Factor_I	50.8408	1	50.8408	101.51	0.0000
J:Factor_J	521.401	1	521.401	1041.07	0.0000
K:Factor_K	1516.5	1	1516.5	3027.96	0.0000
Total error	18.03	36	0.500833		
Total (corr.)	23059.8	47			

R-squared = 99.9218 percent; R-squared (adjusted for df) = 99.8979 percent; Standard Error of Est. = 0.707696.
Mean absolute error = 0.50625; Durbin-Watson statistic = 3.14504 (P = 0.9982).
Lag 1 residual autocorrelation = −0.573073.

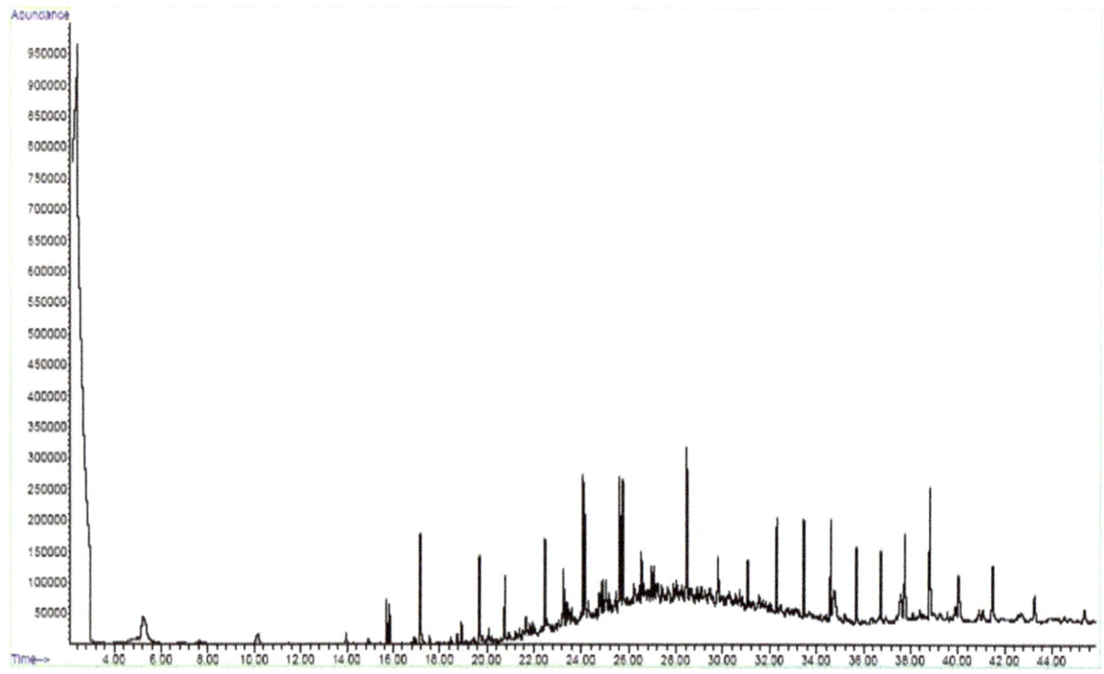

Figure 3 **The GC-MS analysis of naturally contaminated soil.**

as macronutrient has been reported before [28]. Da Silva et al. (2009) reported the improvement in bioremediation of coastal sand contaminated to crude oil by using commercial mineral NPK fertilizer [29]. Borresen and Rike (2007) showed that the concentration of phenanthrene in the soil amended with NP and biosolid was 1.7 and 2.9 times lower, respectively [30].

The relevant effects of eleven factors were sorted from the highest to the lowest in the Pareto chart presented in Figure 2a. The Pareto chart showed that A, B, C, D, G, H, J and K had positively affected phenanthrene degradation, whereas E, F and I had negative effects. All of these factors are in the right side of the t-value line in this chart showing the significant effect of them.

The positive effect of factors shows that the higher concentration of them has more efficiency in biodegradation. Among the eleven, eight factors had the positive effect on phenanthrene removal which confirms the biostimulation of phenanthrene biodegradation by nutrient addition, but Braddock et al. (1997) in their research for bioremediation of hydrocarbon-contaminated arctic soils showed the most biostimulation in the less concentration of nutrients [31]. It may be related to growth inhibition in N and P rich soils [32]. An initial inhibition of bacterial growth by magnitude of 2 Log was seen in the population of consortium in the runs with high concentration of N and P, but it didn't take too long. Besides in the other of our research with pure culture of bacteria, this reduction was about 3 Log. Lower inhibition of bacterial growth in the mix culture showed different responses for the diverse bacterial populations against to the environmental condition [32] and a need for optimum composition of macro, micro and trace nutrients. If the suitable ratio of nutrients does not supply,

Table 5 The PAHs removal efficiency in naturally contaminated soil

PAH	Initial concentration (mg/kg)	Final concentration (mg/kg)	Removal (%) in run 1	Removal (%) in run 3	Removal (%) in run 4
Phenanthrene	72	31	56.9 ± 2.3	41.2 ± 1.4	30.5 ± 1.6
Pyrene	61	29	52.4 ± 2.6	37.5 ± 3.6	33.6 ± 2.8
Anthracene	92	26	71.7 ± 1.9	68.7 ± 2.5	50.3 ± 3.4
Flourene	46	27	41.3 ± 3.3	25.2 ± 4.2	22.8 ± 2.7

the key degrader microorganisms may become ineffective which leading to less progress in biodegradation [33]. After a short time the population of bacterial consortium increased and the maximum density was observed in the runs with higher removal efficiency which may be resulted from a more need of degrader bacteria to the nutrients composition provided in these runs.

Also in the Pareto chart, C was the most effective factor followed by showing the high effect of nitrogen and phosphorus respectively. Olaniran et al. (2006) and Margesin and Schinner (1999) reported the more biotransformation was seen in using nitrogen and phosphorus as fertilizer [34,35]. The main effect plot (Figure 2b) confirms the most effect of macronutrients of C and A too. The slope of NH_4NO_3 effect shows that the response of phenanthrene removal was sensitive to this factor regarding to dominant effect of nitrogen in biodegradation. Nitrogen in the form of NH_4^+ or NO_3^- is readily assimilated in bacterial metabolism [17]. Ferandez-Luqueno et al. (2009) showed that degradation of poly acryl amide caused to release of nitrogen which leading to increment of the concentration of it and promotion in PAH removal [36]. But Liebeg and Cutright (1999) in their investigation of the effect of macro/micro nutrient, reported that phosphorus was the dominant nutrient in PAH bioremediation [17]. The composition of the best run in their report was consisted of 3% nitrogen, 11% phosphorus and 75% sulfur. In our experiment the runs with 3-12% phosphorus on a dry weight basis had the higher efficiency but phosphorus was not the dominant nutrient. The amount of nitrogen in the runs with more biodegradation of phenanthrene (runs: 1, 3, 4, 6 and 7) was in the range of 67-87%. The analysis of variance confirms that factor C (as NH_4NO_3) with the effect of 59% was the most effective nutrient in phenanthrene biodegradation. According to Cookson (1995) the concentration of 150 mg of nitrogen and 30 mg phosphorus has been required for degradation of one gram of a theoretical hydrocarbon into cellular material [37]. Betancur – Galvis et al. (2006) used nitrogen in the concentration of tenfold of phosphorus in biostimulation of PAH contaminated saline–alkaline soils [14]. Atagana et al. (2003) showed more removal of creosote in biostimulation of contaminated soil with lower amount of nitrogen but the more microbial growth was in the higher amount of it [38].

The negative effect of iron and copper (factors E and I) may be related to the lack of requirement of them by the dominant biodegrader bacteria in this experiment [17]. The need for micronutrient or trace element is very different in diverse microorganisms. In our study nutrient solution with composition of: 75% N, 10% P and 14% K was the best mineral salt medium for phenanthrene biodegradation.

Analysis of variance showed that after the factors A and B with the effect of 12.3 and 9.2%, factor C is the most effective factor among the cosidred variables. The total effect of macro, micro and trace nutrients were 80.5, 12.9 and 6.6%, respectively.

The slope of the parameters in the main effect plot showed that all of them had significant effect in the process and the analysis of ANOVAs indicating that they are significantly different from zero at the 95.0% confidence level too. The R-Squared statistic indicates that the model as fitted, explains 99.9218% of the variability in removal. The optimal settings of the experimental factors have been determined and are displayed in the summary in Table 6. Plackett-Burman design has great potential for screening of several variables by assessing the relative importance of these parameters.

Conclusion

Biostimulation of PAHs contaminated soils through nutrient addition enhance the biodegradation rate in the process. Our result on statistical screening of media components by Plackett–Burman design proved the advantages of selecting significant media components while phenantrene biodegradation with a bacterial consortium was investigated. The suitable conditions for phenanthrene removal were: as g/L 6.98 NH_4NO_3, 2.99 KH_2PO_4, 0.6868 K_2HPO_4, 0.1007 $MgSO_4$,0.0103 $FeCL_3$, 0.011556 NaCl, and as mg/ L 12.998 H_3BO_3, 3.96858 $MnSO_4.H_2O$,0.37718 $CuSO_4.5H_2O$, 0.2 $NiCl_2$, 0.1 $CoCl_2$, 0.0499576 $ZnSO_4.H_2O$, 0. 0305687 $FeSO_4.7H_2O$ and 0.999×10^{-2} Na_2MoO_4. Plackett–Burman design has good potential for preliminary optimization and more accurate quantitative analysis of the effect of great number of variables for phenanthrene biodegradation.

Table 6 Factor settings at optimum conditions determines by Plackett-Burman design

Factor		Setting	
Factor_A	KH_2PO_4	0.996965	2.99 (g/L)
Factor_B	K_2HPO_4	−0.850538	0.6868 (g/L)
Factor_C	NH_4NO_3	0.997797	6.98 (g/L)
Factor_D	$MgSO_4$	−0.995396	0.1007 (g/L)
Factor_E	$FeCL_3$	−0.996413	0.0103 (g/L)
Factor_F	NaCl	−0.99606	0.011556 (g/L)
Factor_G	$ZnSO_4.H_2O$	0.997174	0.0499576 (mg/L)
Factor_H	$MnSO_4.H_2O$	0.984212	3.96858 (mg/L)
Factor_I	$CuSO_4.5H_2O$	0.879905	0.37718 (mg/L)
Factor_J	$FeSO_4.7H_2O$	−0.958447	0. 0305687 (mg/L)
Factor_K	H_3BO_3	0.999567	12.998 (mg/L)
	Na_2MoO_4		0.999×10^{-2} (mg/L)
	$CoCl_2$		0.1 (mg/L)
	$NiCl_2$		0.2 (mg/L)

Competing interests
The authors declare that they have no competing interests.

Authors' contributions
SJ and RRK carried out the scientific design and proposed the idea of research, participated in analyzing the findings and drafted the paper. AMB and SN participated as scientific consulter for analyzing the findings, designed the microbial experiments and also participated in paper preparation. AE carried out the statistical design and analysis. MJ and FR carried out the implementation of experiments in laboratory and any other activity in field kike sampling and etc. All authors read and approved the final manuscript.

Acknowledgements
The authors would like to thank Tehran University of Medical Sciences and Tarbiat Modares University and Ahvaz jondishapour University of Medical Sciences for their kind support.

Author details
[1]Department of Environmental Health Engineering, School of Public Health, Iran University of Medical Sciences, Tehran, Iran. [2]Department of Environmental Health Engineering, School of Public Health, Shahid Beheshti University of Medical Sciences, Tehran, Iran. [3]Department of Environmental Health Engineering, School of Public Health and Center for Water Quality Research, Institute forEnvironmental Research, Tehran University of Medical Sciences, Tehran, Iran. [4]Environmental Technology Research Center, Ahvaz Jondishapour University of MedicalSciences, Ahvaz, Iran. [5]School of Public Health, Ahvaz Jondishapour University of MedicalSciences, Ahvaz, Iran. [6]Petroleum University of technology, Abadan faculty of petroleum engineering, Abadan, Iran.

References

1. Gan S, Lau E, Ng H: **Remediation of soil contaminated with PAHs.** *J Hazard Mater* 2009, **172:**532–549.
2. Jorfi S, Rezaee A, Moheb-ali G, Jaafarzadeh HN: **Pyrene removal from contaminated soils by modified Fenton oxidation using iron nano particles.** *J Environ Health Sci Eng* 2013, **11:**17–25.
3. Que C, Zhou H, Wong Y, Tam N: **Isolation of PAH-degrading bacteria from mangrove sediments and their biodegradation potential.** *Mar Pollut Bull* 2005, **51:**1054–1061.
4. Ni H, Zhou W, Zhu L: **Enhancing plant-microbe associated bioremediation of phenanthrene and pyrene contaminated soil by SDBS-Tween 80 mixed surfactants.** *J Environ Sci* 2014, **26:**1071–1079.
5. Huang XD, El-Alawi Y, Penrose DM, Glick BR, Greenberg BM: **A multi-process phytoremediation system for removal of polycyclic aromatic hydrocarbons from contaminated soils.** *Environ Pollut* 2004, **130:**465–476.
6. Ferrarese E, Andreottola G, Oprea L: **Remediation of PAH-contaminated sediments by chemical oxidation.** *J Hazard Mater* 2008, **152:**128–139.
7. Zhang L, Li P, Gong Z, Li X: **Photocatalytic degradation of polycyclic aromatic hydrocarbons on soil surfaces using TiO_2 under UV light.** *Hazard Mater* 2008, **158:**478–484.
8. Yeung A, Gu Y: **A review on techniques to enhance electrochemical remediation of contaminated soils.** *J Hazard Mater* 2011, **15:**11–29.
9. Pope C, Peters W, Howard J: **Thermodynamic driving forces for PAH isomerization and growth during thermal treatment of polluted soils.** *J Hazard Mater* 2000, **79:**189–208.
10. Berezina N, Yada B, Lefebvre R: **From organic pollutants to bioplastics: insights into the bioremediation of aromatic compounds by Cupriavidus necator.** *New Biotechnol* 2015, **32:**47–73.
11. Yu X, Wu S, Wu F, Wong M: **Enhanced dissipation of PAHs from soil using mycorrhizal ryegrass and PAH-degrading bacteria.** *J Hazard Mater* 2011, **186:**1206–1217.
12. Jorfi S, Rezaee A, Moheb-ali G, Jaafarzadeh N: **Application of biosurfactants produced by pseudomonas aeruginosa SP4 for bioremediation of soils contaminated by pyrene.** *Soil Sediment Contam* 2013, **22:**890–911.
13. Nikolopoulou M, Kalogerakis N: **Enhanced bioremediation of crude oil utilizing lipophilic fertilizers combinedwith biosurfactants and molasses.** *Mar Pollut Bull* 2008, **56:**1855–1861.
14. Betancur-Galvis LA, Alvarez-Bernal D, Ramos-Valdivia AC, Dendooven L: **Bioremediation of polycyclic aromatic hydrocarbon-contaminated saline–alkaline soils of the former Lake Texcoco.** *Chemosphere* 2006, **62:**1749–1760.
15. Nam K, Kim J: **Role of loosely bound humic substances and humin in the bioavailability of phenanthrene aged in soil.** *Environ Pollut* 2002, **118:**427–433.
16. Bailey JE, Ollis DF: *Biochemical Eng. fundamentals.* 2nd edition. New York: McGraw-Hill; 1986.
17. Liebeg E, Cutright T: **The investigation of enhanced bioremediation through the addition of macro and micro nutrients in a PAH contaminated soil.** *Int Biodeterior Biodegrad* 1999, **44:**55–64.
18. Chauhan K, Trivedi U, Patel K: **Statistical screening of medium components by Plackett–Burman design for lactic acid production by Lactobacillus sp. KCP01 using date juice.** *Bioresour Technol* 2007, **98:**98–103.
19. Zhou J, Yu X, Ding C, Wang Z, Zhou Q, Pao H, Cai W: **Optimization of phenol degradation by Candida tropicalisZ-04 using Plackett-Burman design and response surface methodology.** *J Environ Sci* 2011, **23:**22–30.
20. Nasseri S, RezaeiKalantary R, Nourieh N, Naddafi K, Mahvi A, Baradaran N: **Influence of bioaugmentation in biodegradation of PAHs-contaminated soil in bio-slurry phase reactor.** *Iran J Environ Health Sci Eng* 2010, **7:**199–208.
21. RezaeiKalantary R, Badkoubi A, Mohseni-Bandpi A, Esrafili A, Jorfi S, Dehghanifard E: **Modification of PAHs biodegradation with humic compounds.** *Soil Sediment Int J* 2013, **22:**185–198.
22. Khosravi-Darani K, Zoghi A: **Comparison of pretreatment strategies of sugarcane baggase: experimental design for citric acid production.** *Bioresour Technol* 2008, **99:**6986–6993.
23. Kim I, Park J, Kim K: **Enhanced biodegradation of polycyclic aromatic hydrocarbons using nonionic surfactants in soil slurry.** *Appl Geochem* 2001, **16:**1419–1428.
24. USEPA: **Polycyclic Aromatic Hydrocarbons (PAHs).** 2008 http://www.epa.gov/osw/hazard/wastemin/priority.htm.
25. Mills MA, Bonner JS, Page CA, Autenrieth RL: **Evaluation of bioremediation strategies of a controlled oil release in a wetland.** *Mar Pollut Bull* 2004, **49:**425–435.
26. Coulon F, Pelletir E, Gourhant L, Delille D: **Effects of nutrient and temperature on degradation of petroleum hydrocarbons in contaminated sub-Antarctic soil.** *Chemosphere* 2005, **58:**1439–1448.
27. Yu K, Wong A, Yau K, Wong Y, Tam N: **Natural attenuation, biostimulation and bioaugmentation on biodegradation of polycyclic aromatic hydrocarbons (PAHs) in mangrove sediments.** *Mar Pollut Bull* 2005, **51:**1071–1077.
28. Teng Y, Luo Y, Ping L, Zou D, Li Z, Christie P: **Effects of soil amendment with different carbon sources and other factors on the bioremediation of an aged PAH-contaminated soil.** *Biodegradation* 2010, **21:**167–178.
29. Da Silva AC, de Oliveira S, Bernardes DS, de França FP: **Bioremediation of marine sediments impacted by petroleum.** *Appl Biochem Biotechnol* 2009, **153:**58–66.
30. Borresen MH, Rike AG: **Effects of nutrient content, moisture content and salinity on mineralization of hexadecane in an Arctic soil.** *Cold Reg Sci Technol* 2007, **48:**129–138.
31. Braddock J, Ruth M, Catteral P, Walworth J, Mc Carthy K: **Enhancement and inhibition of microbial activity in hydrocarbon-contaminated arctic soils: implications for nutrient-amended bioremediation.** *Environ Sci Technol* 1997, **3:**2078–2084.
32. Ruberto L, Vazquez S, Mac Cromack W: **Effectiveness of the natural bacterial flora, biostimulation and bioaugmentation on the bioremediation of a hydrocarbon contaminated Antarctic soil.** *Int Biodeterior Biodegrad* 2003, **52:**115–125.
33. Smith V, Graham D, Cleland D: **Application of resource-ratio theory to hydrocarbon biodegradation.** *Environ Sci Technol* 1998, **32:**3386–3395.
34. Olaniran A, Pillay D, Pillay B: **Biostimulation and bioaugmentation enhances aerobic biodegradation of dichloroethenes.** *Chemosphere* 2006, **63:**600–608.
35. Margesin R, Schinner F: **Biological decontamination of oil spills in cold environments.** *J Chem Technol Biotechnol* 1999, **74:**381–389.
36. Fernandez-Luqueno F, Thalasso F, Luna-Guido M, Ceballos-Ramirez J, Ordoñez-Ruiz I, Dendooven L: **Flocculant in wastewater affects dynamics**

of inorganic N and accelerates removal of phenanthrene and anthracene in soil. *J Environ Manage* 2009, **90**:2813–2818.

37. Cookson J: *Bioremediation engineering: design and application*. New York: McGraw-Hill; 1995.
38. Atagana H, Haynes R, Wallis F: Optimization of soil physical and chemical conditions for the bioremediation of creosote-contaminated soil. *Biodegradation* 2003, **14**:297–307.

18

Decolorization and detoxification of sulfonated toxic diazo dye C.I. Direct Red 81 by *Enterococcus faecalis* YZ 66

Madhuri M Sahasrabudhe[1*], Rijuta G Saratale[2], Ganesh D Saratale[3,4*] and Girish R Pathade[5]

Abstract

Isolated *Enterococcus faecalis* YZ 66 strain shows ability to decolorize various industrial dyes among which, it showed complete decolorization and degradation of toxic, sulfonated recalcitrant diazo dye Direct Red 81 (50 mg/L) within 1.5 h of incubation under static anoxic condition. The optimum pH and temperature for decolorization was 7.0 and 40°C, respectively. Significant induction in the activity of intracellular oxidoreductive enzymes suggested its involvement in the decolorization of Direct Red 81. The biodegradation of Direct Red 81 was monitored by UV-Visible, FT-IR spectroscopy and HPLC. The final products were characterized by GC-MS and possible pathway of the degradation of the dye was proposed. The phytotoxicity assay (with respect to plants *Sorghum vulgare* and *Phaseolus mungo*) revealed that the degradation of Direct Red 81 produced nontoxic metabolites. Finally *E. faecalis* was employed to decolorize actual industrial effluent showing decolorization (in terms of ADMI value) with moderate COD and BOD reduction. Moreover the result increases the applicability of the strain for the treatment of industrial wastewaters containing dye pollutants.

Keywords: Direct Red 81, *Enterococcus faecalis*, ABTS, Lignin peroxidase, Azoreductase, ADMI, GC-MS

Background

Azo dyes are xenobiotic compounds characterized by the presence of one or more azo linkages and aromatic rings [1]. They are the largest class of dyes with the greatest variety of colour. At least 3000 different varieties of azo dyes are extensively used in textile, paper, food, cosmetics and pharmaceutical industries [2]. Among various applications of synthetic dyes about 30,000 tons of different dyestuffs are produced per year worldwide [3]. Among these synthetic dyes, azo dyes are the most widely used which account for over 60% of the total number of dyes known to be manufactured [4,5]. Some investigators reported that azo dyes and their metabolites are toxic, carcinogenic and mutagenic in nature which leads to the formation of tumors and allergies besides growth inhibition of bacteria, protozoan, algae, plants and different animals [6,7].

In India textile industry is one of the greatest generators of liquid effluent pollutants which are often contaminated with harmful or poisonous substances. An estimate shows that textiles account for 14% (about 2200 dyeing industries) of India's industrial production and around 27% of its export earnings [6]. However in India particularly for small scale textile industries, where working conditions and economic status do not allow them to treat their wastewater before disposal and they have no choice other than dumping all effluent into the main stream of water resources. Thus dyes released from the textile processing and dye stuff manufacturing industries results in increase in organic load of natural reservoirs. Pollution caused by dye effluents is mainly due to durability of dyes in wastewater, colour fastness, stability and resistance of dyes to degradation [8]. During industrial processing up to 40% of the used dyestuff are released into the process water producing highly coloured wastewater that affect aesthetics, water transparency and gas solubility in water bodies [9,10].

Several physical and chemical methods have been suggested for the treatment of dye contaminated wastewater

* Correspondence: sahasrabudhemadhuri@gmail.com; gdsaratale28@gmail.com
[1]Department of Microbiology, Maulana Azad College, Aurangabad, MS, India
[3]Department of Environmental Biotechnology, Shivaji University, Kolhapur, MS, India
Full list of author information is available at the end of the article

but not widely used because of high cost, secondary pollution that can be generated by excessive use of chemicals [11]. In contrast, microbial degradation of dyes does not have similar problems so it is necessary to establish biological wastewater treatment of azo dyes [12]. Currently microbial biodegradation became a promising approach of dye treatment because of its cheaper, effective and more ecofriendly in nature [2]. Varieties of microorganisms including bacteria, fungi, yeasts, actinomycetes, algae and plants are capable of removing dyes from dye effluent, and become an inexpensive and promising tool for the removal of various dyes from textile dye effluents [6]. Among which bacterial cells represent an inexpensive and promising tool for the removal of various azo dyes from textile dye effluents. Recently a substantial amount of research has been carried out using single bacterial cultures like; *Lysinibacillus* sp. RGS; *Pseudomonas luteola, Bacillus fusiformis* KMK5, *Micrococcus glutamicus* NCIM-2168, and *Aeromonas hydrophila* shows very promising results for the azo dye decolorization [4,5,11,13].

Enterococcus faecalis is a nonmotile, facultatively anaerobic coccus and can survive very harsh environments. *Enterococcus* sp. was found to be catabolically versatile with the ability to utilize a wide range of unusual substrates such as; chlorpyrifos, pentaerythritol tetranitrate, 2,4,6-trinitrotoluene and phosphonate [14]. Biological neutralization of an alkaline effluent by an alkaliphile, *Enterococcus* faecium strain R-5 was reported earlier [15]. Recently Pingitore et al. (2012), reported the importance of *Enterococcus* sp. in dairy industries increases the biotechnological value of this strain [16]. A bacterial consortium, NAR-2 consisting of *Citrobacter freundii* A1, *Enterococcus casseliflavus* C1 and *Enterobacter* cloacae L17 was investigated for biodegradation of Amaranth azo dye (100 mg/l) in 30 min under sequential microaerophilic–aerobic condition [17]. Recently expression and characterization of an aerobic FMN-dependent azoreductase from *Enterococcus faecalis* was also reported [18].

The mechanism of microbial degradation of azo dyes involves the reductive cleavage of azo bonds (–N = N–) with the help of azoreductase under anaerobic conditions resulted into the formation of colorless solutions [11]. For the reduction of azo dyes, reduction to the anion radical occurs by a fast one-electron transfer reaction, followed by a second, slower electron transfer event to produce the stable dianion [19]. Thus the functional group of azo dye with higher electronic density might be unfavorable to this second electron transfer to form the dianion and leads to low or no decolorization [12]. Due to this reason sulfonated reactive group of azo dyes are normally considered to be more recalcitrant than carboxylated azo dyes. Some investigators reported that the rate limiting step during bacterial decolorization of sulfonated azo dyes is the permeation through the bacterial cell membrane [4,20].

In this study, we have used isolated *Enterococcus faecalis* YZ 66 to decolorize Direct Red 81 (DR 81). The isolated strain could decolorize DR 81 completely up to 500 mg/L. We have optimized various physicochemical parameters and studied enzymatic status during decolorization. Supplementation of carbon and nitrogen source on decolorization performance, identification of metabolites formed after decolorization using analytical techniques and toxicity study of metabolites formed after decolorization were systematically investigated.

Material and methods

Microorganism and culture conditions

Enterococcus faecalis YZ 66 was isolated from dye industry effluent contaminated soils [21] this strain was acclimatized with dye waste obtained from Spectrum Dyes and Chemicals Industry, Surat, India, containing DR81 in higher proportion by a method of Peppler (1979) [22]. The method included two steps: treatability test was carried out by enrichment culture technique and toxicity test in which toxicity was assessed in a series of parallel flasks with geometrically increasing concentration of DR81. Microbial growth was measured daily by turbidity. The concentration at which the toxic components inhibited growth was noted and was used as a warning of an upper concentration limit. Pure culture was maintained on the nutrient agar slants. Composition of nutrient broth and agar used for decolorization is (g/L) peptic digest of animal tissue 5, NaCl 5, beef extract 1.5, yeast extract 1.5 and pH 7.4 ± 0.2.

Dyestuff and chemicals

The dye Direct Red 81 (DR 81) and actual dye effluent were obtained from Spectrum Dyes and Chemicals Industry, Surat, India. 2′-2′Azinobis-(3 ethylbenthiazoline-6 sulphonate) (ABTS) was purchased from Sigma Aldrich, USA. Nutrient broth dehydrated was purchased from Hi-Media, Mumbai, India. Tartaric acid, n-propanol was purchased from Qualigenes, India. All chemicals used were of the highest purity and of analytical grade.

Decolorization studies

Enterococcus faecalis YZ 66 was grown for 24 h at 37°C on nutrient agar [21]. 10% inoculum ($O.D_{600}$ 1.0) was used throughout the study [8,23]. The isolated strain was inoculated in nutrient broth to study the decolorizing ability of the culture. The dye was filter sterilized by using 0.2 μm filter (Sartorius Biolab, Germany) and added after sterilization of medium throughout the study. The dye (50 mg/L) was added immediately and incubated under static condition at 37°C. Aliquot (3 mL) of culture media was withdrawn at different time intervals and centrifuged at 6000×g for 20 min. Decolorization was monitored by measuring the absorbance of the culture at λ_{max} of the dye i.e. 511 nm and change in pH was also recorded. Sterile nutrient broth of different pH 3, 4, 5, 6, 7 and 8 was inoculated with 10% inoculum and incubated at 37°C under static condition. The dye concentration was 50 mg/L. For temperature studies sterile nutrient broth of pH 7.0 was inoculated with 10% inoculum and filter sterilized dye at 50 mg/L was added aseptically. The broth was incubated at 25°C, 30°C, 37°C, 40°C, 45°C and 50°C. All decolorization experiments were performed in triplicates. Abiotic control (without microorganism) was always included in each study.

In order to examine the effect of initial dye concentration on decolorization 50–700 mg/L of DR81 was added to the sterile nutrient broth inoculated with 10% inoculum of *Enterococcus faecalis* YZ 66 ($O.D_{600}$ 1.0) and incubated at 37°C under static condition. The % decolorization was measured. All decolorization experiments were performed in triplicates. In each study abiotic control (without culture) was always included. The % decolorization and average decolorization rate was measured [5] as follows:

$$\%\ \text{Decolorization} = \frac{\text{Initial absorbance} - \text{Observed absorbance}}{\text{Initial absorbance}} \times 100\%$$

Effect of supplementation of carbon and nitrogen sources on decolorization

To study the effect of carbon and nitrogen sources on decolorization of DR81, semi synthetic medium [8] having following composition was used (g/L); $(NH_4)_2SO_4$;0.28, NH_4Cl; 0.23, KH_2PO_4; 0.067, $MgSO_4.7H_2O$; 0.04, $CaCl_2.2H_2O$; 0.022, $FeCl_3.6H_2O$; 0.005, yeast extract; 0.2, NaCl; 0.15, $NaHCO_3$; 1.0 and 1 ml/L of trace element solution containing (g/L) $ZnSO_4.7H_2O$; 0.01, $MnCl_2.4H_2O$; 0.1, $CuSO_4.5H_2O$; 0.392, $COCl_2.6H_2O$; 0.248, $NaB_4O_7.7H_2O$; 0.177 and $NiCl_2.6H_2O$; 0.02. It was further incorporated with different carbon and nitrogen sources (1% each) such as glucose, sucrose, lactose and starch, yeast extract, peptone, malt extract, meat extract and urea respectively. Filter sterilized dye 50 mg/L of the DR81 was added after inoculation of *Enterococcus faecalis* YZ 66 in sterilized media.

Repeated dye decolorization in fed batch process

Decolourization medium containing 50 mg/L DR 81 was inoculated with 24 h grown cells of *Enterococcus faecalis* YZ 66. The resulting solution was then statically incubated at 37°C for the decolorization. After complete colour removal, the cells were collected, rinsed twice with sterile deionized water and transferred into a fresh decolorization medium for the second decolorization batch experiment. The same procedures were repeated seven times. All steps were done under aseptic conditions. For comparison, the repeated batch experiments were also conducted using free cells under identical experimental procedures.

Preparation of the cell free extract

Enterococcus faecalis YZ 66 was grown in nutrient broth at 37°C for 24 h and centrifuged at 10,000 rpm for 20 minutes. The cell pellet was suspended in the potassium phosphate buffer (50 mM, pH 7.4) keeping sonifier output at 50 amp and giving 7 strokes each of 30 seconds with a 2 min interval at 4°C. The homogenate was centrifuged and supernatant was used as a source of enzymes. A similar procedure was followed to the cells of *Enterococcus faecalis* YZ 66 obtained after complete decolorization.

Oxidative and reductive enzyme assays

The activities of laccase and lignin peroxidase were assayed spectrophotometrically in the cell free extract. Laccase activity was determined in a 2 mL mixture containing ABTS (10%) in 0.1 M acetate buffer pH 4.9 and measured as an increase in optical density at 420 nm [21]. Lignin peroxidase (LiP) activity was determined by monitoring the propanaldehyde formed at 300 nm in a reaction mixture of 2.5 mL containing 100 mM n-propanol, 250 mM tartaric acid and 10 mM H_2O_2 [13]. All enzyme assays were carried out at 37°C with reference blanks that contained all components except the enzyme to be assayed. All enzyme assays were conducted in triplicates and the average rates were calculated to represent the enzyme activity. One unit of enzyme activity was defined as a change in absorbance U/mL/min of the enzyme. NADH-DCIP reductase and azoreductase activity was carried out as per the method reported by Saratale (2013) [13].

Decolorization of dye industry effluent

For the dye wastewater study, anaerobically digested effluent was used for further aerobic treatment by using the selected isolate in pure culture and the effluent was checked for COD and BOD [24]. For color removal efficiencies of dye wastewater the true color level independent of hue was measured using the American Dye Manufacturers' Institute (ADMI 3WL) tristimulus filter method. This method is applicable to colored waters and wastewaters having color characteristic. The decolorization of actual dye wastewater by *Enterococcus faecalis* was determined by measuring ADMI from the aqueous solutions. ADMI removal percent (%) is the ratio between the removal ADMI value at any contact time and the ADMI value at initial concentration was calculated [5]. To understand the degree of biodegradation (mineralization) of dye wastewater reduction in chemical oxygen demand (COD) and biological oxygen demand (BOD) of the culture before and after incubation with *Enterococcus faecalis* YZ 66 was measured [24]. The nutrient medium was used as blank and similar condition was used for test.

Analytical methods

The metabolites produced during the biodegradation of DR81 at 1.5 h i.e. after decolorization of the medium were extracted twice with equal volume of dichloromethane (DCM). The DCM extracts were pooled and evaporated at 40°C in a rotary evaporator and then transferred to a test tube [25]. The extracted residue was dissolved in a small volume of HPLC grade methanol and used for analysis. During UV visible spectral analysis, changes in absorption spectrum in the decolorized medium (400-800 nm) was recorded in comparison with the spectra of the undegraded dye [5]. HPLC analysis was performed in an isocratic system (Shimadzu SCL 10 AVP) equipped with dual absorbance detector using C-18 column with HPLC grade methanol as mobile phase at the flow rate of 1.0 mL/min for 10 min at 511 nm. The mobile phase used for TLC was composed of methanol: ethyl acetate: n-propanol: water: acetic acid (1:2:3:1:0.2 v/v) and the separation was done on precoated silica gel plates 'Merck'. TLC plate was developed using iodine chamber [8]. Metabolites formed after decolorization of DR81 were characterized by using Fourier Transform Infrared Spectroscopy (Perkin Elmer 1000) (FTIR) Analysis was done in the mid IR region of 400-4000/Cm with 16 scan speed, the pellets prepared using spectrophotometric pure KBr (5:95) were fixed in sample holder and analysis was carried out. Extracted metabolites were subjected to FTIR. The metabolites formed after decolorization was identified by using Gas Chromatography-Mass Spectroscopy (Shimadzu GC-MS QP2010). The ionization voltage was 70 eV. Gas chromatography was conducted in temperature programming mode with a Resteck column (0.25 mmX 30 mm). The initial column temperature was 40°C for 4 min, which was increased linearly at 10°C /min up to 270°C and held at 4 min. The temperature of injection port was 275°C and GC-MS interface was maintained at 300°C. The helium was carrier gas; flow rate 1 mL/min and 30 min run time.

Toxicity studies

Phytotoxicity tests were carried out in order to assess the toxicity of DR81 and metabolites formed after decolorization. Phytotoxicity tests were carried out at a final concentration of 400 ppm on two kinds of seeds. One from grains i.e. *Sorghum vulgare* (monocot) and second from pulses i.e. *Phaseolus mungo* (dicot), commonly cultivated in India. Phytotoxicity was conducted at room temperature (10 seeds of each) by watering separately 5 ml sample of control DR81 and its degradation products per day. Control set was irrigated using distilled water at the same time. Germination % as well as the length of plumule and radical was recorded after 7 days of incubation [8].

Statistical analysis

Data was analyzed by one way analysis of variance (ANOVA) with Tukey- Kramer multiple comparison test. Readings were considered significant when P was ≤0.05.

Results and discussion

Decolorization experiment

The isolated *Enterococcus faecalis* YZ 66 was able to decolorize DR 81 within 1.5 h at a dye concentration of 50 mg/L. UV visible scan of the culture supernatant withdrawn at different time intervals indicated the decolorization and decrease in dye concentration from batch culture. Peak obtained at 511 nm disappeared after complete decolorization. The absorbance peak in the visible region disappeared indicating complete decolorization. In the UV spectra, the peak at 511 nm was replaced by new peak at 240 nm (Additional file 1: Figure S1). The absorbance peaks in the visible region disappeared indicating complete decolorization [26]. Decolorization with respect to time showed complete decolorization of the dye in 1.5 hours. There was proportionate increase in wet weight indicating growth of *E. faecalis* in the presence of dye (Additional file 1: Figure S2). There was no abiotic loss of DR 81 within 24 h incubation indicating that the decolorization of DR 81 was due to biological mechanism rather than adsorption. To confirm whether this decolorization is due to the variation in pH, change in pH was recorded which is in the range of 7.0 ± 0.2.

Effect of physicochemical conditions on the decolorization performance

The effect of various physiochemical conditions such as pH, temperature, dye concentration, carbon and nitrogen sources on decolorization of DR 81 by *E. faecalis* was studied in detail. All parameters were studied at 37°C under static condition. 10% inoculum A_{600} 1.0 was used at a dye concentration 50 mg/L.

Effect of pH

It was observed that pH of the media, affects the colour of the solution and the solubility of the dye and the enzymatic activity related to decolorization is also dependent on the pH. Generally bacterial cultures exhibit maximum decolorization at pH values near 7 or slightly alkaline pH values and the rate of colour removal tends to decrease rapidly at strongly acidic or slightly alkaline pH values [12,27]. *E. faecalis* showed complete decolorization of DR 81 at pH 7.0 within 1.5 h. It showed decolorization in the pH range of 5–8 while at pH 3 and 4 (about 40%) and at pH 9 and 10 about (about 30%) decolorization was observed after 24 hour of incubation (Additional file 1: Figure S3). Similar results were observed in *Micrococcus* sp. in the decolorization of 300 mg/L of Orange MR [28].

Effect of temperature

Pearce et al., (2003) [12] reported that the rate of colour removal increases with increasing temperature within a defined range that depends upon the system. The temperature required to produce the maximum rate of colour removal tends to correspond with the optimum cell culture growth temperature which is in the range of 35-45°C. Temperature affects microbial growth, enzymes production and consequently, the percentage of decolouration. It was reported by Mathew and Madamwar, 2004 [23] that various microorganisms showed their survival at various temperatures ranging from 25-50°C. The decline in colour removal activity at higher temperature can be attributed to the loss of cell viability or to the denaturation of azoreductase enzyme. However, it has been shown that with certain whole bacterial cell preparation, azo reductase enzyme is relatively thermostable and can remain active up to temperature of 60°C over short period of time [29]. *E. faecalis* YZ 66 decolorized the dye under study in the range of 96-99% within a temperature of 30-40°C. At 30°C, 99.35% decolorization was observed while at 40°C, 98.54% decolorization was seen, thus showing negligible difference in percent decolorization at both the temperatures (Additional file 1: Figure S3). At 45°C and 50°C, 17.85 and 14.81% decolorization was observed, respectively (Additional file 1: Figure S3). Similar results was observed in *Pseudomonas aeruginosa* degrades 97% of Remazol Red (50 mg/L) at 40°C, 72% at 10°C and 82% at 30°C, respectively [30].

Effect of initial dye concentration

Decolorization of different initial concentrations of the dye from 50–700 mg/L was studied under static anoxic condition. Rate of decolorization of dye increased with increase in concentration of the dye up to 300 mg/L but the time required for decolorization was more. The *E. faecalis* showed faster decolorizing ability up to 300 mg/L after which the rate of decolorization falls decreasing (Figure 1). Fifty four hours are required to decolourize 85.74% of the dye at 500 mg/L concentration. The activity was lower at dye concentration 600 mg/L and above which decolorization was strongly inhibited at dye concentration at 700 mg/L (Figure 1). It has been proposed that dye concentration can influence the efficiency of microbial decolorization through combination of factors imposed by dye at high dye concentration [31]. Similar results were observed in *Lysinibacillus* sp. (for Metanil Yellow) [32], *Sphingomonas paucimobilis* (for Methyl Red) [33], and in *Lysinibacillus* sp. RGS (for Remazol Red) [13].

Effect of supplementation of carbon and nitrogen sources on the decolorization performance

Dyes are deficient in carbon and thus biodegradation without supplying extra carbon or nitrogen source is very difficult [34]. Carbon and nitrogen sources have an important influence on the extent of decolouration using microorganisms. In order to enhance the decolorization

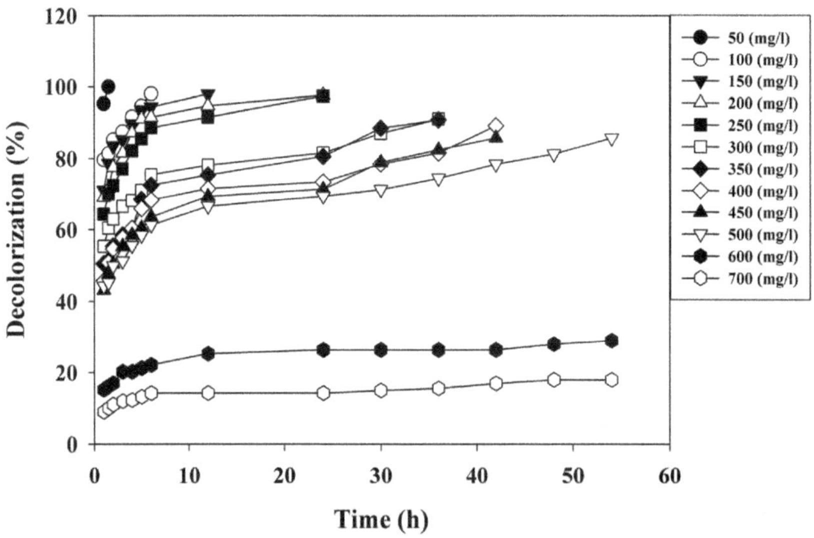

Figure 1 Effect of initial dye concentration of C.I. Direct Red 81on decolorization performance by using *Enterococcus faecalis* YZ 66.

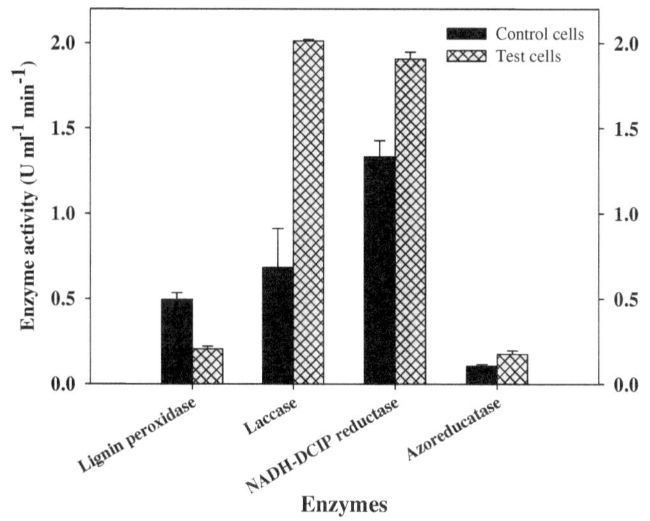

Figure 2 Oxidative (lignin peroxidise and laccase) reductive (azoreductase and NADH-DCIP reductase) enzyme activity profile in control cells of *Enterococcus faecalis* YZ 66 at (0 h) and the induced cells obtained after complete decolorization of C.I. Direct Red 81 (1.5 h).

performance of the DR-81, an extra carbon and nitrogen source was supplied in semi synthetic medium. There was no decolorization observed in semi synthetic medium. In the presence of lactose 98.12% decolourization was observed followed by 96.16, 95.25, 95.61 and 93.76% in the presence of meat extract, peptone, glucose and starch, respectively while less decolourization with other supplements of carbon and nitrogen source within 24 h of incubation (Additional file 1: Figure S4). In addition, supplying urea as a nitrogen source did not enhance decolorizing ability. Different microbial metabolic characteristics lead to differences in the uptake of sources, thus affecting azo dye decolorization. Addition of carbon source found to be less effective to promote the decolorization performance probably due to the preference of the cells in assimilating the added carbon sources over using the dye compound as the carbon source [6,35]. Nitrogen sources are found important for microbial decolouration since it was observed that this source is essential for the regeneration of NADH [6,35].

Decolorization with repeated addition of dye aliquots

An ability of *E. faecalis* YZ 66 to decolorize repeated addition of DR 81 dye aliquot (50 mg/L) was studied under static condition. The isolate have an ability to decolorize 100% dye up to seventh aliquot and after that subsequent cycle showed no decolorization (Additional file 1: Figure S5). The eventual cessation of decolorization was likely due to nutrient depletion [5]. Thus *E. faecalis* YZ 66 showed the ability to decolorize repeated addition of the dye aliquots is noteworthy for its commercial application.

Enzymes involved in dye decolorization

The use of microbial techniques to deal with pollution is a key research area in the environmental sciences. In these processes microbes acclimatize themselves to the toxic wastes and resistant strains develop naturally,

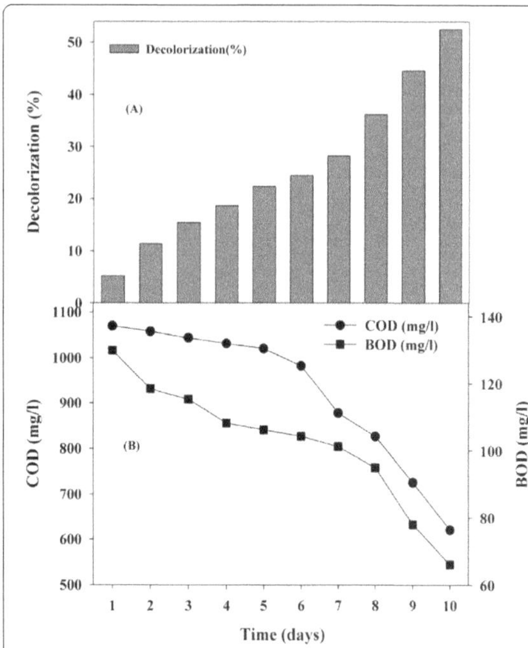

Figure 3 Color removal of (A) dye industrial effluent in terms of ADMI removal ratio values (about 52%); (B) COD and BOD reduction after 10 days of incubation time by *Enterococcus faecalis* YZ 66.

which then transform various toxic chemicals into less harmful forms. The mechanism behind the biodegradation of recalcitrant compounds (azo dyes) in the microbial system is based on the action of the biotransformation enzymes [36]. Besides uptake, the presence and activity of a network of detoxification enzymes is crucial for the metabolism and eventually the degradation of chemicals. To understand the decolorization mechanism, enzyme activities of laccase, lignin peroxidase, NADH-DCIP reductase, and azo reductase were monitored over time. The enhanced activities of enzymes were noted in induced cells (after decolorization) (Figure 2). The enzymatic profile presumably indicates communal action of oxidoreductive enzymes for the degradation of DR 81 into simple metabolites by *E. faecalis* (Figure 2). No enzyme activities were observed in cell free supernatant. The role of oxidoreductive enzymes in the decolorization of azo dyes have been characterized in various bacteria are well documented in recent reviews [6,37].

Decolorization studies of dye wastewater by *E. faecalis*

Most of the microbial decolorization studies in several laboratories showed the ability of bacteria, fungi, and algae in removing the color of textile dyes, but they do not find much application in treatment system for industrial effluent because of heterogeneity of the components in effluent depending upon production schedule. However, it is very important to test decolorization in real textile effluents, which are complex systems having strong colors, large amounts of suspended solids, broadly fluctuating pHs, high temperatures, high COD and high salt concentrations that can be inhibitory to microorganisms [13]. Considering this perspective we have checked the efficiency of *E. faecalis* to decolorize actual textile

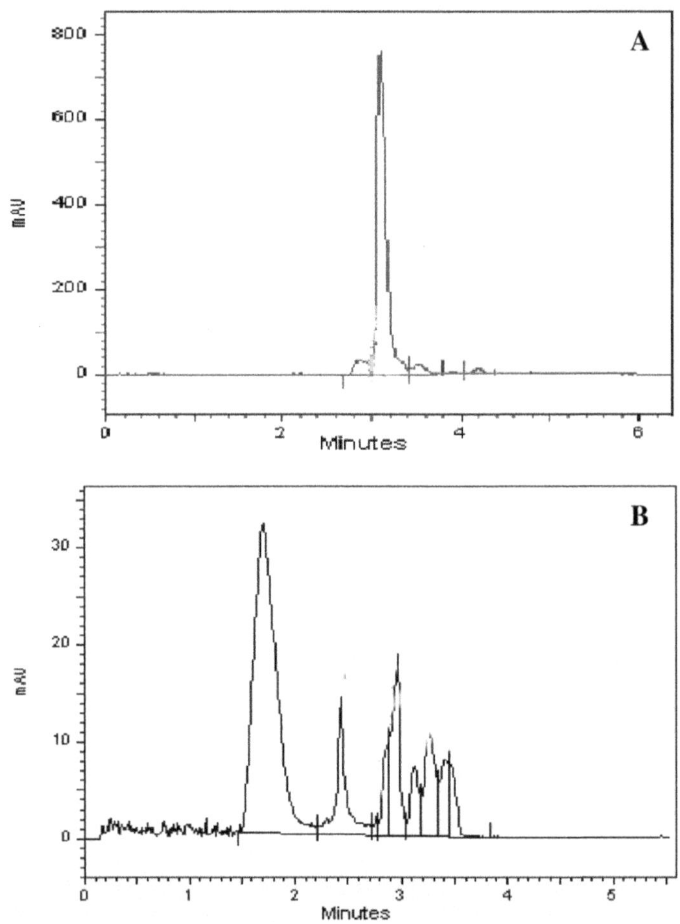

Figure 4 HPLC analysis of products (extracted with ethyl acetate) formed by degradation of C.I. Direct Red 81: (A) at 0 h (control with peak at 1.71. min), (B) metabolites formed by *Enterococcus faecalis* YZ 66 after complete decolorization (1.5 h with peak at 3.008, 3.861 & 4.021 min) indicating degradation of DR 81 into different metabolites.

wastewater. The true color of textile wastewater measured by using ADMI 3WL suggesting that *E. faecalis* could achieve higher color removal value (52%) with moderate reduction in COD (about 42%) and BOD (about 48%) after 10 days of incubation (Figure 3). Decolorization performance of dye wastewater by *E. faecalis* is comparable with

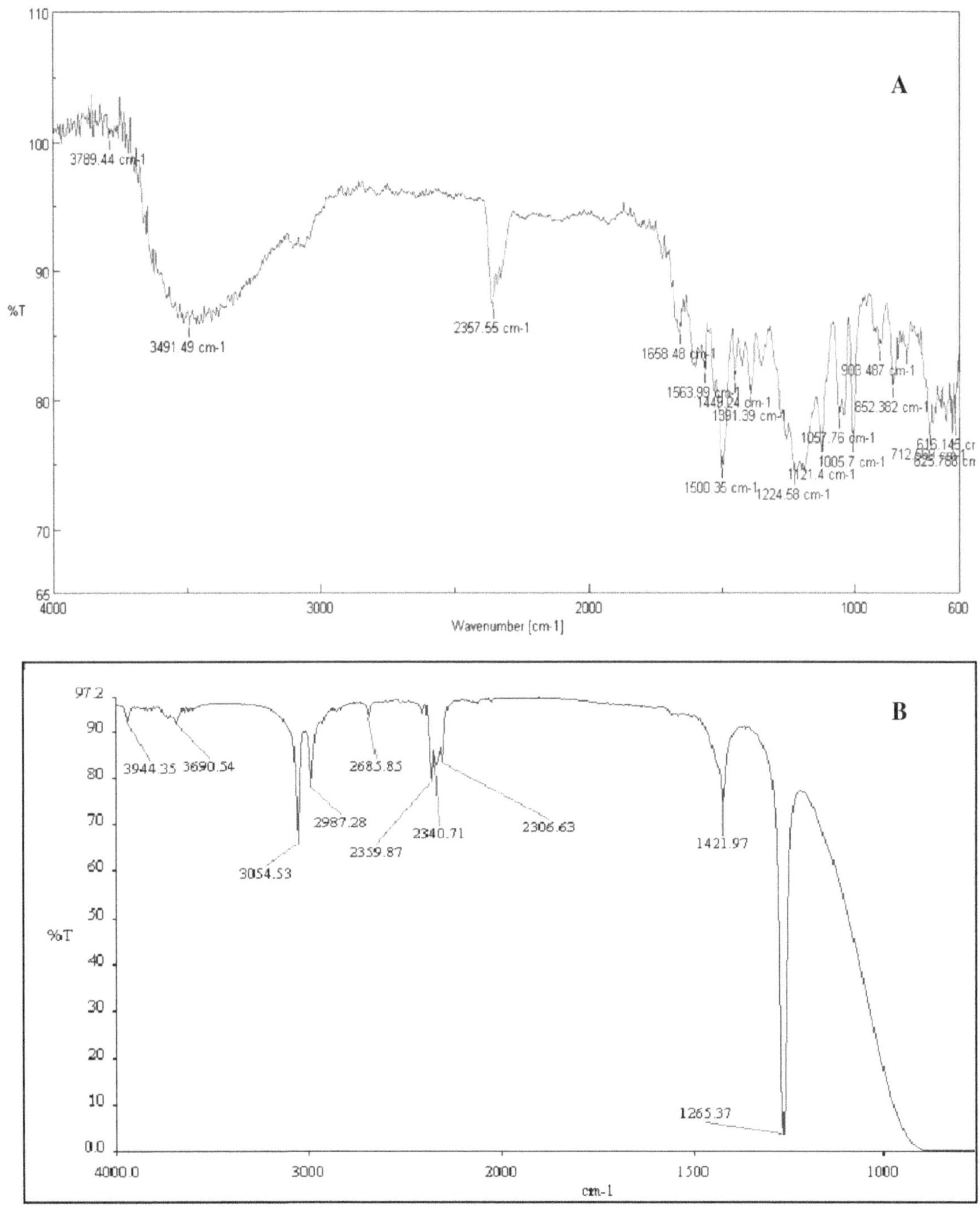

Figure 5 FTIR analysis of products (extracted with diicholoromethane) formed by degradation of C.I. Direct Red 81: (A) at 0 h (control), (B) metabolites formed by *Enterococcus faecalis* YZ 66 after complete decolorization (1.5 h).

Citrobacter sp. strain KCTC 18061P strain removed 70% of effluent color within 5 days with 35% COD reduction [38]. Untreated dye effluents cause serious environmental and health hazards whereas in aqueous ecosystems is aesthetically unpleasant and leads to a reduction in sunlight penetration, dissolved oxygen concentration and had acute toxic effects on aquatic flora and fauna. This study is of particular relevance since the Panchganga river and Ichalkaranji area near Kolhapur, India are heavily industrialized, with significant wastewater discharge from textile and dye manufacturing industries which causes the harmful impacts to the environment. Our strain *E. faecalis* showed better colour removal of actual dye wastewater with significant reduction in COD and could be a potential strain for the treatment of textile dyestuffs and textile and dye industry effluent via appropriate bioreactor operations and will be useful to small textile industries in an ecoefficient and economically feasible that could effectively decolorize and detoxify dye containing wastewater.

Analysis of metabolites resulting from decolorization

To understand and confirm the possible mechanism of dye decolorization, analysis of products of biodegradation of DR 81 were studied by TLC, HPLC, FTIR and

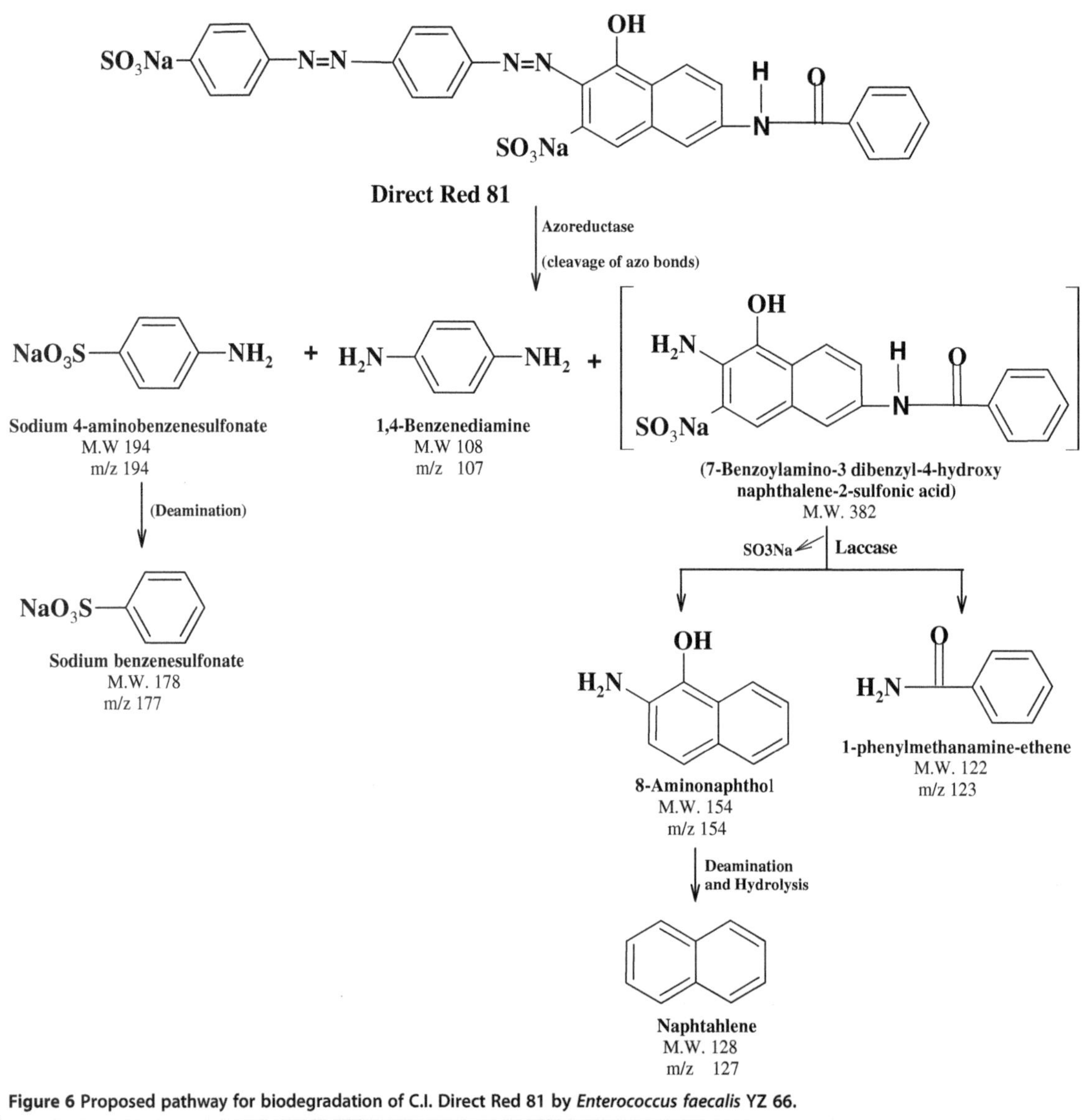

Figure 6 Proposed pathway for biodegradation of C.I. Direct Red 81 by *Enterococcus faecalis* YZ 66.

Table 1 Mass spectrum data of degraded products of C.I. Direct Red 81 by Enterococcus faecalis YZ 66

Mass spectrum	Name of the product
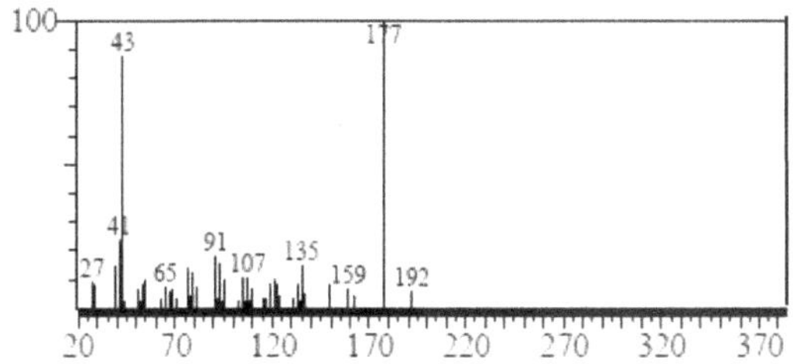	Sodium-4-aminobenzenesulfonate (M.W.:194 m/z:192)
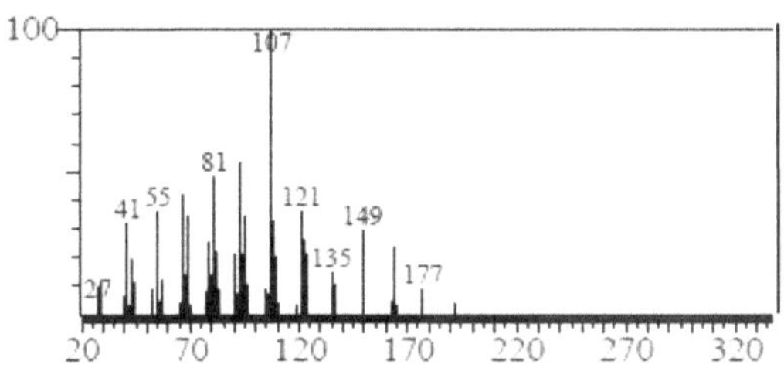	1,4-benzenediamine (M.W.:108 m/z:107)
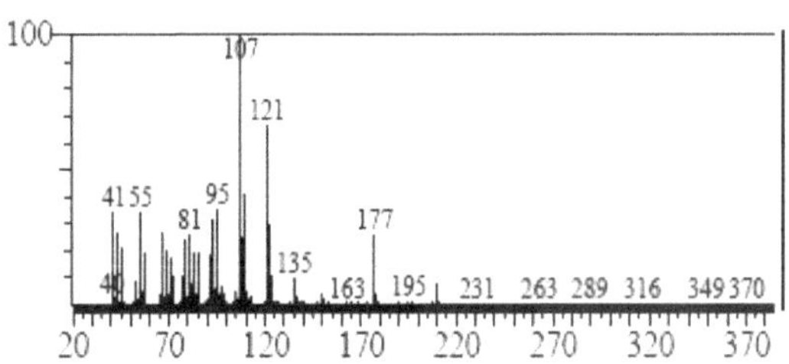	Sodium benzenesulfonate (M.W.:178 m/z:177)
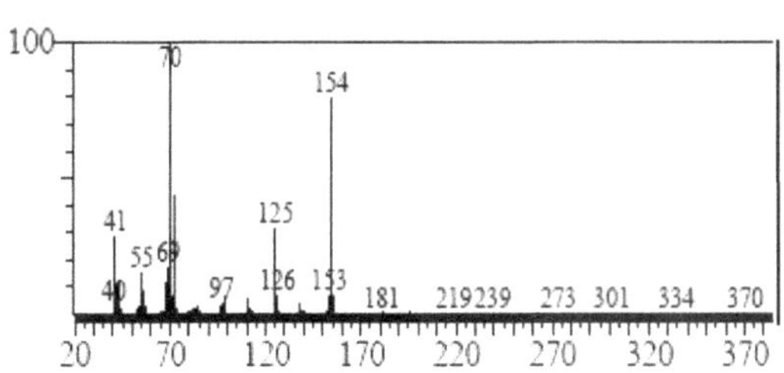	8-aminonaphthol (M.W.:154 m/z:154)
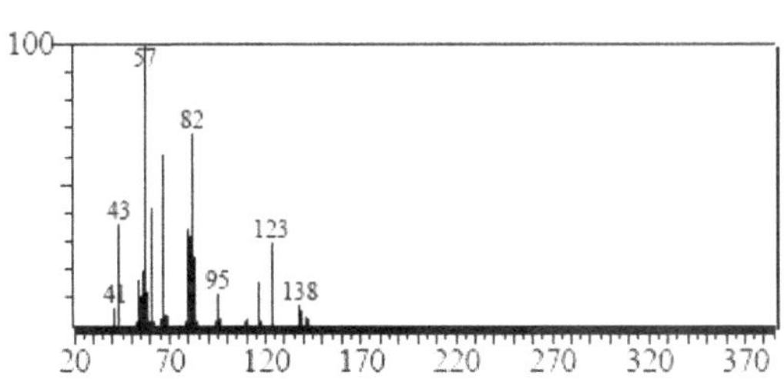	1-phenylmethanamine-ethene (M.W.:122 m/z:123)

Table 1 Mass spectrum data of degraded products of C.I. Direct Red 81 by Enterococcus faecalis YZ 66 *(Continued)*

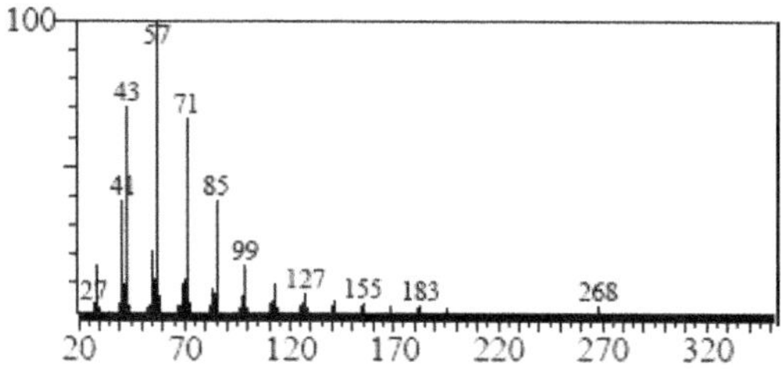	Naphthalene (M.W.:128 m/z:127)

GC-MS. TLC analysis showed the appearance of one spot in the sample containing the extracted metabolites of completely decolorized medium with R_f value 0.71 where as R_f value of DR 81 was noted as 0.97 confirming the biodegradation of DR 81 by *E.faecalis* YZ 66. HPLC elution profile of DR81 showed a distinct single peak at retention time of 1.71. min. Three peaks at retention time of 3.008, 3.861 and 4.021 min were showed that the degradation of DR 81 into different products by *E. faecalis* YZ 66. Disappearance of a distinct peak of DR81 confirmed the degradation of the dye. HPLC analysis of metabolites formed after biodegradation of DR 81 showed the peaks with different retention times than the original dye which indicates the biodegradation of DR 81 into different metabolites (Figure 4A and B).

The FTIR spectrum of a control dye and metabolites was compared. The spectrum of the control dye displayed a peak at 3789.44 cm^{-1} and 3491.49 cm^{-1} N-H stretching. The peak at 1658.48 cm^{-1} represents N = N symmetric stretch. A peak at 1563.99 cm^{-1} represents N-H bending. A peak at 1224.58 cm^{-1} represents C-O stretch band of phenol. The peaks at 1121.4 and 1057.76 cm^{-1} represents C-N stretch along with O = S = O symmetric stretch. The peaks at 616, 712 and 852.382 cm^{-1} represents C-H of substituted aromatics. FTIR spectrum of metabolites obtained after decolorization showed peaks at 3994.35 cm^{-1} and 3690.54 cm^{-1} represents phenolic –OH group, 3054.53 cm^{-1} represents = C-H stretch, 2987.28 cm^{-1} showed –C-H stretch and 1265.37 cm^{-1} represented –C-O stretching vibrations (Figure 5 A and B).

To verify the degradation products formed during dye decolorization by *E. faecalis*, GC–MS analysis was carried out. The low molecular weight aromatic compounds were produced from the degradation of Direct Red 81 by *E. faecalis*. Accordingly, the pathway for the degradation of Direct Red 81 is proposed as depicted in Figure 6, showing various steps involved in the degradation mechanism. However, very little is known about the nature of the degradation products formed in these reactions (Table 1) and the reaction mechanism about oxidoreductive enzymes. We propose that initially primary reductive cleavage in azo bond of Direct Red 81 results in the product such as, sodium-4-aminobenzenesulfonate, 1,4-benzenediamine and 7-benzylamino-3-dibenzyl-1-4-hydroxy naphthalene-2-sulfonic acid. Further deamination of sodium-4-aminobenzenesulfonate results into sodium benzenesulfonate with a mass peak of 178. Whereas the asymmetric cleavage of product 7-benzylamino-3-dibenzyl-1-4-hydroxy naphthalene-2-sulfonic acid by oxidative enzymes (laccase) resulted in the formation of

Table 2 Phytotoxicity studies of C.I. Direct Red 81 and its metabolites formed after biodegradation on *Phaseolus mungo* and *Sorghum vulgare*

	For Direct Red 81					
	Phaseolus mungo			*Sorghum vulgare*		
Parameters studied	Water	Direct red 81[a]	Extracted dye metabolites[a]	Water	Direct red 81[a]	Extracted dye metabolites[a]
Germination (%)	100	70	100	100	70	100
Shoot length (cm)	10.3	8.18	11.54	10.46	8.31	10.77
	±1.91	±1.70	±1.11	±1.12	±1.437	±1.31
Root length (cm)	5.11	4.55	7.23	6.64	5.042	8.0
	±1.35	±0.87	±1.13	±0.512	±0.692	±1.011

[a]400 ppm concentration.
Values are mean of three experiments, SEM (±), significantly different from the control (seeds germinated in distilled water) at $*P < 0.05$, $**P < 0.001$, by one-way analysis of variance (ANOVA) with Tukey–Kramer multiple comparisons test.

1-phenylmethanamine-ethene and 8-aminonaphthol as a products. Further deamination reaction resulting in the formation of low molecular weight compound such as naphthalene as a final product (Figure 6). Therefore, analytical studies confirmed the biodegradation of Direct Red 81 dye, in which the smaller molecular weight intermediates are formed by the consecutive action of oxidoreductive enzymes present in *E. faecalis*.

Phytotoxicity studies

Untreated or partially treated effluent may be disposed off in the water bodies and this water can be used for irrigation purpose. Thus it was found necessary to study phytotoxicity of the dye before and after degradation. The relative sensitivities towards the dye DR 81 and its degradation products in relation to *Sorghum vulgare* and *Phaseolus mungo* seeds were represented in the Table 2. There was no significant difference in the root and shoot length in case of the selected plants irrigated with the dye but in case of metabolites irrigated selected plants root and shoot length was significantly increased ($P \leq 0.05$) as compared to control. Phytotoxicity study showed good germination rate as well as significant growth in the plumule and radical for both the plants ($P \leq 0.05$) in the metabolites extracted after decolorization as compared to dye sample. This indicates the detoxification of DR 81 by *E. faecalis*. Hence this indigenous bacterial strain could be a good biocatalyst for the treatment of textile dyes and effluent containing dyes.

Conclusions

This study demonstrates that isolated *Enterococcus faecalis* YZ 66 was able to degrade and detoxify the toxic sulfonated azo dye Direct Red 81 under static condition. Enzyme analysis indicated prime involvement of oxidoreductive enzymes in the decolorization process. The COD and BOD measurement showed mineralization of Direct Red 81 and phytotoxicity studies shows nontoxic residual metabolites. Analytical studies of extracted products confirmed the biodegradation of Direct Red 81 by *Enterococcus faecalis* YZ 66. A possible pathway for biodegradation of this dye was proposed with the help of GC-MS analysis. This strain also showed better colour removal of dye industry wastewater with significant reduction in COD and BOD and could be a potential strain for the treatment of textile dyestuffs and dye industry effluent by using appropriate bioreactor.

Additional file

Additional file 1: Figure S1. UV-visible spectral scans of C.I. Direct Red 81 (50 mg/L) after complete decolorization by *Enterococcus faecalis* YZ 66. **Figure S2.** Decolorization and growth performance of C.I. Direct Red 81 by *Enterococcus faecalis* YZ 66. **Figure S3.** Effect of (A) pH and (B) temperature on decolorization performance of C.I. Direct Red 81 (50 mg/L) by using *Enterococcus faecalis* YZ 66. **Figure S4.** Effect of supplementation of different carbon and nitrogen sources on the decolorization of C.I. Direct Red 81 (50 mg/L) by using *Enterococcus faecalis* YZ 66. **Figure S5.** Effect of repeated addition of C.I. Direct Red 81 on decolorization performance by using *Enterococcus faecalis* YZ 66.

Competing interests

The authors declare that they have no competing interests.

Authors' contributions

MMS has carried out the experiments, analyzed data and wrote the manuscript. RGS and GDS carried out GC-MS analysis and proposed pathway of degradation of the selected dye. GRP has conceived the strategies, developed the concept, supervised the study and finalized the manuscript. All authors read and approved the final manuscript.

Author details

[1]Department of Microbiology, Maulana Azad College, Aurangabad, MS, India. [2]Department of Biotechnology, Shivaji University, Kolhapur, MS, India. [3]Department of Environmental Biotechnology, Shivaji University, Kolhapur, MS, India. [4]Department of Biochemistry, Shivaji University, Kolhapur, MS, India. [5]H.V. Desai College, Pune, MS, India.

References

1. Mohana S, Desai C, Madamwar D: **Biodegradation and decolorization of anaerobically treated distillery spent wash by a novel bacterial consortium.** *Bioresour Technol* 2007, **98:**333–339.
2. Jirasripongpun K, Rujikan N, Jongjira N, Boonsiri C: **Decolorization and degradation of C.I. Reactive Red 195 by *Enterobacter* sp.** *Thammasat Int J Sci Technol* 2007, **12:**6–11.
3. Keharia H, Patel H, Madamwar D: **Decolorization screening of synthetic dyes by anaerobic, methanogenic sludge using batch decolourization assay.** *World J Microbiol Biotechnol* 2004, **20:**365–370.
4. Kodam KM, Soojhawon I, Lokhande PD, Gawai KR: **Microbial decolorization of reactive azo dyes under aerobic conditions.** *World J Microbiol Biotechnol* 2005, **21:**367–370.
5. Saratale RG, Saratale GD, Chang JS, Govindwar SP: **Ecofriendly degradation of sulphonated diazo dye Reactive Green 19A using *Micrococcus glutamicus* NCIM 2168.** *Bioresour Technol* 2009, **100:**3897–3905.
6. Saratale RG, Sartale GD, Chang JS, Govindwar SP: **Bacterial decolorization and degradation of azo dyes: A review.** *J Taiwanese Inst Chem Eng* 2011, **42:**138–157.
7. Gottlieb A, Shaw C, Smith A, Wheatley A, Forsythe S: **The toxicity of textile reactive azo dyes after hydrolysis and decolourisation.** *J Biotechnol* 2003, **101:**49–56.
8. Parshetti GK, Saratale GD, Telke A, Govindwar SP: **Biodegradation of hazardous triphnylmethane dye methyl violet by *Rhizobium radiobacter* MTCC 8161.** *J Basic Microbiol* 2009, **49:**179–196.
9. Fu V, Viraraghavan T: **Fungal decolorization of dye wastewater -A review.** *Bioresour Technol* 2001, **79:**251–262.
10. Manu B, Chaudhari S: **Anaerobic decolorization of simulated textile wastewater containing azo dyes.** *Bioresour Technol* 2002, **82:**225–231.
11. Chang JS, Chou C, Lin Y, Ho J, Hu TL: **Kinetic characteristics of bacterial azo-dye decolorization by *Pseudomonas luteola*.** *Water Res* 2001, **35:**2841–2850.
12. Pearce CI, Lloyed JR, Guthrie JT: **The removal of colour from textile wastewater using whole bacterial cells: a review.** *Dyes Pigments* 2003, **58:**41–45.
13. Saratale RG, Purankar M, Gandhi S, Kurade M, Oh SE, Govindwar SP, Saratale GD: **Decolorization and degradation of C.I. Remazol Red and textile effluent by isolated *Lysinbacillus fusiformis* strain IGI.** *J Biosci Bioeng* 2013, **115:**658–667.

14. Singh BK, Walker A, Morgan JAW, Wright DJ: **Biodegradation of chlorpyrifos by *Enterobacter* Strain B-14 and its use in bioremediation of contaminated soils.** *Appl Environ Microbiol* 2004, **70:**4855–4863.
15. Jain RM, Mody KH, Keshri J, Jha B: **Biological neutralization of chlor-alkali industry wastewater.** *Mar Pollut Bull* 2011, **62:**2377–2383.
16. Pingitore EV, Todorov SD, Sesma F, Bernadette DGMF: **Application of bacteriocinogenic *Enterococcus mundtii* CRL35 and *Enterococcus faecium* ST88Ch in the control of *Listeria monocytogenes* in fresh Minas cheese.** *Food Microbiol* 2012, **32:**38–47.
17. Giek FC, Noor AAR, Lee SC, Norzarini A, Rozita N, Mohamed R, Mohamad I: **Communal microaerophilic–aerobic biodegradation of Amaranth by novel NAR-2 bacterial consortium.** *Bioresour Technol* 2012, **105:**48–59.
18. Chen H, Wang RF, Cerniglia CE: **Molecular cloning, overexpression, purification, and characterization of an aerobic FMN-dependent azoreductase from *Enterococcus faecalis*.** *Protein Expr Purif* 2004, **34:**302–310.
19. Zimmermann T, Kulla H, Leisinger T: **Properties of purified Orange II azoreductase, the enzyme initiating azo dye degradation by *Pseudomonas* KF46.** *Eur J Biochem* 1982, **129:**197–203.
20. Lourenco ND, Novais JM, Pinheiro HM: **Reactive textile dye colour removal in a sequencing batch reactor.** *Water Sci Technol* 2000, **42:**321–328.
21. Mate MS, Pathade G: **Biodegradation of C.I. reactive Red 195 by *Enterococcus faecalis* strain YZ 66.** *World J Microbiol Biotechnol* 2012, **28:**815–826.
22. Peppler HJ, Perlmani D: *Microbial Technology*, Volume II. 2nd edition. London: Academic; 1979:211–222.
23. Mathew S, Madamwar D: **Decolorization of Ranocid fast Blue dye by bacterial consortium SV 5.** *Appl Biochem Biotechnol* 2004, **118:**371–381.
24. APHA, WEF, AWWA: *Standard Method for the examination of water and Wastewater*, 18th Edition. Washington, DC, USA: 1992
25. Moutaouakkil A, Zeroual Y, Dzayri FZ, Talbi M, Lee K, Blaghen M: **Bacterial degradation of azo dye Methyl red by *Entrobacter agglomerans*.** *Ann Microbiol* 2003, **53:**161–169.
26. Elisangela F, Andrea Z, Dias GF, Cristiano R, Durrant R, Artur CP: **Biodegradation of textile azo dye by a facultative *Staphylococcus arlettae* VN-11 using a sequential microaerophilic/aerobic process.** *Int Biodeterior Biodegrad* 2009, **63:**280–288.
27. Junnarkar N, Murty DS, Bhatt N, Madamwar D: **Decolorization of diazo dye Direct Red 81 by a novel bacterial consortium.** *World J Microbiol Biotechnol* 2006, **22:**163–168.
28. Rajee O, Patterson J: **Decolorization of azo dye (Orange MR) by an autochthonous bacterium *Micrococcus* sp. DBS 2.** *Indian J Microbiol* 2011, **51:**159–163.
29. Jadhav SU, Kalyani DC, Telke AA, Phugare SS, Govindwar SP: **Evaluation of the efficiency of a bacterial consortium for the removal of colour, reduction of heavy metal, and toxicity from textile dye effluent.** *Bioresour Technol* 2010, **101:**165–173.
30. Jadhav SB, Phugare SS, Patil PS, Jadhav JP: **Biochemical degradation pathway of textile dye Remazol Red and subsequent toxicological evaluation by cytotoxicity, genotoxicity and oxidative stress studies.** *Int Biodeterior Biodegrad* 2011, **65:**733–743.
31. Bhatt N, Patel KC, Keharia H, Madamwar D: **Decolorization of diazo dye Reactive blue 172 by *Pseudomonas aeruginosa* NBAR 12.** *J Basic Microbiol* 2005, **45:**407–418.
32. Anjaneya O, Souche SY, Santoshkumar M, Karegoudar TB: **Decolorization of sulfonated azo dye Metanil Yellow by newly isolated bacterial strains: *Bacillus* sp. strain AK1 and *Lysinibacillus* sp. strain AK2.** *J Hazard Mater* 2011, **190:**351–358.
33. Ayed L, Mahdhi A, Cheref A, Bakhrouf A: **Decolorization and degradation of azo dye Methyl Red by an isolated *Sphingomonas paucimobilis*: biotoxicity and metabolites characterization.** *Desalination* 2011, **274:**272–277.
34. Tony BD, Goyal D, Khanna S: **Decolorization of textile azo dyes by aerobic bacterial consortium.** *Int Biodeter Biodegrad* 2009, **63:**462–469.
35. Khelifi E, Ayed L, Bouallagui H, Touhami Y, Hamdi M: **Effect of nitrogen and carbon sources on Indigo and Congo red decolorization by *Aspergillus alliaceus* strain 121C.** *J Hazard Mater* 2009, **163:**1056–1062.
36. Saratale G, Kalme S, Bhosale S, Govindwar S: **Biodegradation of kerosene by *Aspergillus ochraceus* NCIM-1146.** *J Basic Microbiol* 2007, **47:**400–405.
37. Pandey A, Singh P, Iyengar L: **Bacterial decolorization and degradation of azo dyes.** *Int Biodeterior Biodegrad* 2007, **59:**73–84.
38. Jang MS, Jung BG, Sung NC, Lee YC: **Decolorization of textile plant effluent by *Citrobacter* sp. strain KCTC 18061P.** *J Gen Appl Microbiol* 2007, **53:**339–343.

A novel technique for detoxification of phenol from wastewater: Nanoparticle Assisted Nano Filtration (NANF)

L. D. Naidu[1*], S. Saravanan[1], Mukesh Goel[2], S. Periasamy[3] and Pieter Stroeve[4]

Abstract

Background: Phenol is one of the most versatile and important organic compound. It is also a growing concern as water pollutants due to its high persistence and toxicity. Removal of Phenol from wastewaters was investigated using a novel nanoparticle adsorption and nanofiltration technique named as Nanoparticle Assisted Nano Filtration (NANF).

Methods: The nanoparticle used for NANF study were silver nanoparticles and synthesized to three distinct average particle sizes of 10 nm, 40 nm and 70 nm. The effect of nanoparticle size, their concentrations and their tri and diparticle combinations upon phenol removal were studied.

Results: Total surface areas (TSA) for various particle size and concentrations have been calculated and the highest was 4710×10^{12} nm^2 for 10 nm particles and 180 ppm concentration while the lowest was for 2461×10^{11} for 70 nm and 60 ppm concentrations. Tri and diparticle studies showed more phenol removal % than that of their individual particles, particularly for using small particles on large membrane pore size and large particles at low concentrations. These results have also been confirmed with COD and toxicity removal studies.

Conclusions: The combination of nanoparticles adsorption and nanofiltration results in high phenol removal and mineralization, leading to the conclusion that NANF has very high potential for treating toxic chemical wastewaters.

Keywords: Phenol, Nanoparticle, Nanofiltration, NANF, COD, Nanoporous membranes

Background

Water is the central element of all vital social and economic processes. Because of the development of consumer society, harmful chemicals are being generated in huge quantities throughout the world. The problems derived from the toxicological effects of these organic compounds must be resolved for the benefits of entire society. The problem is certainly complex and it is imperative that novel procedures are required to deal with this extensive range of tribulations. There are ample treatment technologies for sewage, distillery effluents and so on, which contain biodegradable organics, but not so much for toxic effluents containing xenobiotic compounds, which are often non-biodegradable or only partially biodegradable. Biological treatment though promises much in this regard is handicapped by its slow oxidation characteristics and incomplete mineralization of toxic chemicals [1–5].

Nanotechnology has also attracted the versatile membrane filtration process for water treatment. Nanofiltration and Reverse Osmosis (RO), are two common membrane filtration processes for toxic chemical removal from wastewaters and have been successfully applied in removing BOD/COD from many wastewaters [6–9]. The powerful membrane technology is however, limited by high costs associated with it and other operational considerations. One of the significant ways to overcome this limitation is to couple membrane separation with another technology. An interesting solution is proposed by some researchers; polymer enhanced ultrafiltration (PEUF) [10, 11]. This was especially found to be more relevant in removing heavy metals from

* Correspondence: ldnaidu@gmail.com
[1]Department of Chemical Engineering, National Institute of Technology, Tiruchirapalli 15, India
Full list of author information is available at the end of the article

aqueous solutions. It combines UF with metal complexation using water-soluble polymers. The formed complexes are sufficiently large to be retained by a UF membrane. There has been several reports on the use of PEUF for treatment of wastewaters. This is however, limited to metal removal and has not found applications in removing toxic chemicals from water and wastewater [12, 13].

A modified version of this technology is called Nanoparticle Assisted Nanofiltration (NANF); nanofiltration in conjunction with adsorption using nanoparticles. The large surface area of nanoparticles makes them a potent adsorbent for toxic chemical removal. The first step in the process is adsorption of toxic organic chemicals on nanoparticles, followed by filtration with a NF membrane, which permits the passage of water and other smaller compounds, but retaining the toxic organic compounds. For polluted water, the full treatment steps conventionally adopted in practice includes pre-oxidation, enhanced coagulation, sedimentation, sand filtration, main oxidation, GAC filtration, etc. Such a treatment chain is however too long to be afforded for developing countries. Therefore NANF, using nano-adsorbents, is a suitable alternative toconventional wastewater treatment plants in dealing with toxic chemicals. In this work, silver nano-particles have been tested for their ability to adsorb organic compounds, which are then retained with a NF membranes. Phenol is used as the model organic compound for these experiments because it is one of the major waste-products in a wide range of manufacturing industries, e.g., chemical and pharmaceutical, paintsand textiles, paper and pulp, plastics and polymer, oil and gasoline as well as coking ovens and metallurgical furnaces [14, 15]. Three different nanoparticles sizes, 10, 40 and 70 nm, and four different NF membranes of diameter, 10, 30, 50 and 80 nm were used for the experiments.

Methods

Adsorbent

Silver nanoparticles of different sizes were synthesized using a chemical reduction method [16]. Dynamic light scattering (DLS, Brookhaven Instruments Corporation, USA) was used to analyze the size of the nanoparticles.

Adsorbate

Phenol was obtained from Merck, and its 500 ppm stock solution was prepared in double distilled water. Solution of 200 ppm phenol used in the experiments was prepared from stock solution and double distilled water was used for necessary dilutions. All reagents used in the investigation were of analytical grade.

Analytical measurements

Phenol was analyzed using the Amino-Antipyrine method [17]. For phenol analysis, 3 ml of distilled water was added with 1 ml of centrifuged sample from each reactor to make the sample to 4 ml. The sample vial was added with 0.60 ml of K_3 Fe CN_6, 0.20 ml of NH_4OH and 0.60 ml of 4-amino antipyrine. The absorbance of the sample was measured with UV–VIS spectrophotometry (Lab India). Theoretical density values were measured at 406 nm to yield the concentration of phenol. Chemical oxygen demand (COD) was measured using a HACH Colorimeter (DR 890). The COD solution (HR grade – 0–1500 ppm) was prepared by mixing 0.25 ml of COD solution A and 2.8 ml of COD solution B. To this solution, 2 ml of centrifuged sample (include dilution) was added. The digestion was done at 1500 C in a HACH COD digester for two hours using HACH COD vials. Final COD value after air cooling was measured in a HACH-DR/890 colorimeter. The toxicity of phenols before and after treatments was realized using Resazurin reduction method [18]. Pre-cultivated *Bacillus cereus* culture was used as the test organisms and was cultivated on nutrient broth medium. Resazurin changes color in the presence of dehydrogenase enzyme activity resulting from microorganisms actively growing in a culture medium. In the presence of an active bacterial culture, resazurin changes color from blue to pink. If bacterial growth is inhibited, no reduction of the resazurin occurs, and such a sample would remain blue. The reaction time for assay was 20 min and would be inhibited by the addition of 50 μl of $HgCl_2$. The colour intensity of centrifuged samples is determined using the UV–VIS spectrophotometer at 610 nm. Resazurin solution was prepared by dissolving 1 tablet of resazurin (5 mg in 100 ml) in 100 ml of distilled water. The resazurin solution was stored in a brown bottle and kept in a temperature of 4 °C.

NANF experiments

All experiments were conducted at 200 ppm phenol concentration and at pH 7 by the addition of NaOH or HCl. Experiments were conducted in a stirred membrane reactor. The schematic of the reactor is shown in Fig. 1. The details of the reactor is given elsewhere [19]. The effective permeation area of the NF membrane was 1.77 cm^2, and the volumes of both compartments were 35.0 mL. Mixing was obtained using magnetic stirrers, which were present in both compartments.

Experimental procedure

Commercial PCTE membranes (Poretics, Inc.) were used as NF membranes [20]. Membranes are numbered according to their diameter. Membrane with 10 nm pore size was referred as M1, whereas, membranes with 80 nm pore size

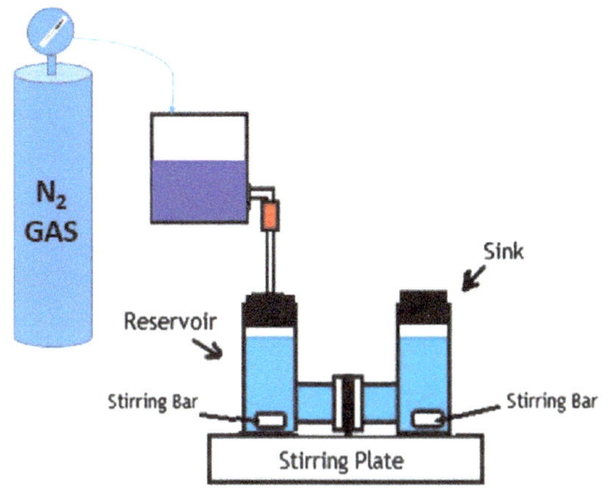

Fig. 1 Schematic diagram of Experimental Set up

was referred as M4. The intermediate membranes were referred as M2 and M3 respectively. The pressure is obtained through compressed N_2 gas. One hour before experiment, 0.5 mL of nanoparticles solution was mixed with 35 mL of neutral solution containing 200-ppm phenol. The solution was then moved to the membrane reactor. Additionally, each experiment was conducted without any nanoparticles to comprehensively evaluate the potential adsorption and filtration performance of NANF.

Theoretical surface area calculation

The surface area calculation was performed as given below.

a) Surface area of a single nanoparticles of 10 nm diameter (5 nm radius) is given by

$$= 4 \times 3.14 \times 5^2 \text{ nm}^2$$
$$= 314 \text{ nm}^2$$

b) Number of particles is obtained from
   i) Concentration of the nanoparticles solution, i.e. 60, 120 and 180 ppm
   ii) Nanoparticle average size
   iii) Total volume of nanoparticle solution used, i.e. 0.5 ml
   iv) Density of silver metal (10.49 g/cc)

From i), we calculate the total weight of the nanoparticles used for adsorption, for example60 ppm means 60 g in 10^6 ml of water, so for 0.5 ml of water

$$= 60 \times 0.5/10^6$$
$$= 30 \times 10^{-6} \text{ g-------x}$$

From ii) we can calculate the volume of individual particles, for example for 10 nm particle

$$= 1.33 \times 3.14 \times 5^3 \times (10^{-7})^2 \text{cm}^3$$
$$= 523 \times 10^{-21} \text{cm}^3$$

Weight of an individual particle can be calculated from the density value of silver metal (10.49 g/cc)

$$= 10.49 \times 523 \times 10^{-21} / 1$$
$$= 548 \times 10^{-20} \text{g ------- y}$$

Hence, total number of particles can be obtained from

$$= x/y$$
$$= 30 \times 10^{-6}\text{g} / 548 \times 10^{-20}\text{g}$$
$$= 3000 \times 10^{12} / 548$$
$$= 5.47 \times 10^{12}$$
$$\approx 5 \times 10^{12}$$

c) Total surface area is obtained by

$$= \text{Surface area of individual particle}$$
$$\times \text{ Total number of particles}$$
$$= 314 \text{ nm}^2 \times 5 \times 10^{12}$$
$$= 1570 \times 10^{12} \text{nm}^2$$

Results and discussion

The current study combines two materials, i.e. the membrane filters and the nanomaterials, to exploit their inherent characteristics for the removal of toxic materials from water. For this purpose, membranes of various pore sizes are used along with silver nanoparticles with various average particle sizes.

Silver nanoparticles

Figure 2 shows the transmission electron micrograph of the silver nanoparticles, which were prepared using three distinct combinations of the bottom-up nanoparticles synthesis process for obtaining nanoparticles of three different average nanoparticles sizes i.e. 10 nm (Fig. 2a), 40 nm (Fig. 2b) and 70 nm (Fig. 2c). Silver nanoparticles with average size of 10 nm were obtained with 10 °C synthesis temperature, Polyvinyl Pyrolidone (PVP) as the dispersing and stabilizing agent and hydrazine as the reducing agent; 40 nm were obtained with 40 °C synthesis temperature, PVP and hydrazine; and 70 nm were obtained with 40 °C synthesis temperature, oligo condensate of naphthalene sulfonic acid (OCNS) and D-glucose combinations. It can be observed from the transmission electron micrographs (Fig. 2) that the particles have been formed with proper geometrical shape, almost spherical for most particles. It can be observed that the particles prepared with PVP have less variation in the particle size distribution (narrow) while those prepared with OCNS comparatively have a wide particle

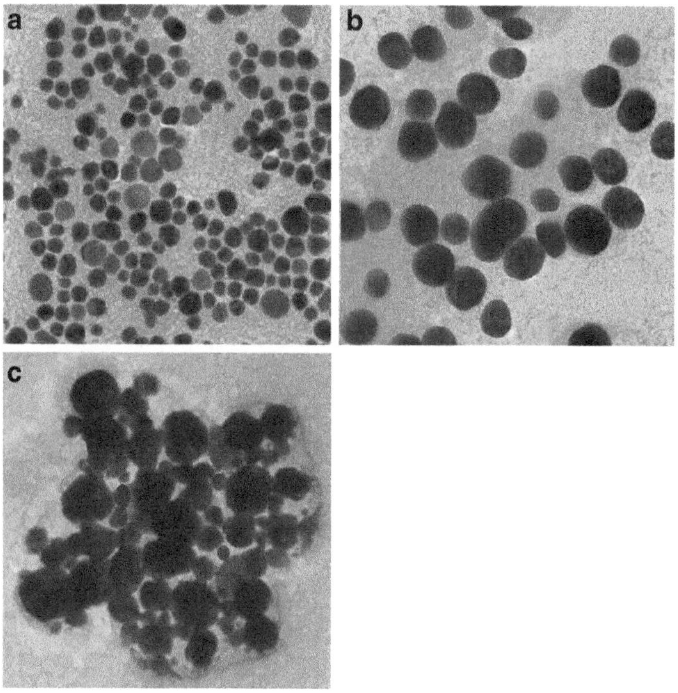

Fig. 2 Transmission electron micrograph of silver nanoparticles with average particle size of **a**) 10 nm, **b**) 40 nm and **c**) 70 nm

size distribution (broad). The mechanism of such particles formation with varying average particle size and distribution are discussed elsewhere [13].

Nanoparticle assisted nano filtration (NANF) effect on phenol removal from water

The work started with adsorption of phenols using silver nanoparticles. Since three distinct average size particle samples were prepared, the three particles batches are used in equal proportion, with the total particle concentrations at 60 ppm. Gulrajani et al. [16] have shown that at 60 ppm particle concentration good particle performance was obtained in their study. The effects of nanoparticle size and membrane pore size on filtration of the phenol are presented in Table 1. It can be observed from Table 1 that for all particle sizes the presence of the nanoparticles show a significant improvement in filtering the phenol with reference to the respective counterpart, i.e. in the absence of the particles for the respective membrane. The results clearly show the effectiveness of the NANF process for phenol removal. As high as 675 % improvement is observed for NANF process (membrane filtration with nanoparticles), compared to the process without nanoparticles. These results can be explained based on the mechanism of phenol adsorption followed by agglomeration of the nanoparticles, which results in a net increase in the particle size, which then cannot pass through the membrane pores and hence are filtered. Such effect is schematically shown in Fig. 3. In the absence of nanoparticles (Fig. 3a) the molecular phenolic compounds can easily pass through the pores of the membranes by diffusion. However, in the presence of nanoparticles (Fig. 3b) whose surface area is very large, there exists a spontaneous attraction of phenolic compounds towards the surface of the particles due to i) electrostatic charge attraction between the nanoparticles and the phenolic compounds (Fig. 3c) and ii) aggregate formation to obtain a lower energy state of the solute and the squeezing out of solvent in support of solute separation [21].

Table 1 Phenol removal percentage with and without nanoparticles for mixtures of nanoparticles

Membrane size (nm)	Phenol Removal (%)		% improvement
	Nanoparticles	No Nanoparticle	
10	68	27	152
30	57	18	216
50	42	11	281
80	31	4	675

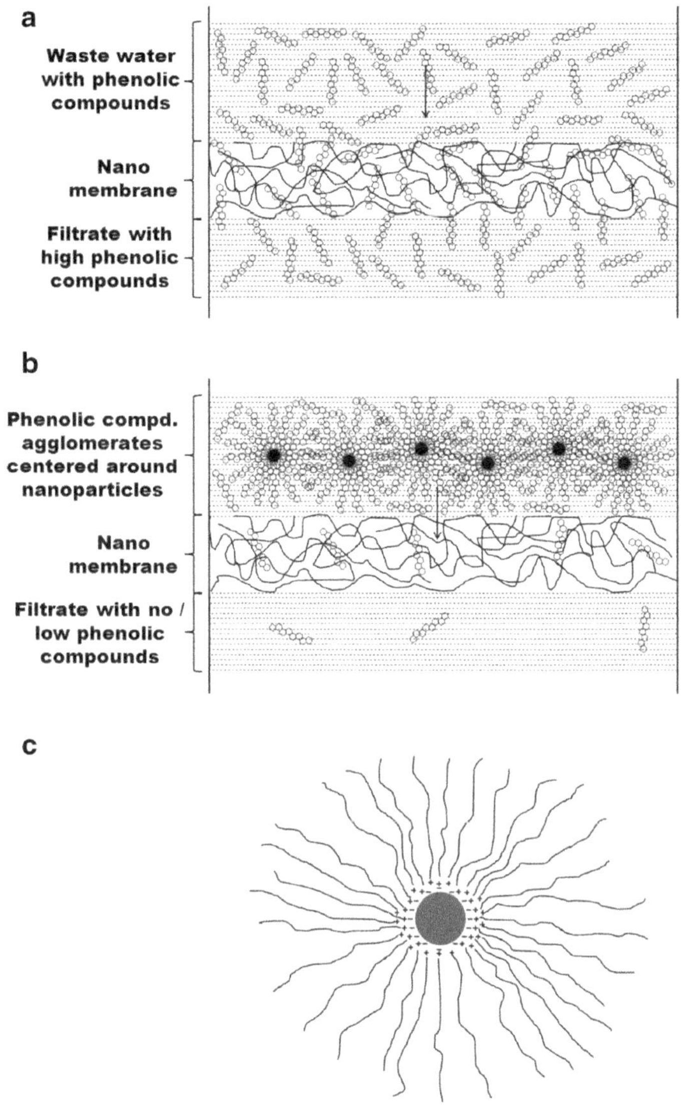

Fig. 3 Schematic of filtration of model phenolic compound through nano membrane **a**) without nano particles, **b**) with nanoparticles (NANF) and **c**) electrostatic adsorption

The effect of membrane pore size on NANF is observed to be significant. Filtration effects are large for the 80 nm pore size membranes (M4), i.e. up to a 675 % improvement (obtained from the phenol removal % with the presence and absence of nanoparticles [((31–4) × 100)/4]). It is surprising to note that such % improvement is low for 10 nm pore size membranes (M1), i.e. only a 151 % improvement. This may be explained to the fact that for large pore size almost all the phenolic compounds pass through the pores in the absence of the particles and the phenol removal % is very low; but with the low value in the denominator, for the case of the presence of nanoaparticles, the % improvement becomes large, as shown in the calculation above. However, in contrast with the small pore size membrane, i.e. 10 nm, there is a considerable amount of phenol removal even in the absence of particles, i.e. 27 %. Hence, even though the actual amount of phenol removal is high in the combined nanoparticles and membrane approach, i.e. 68 %, the % improvement in the presence of nanoparticles is only 151 %. It can be noticed that the best results are obtained for the small pore size membrane (10 nm) and the nanoparticles combinations. This can be explained to the fact that the nanoparticles have enormous surface area particularly below the 100 nm size. The specific surface area of the particles increases exponentially with decreasing size (Theodore, 2005). Hence, availability of large surface areas contributes to surface adsorption of the toxic compounds, which are then filtered by membrane filtration. The low phenol removal %, i.e. 31 % with the 80 nm membrane, even in the presence of the nanoparticles, is due to the presence of nanoparticles whose size is less than 80 nm. These particles pass through the pores resulting in lower nanoparticles concentration and a consequent decrease in phenol removal. This indicates that the smaller particles play an important role in adsorption and hence the phenol removal %. The effectiveness of the hybrid process is also reported by Geckeier et al. [22]. They studied polymer enhanced ultrafiltration (PEUF), which was helpful in removing metal from wastewater by complexing of polymers with the metals and subsequent removal from the wastewater. Similar study was reported by Li et al. [23].

Effect of concentration and particle size of silver nanoparticles

Silver nanoparticles with three different average particle sizes, i.e. 10, 40 and 70 nm, each with three different particle concentrations, 60, 120 and 180 ppm, are used for phenol removal study on membranes (M1 – M5). The results are depicted in Fig. 4 and in Fig. 5. Figure 4a shows the effects of the three particle concentrations, for 10 nm silver nanoparticles. Similarly, Figs. 4b and c show the effects of the three particle concentrations for 40 and 70 nm nanoparticles, respectively. Inversely, Fig. 5 shows the effect of particle size for a given concentration, i.e. Figure 5a shows the effect of the three nanoparticles sizes for the 60 ppm particle concentration. Similarly, Figs. 5b and c show the effect of the three nanoparticles size for the 120 and 180 ppm concentrations, respectively.

From Figs. 4-5, it can be inferred that for the 10 nm pore size membrane that about 100 % phenol removal is obtained for 120 and 180 ppm concentrations for all the three nanoparticles sizes. Further, it can be observed that for the 60 ppm particle concentration that, for almost for all particle sizes (Fig. 5a), the results are less than 100 % phenol removal on the 10 nm pore size membrane. Further, it can be observed that the phenol removal % decreases with increase in pore size of the membrane for almost all particle sizes, but with a varying slopes (Fig. 4a to c). The decrease is sharp after the 30 nm membrane and is well noticed for 60 ppm particle concentration (Fig. 4a to c and Fig. 5a). However, for higher particle concentrations, the decrease in phenol removal % with increasing membrane pore size is low. For the 10 nm size nanoparticles, a notable decrease is observed both for 120 (Fig. 5b) and 180 ppm (Fig. 5c)

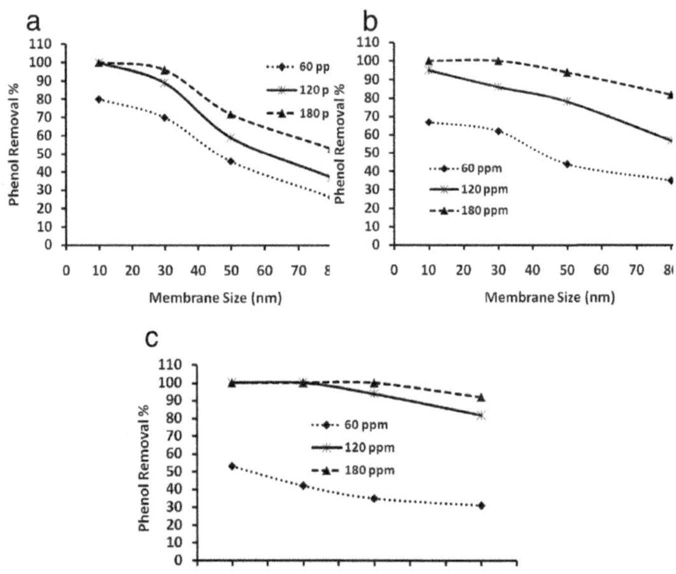

Fig. 4 Phenol removal %, through membranes having various average pore sizes, as an effect of concentration of silver nanoparticles of **a**) 10 nm, **b**) 40 nm and **c**) 70 nm

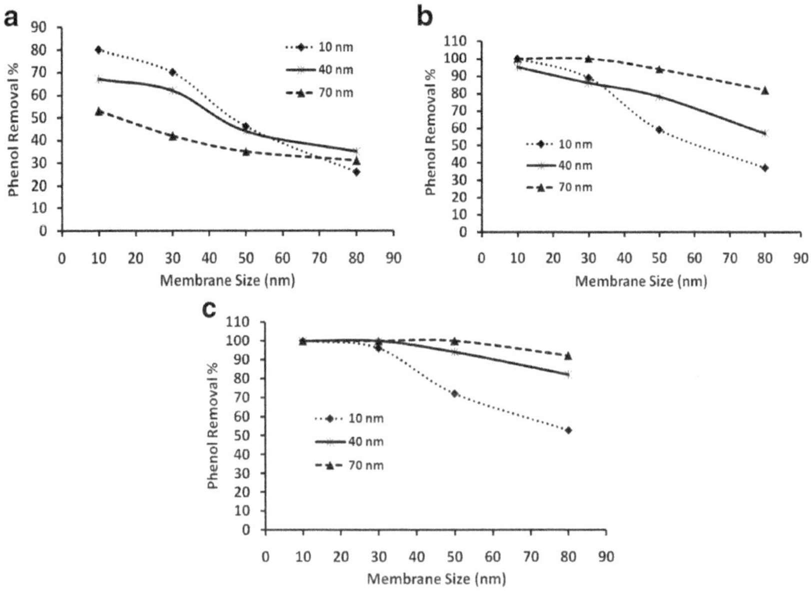

Fig. 5 Phenol removal %, through membranes having various average pore sizes, with silver nanoparticles as an effect of the particles sizes, **a**) 60 ppm, **b**) 120 ppm and **c**) 180 ppm

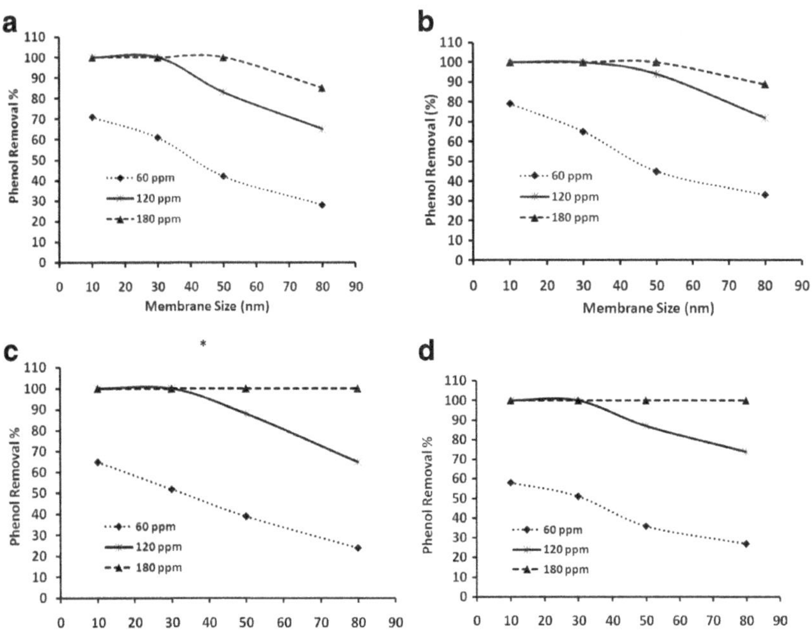

Fig. 6 Phenol removal % through membranes having various average pore sizes with mixture of nanoparticles, **a**) 1:1:1 (10 nm, 40 nm, 70 nm), **b**) 1:1 (10 nm, 40 nm), **c**) 1:1 (10 nm, 70 nm) and **d**) 1:1 (40 nm, 70 nm)

above the 30 nm membrane size. However, the decrease is less for the particle size of 40 nm both for 120 (Fig. 5b) and 180 ppm (Fig. 5c) after the 30 nm membrane size and is even less, almost reaching 100 % for the particle size of 70 nm both for 120 (Fig. 5b) and 180 ppm (Fig. 5c) above the 30 nm membrane size (Fig. 4c).

These observations can be explained based on nanoparticles size, membrane pore size and the surface area available of the nanoparticles for adsorption [24]. The surface area of the individual nanoparticles, total number of nanoparticles available for a given particle size and concentration and hence the total surface area (TSA) available for the adsorption studies have been discussed previously and the values are presented in Table 2. It can be observed from Table 2 that the highest TSA is obtained for the 10 nm particle size at 180 ppm concentration, i.e. 4710×10^{12} nm^2, and the lowest TSA of 2461×10^{11} nm^2 for the particle size of 70 nm at 60 ppm concentration with the total % difference of 1813 [$(4710 \times 10^{12} - 2461 \times 10^{11}) \times 100/2461 \times 10^{11}$]. For the other particle sizes and concentrations, the TSAs of the particles lie in between these two limits. Additionally, it can be observed from Table 2 that for the 60 ppm concentration the TSA for 10, 40 and 70 nm particles are 1570×10^{12}, 4521×10^{11} and 2461×10^{11}, respectively, and so on for the other concentrations. The percentage increases in TSA between the particles with various concentrations are shown separately in Table 3. The general formula for calculating % increase in TSA between various samples is shown as a foot note under Table 3. It can be observed from Table 3 that between samples 1& 4 (between10 and 40 nm both at 60 ppm conc.) the TSA increase is 247 % while between samples 4 & 7 (between 40 and 70 nm both at 60 ppm conc.) the TSA increase is only 84 %. However, the TSA increase between samples 1 & 7 (between 10 and 70 nm both at 60 ppm conc.) is a maximum of 537 %. Further it could be carefully noted that similar comparisons for the higher concentrations, i.e. 120 ppm and 180 ppm, also result in similar TSA % increases i.e. 247 %, 84 % and 537 %. However, in such trials when compared for the smallest particle (10 nm) with highest concentration (180 ppm) (sample 3) and largest particle (70 nm) at the lowest concentration (60 ppm) (sample 7) the TSA increase is the highest of all, about 1813 %.

It can be observed from Figs. 4-6 that the 10 nm particles at 180 ppm concentration with the 10 nm pore membrane result in 100 % phenol removal (Fig. 4a), while the 70 nm at 60 ppm with the 10 nm pore size membrane results in phenol removal of only about 50 % (Fig. 4c), which can be attributed to the large difference in the total surface areas available for adsorption, between these two samples, i.e. 1813 % difference. However, the phenol removal % for 10 nm particles on the larger pore size membranes, shows less than 100 % and is dramatic above the 30 nm pore size for all three concentrations (Fig. 4a), which could be because of the loss of nanoparticles through the large pore size membranes. Contrastingly, with 70 nm particle such decrease is observed to be low for 60 ppm and very low for the higher concentrations i.e. 120 and 180 ppm (Fig. 4c). In fact, at these concentrations almost 100 % phenol removal is observed for 70 nm particles except with the 80 nm pore size membrane, which may be due to some loss of the 70 nm particles with the 80 nm pore size membrane. The observations of phenol removal by the other particle sizes and concentrations lie in between these two limits of smallest particle and highest concentrations with the smallest membrane pore size (Fig. 4a & Fig. 5c) and the largest particle and lowest concentration with the largest membrane pore size (Fig. 4c & Fig. 5a). The results can be attributed to the corresponding TSA along with the loss of particles through the large pore size membrane, i.e., small particles of high concentrations might have large TSA but fail to show 100 % phenol removal with the large pore size membranes, which might be by virtue of their small size because they pass

Table 2 Theoretical calculation of total available surface area of the silver nanoparticles

Sample	Particle size (nm)	Surface Area (nm^2)	Con. in ppm	No. of Particles	Total Surface Area (TSA) (nm^2)	Comments
1	10	314	60	5×10^{12}	1570×10^{12}	Equal to 6
2	10	314	120	10×10^{12}	3140×10^{12}	In between
3	10	314	180	15×10^{12}	4710×10^{12}	Highest
4	40	5024	60	9×10^{10}	4521×10^{11}	Equal to 8
5	40	5024	120	18×10^{10}	9043×10^{11}	In between
6	40	5024	180	27×10^{10}	1356×10^{12}	Close to 1
7	70	15386	60	16×10^{9}	2461×10^{11}	Lowest
8	70	15386	120	32×10^{9}	4923×10^{11}	Close to 4
9	70	15386	180	48×10^{9}	7385×10^{11}	Close to 5

Table 3 Percentage increase in total surface area (TSA) between particles of various sizes and concentrations

Between Sample # and #	Increase in TSA (%)[a]
1 & 4	247
4 & 7	84
1&7	537
2&5	247
5&8	84
2&8	537
3&6	247
6&9	84
3&9	537
3 & 7	1813

[a](TSA of 1 – TSA of 2) × 100/TSA of 2

through the membrane pores and are lost (Fig. 4a & Fig. 5a to Fig. 5c).

The effect of the particle size upon phenol removal can be realized at 60 ppm concentration and with the 10 nm pore size membrane (Fig. 5a) where the loss of small particles should be very low. It can be observed from Fig. 4a that the 10, 40 and 70 nm size particles give about 82 %, 68 % and 52 % phenol removal which might be due to the difference in the TSA i.e. 1570×10^{12} nm^2 (sample 1), 4521×10^{11} nm^2 (sample 4) and 2461×10^{11} nm^2 (sample 7) for 10, 40 and 70 nm size particles, respectively. The corresponding % TSA increase between the samples are found to be 247 %, 84 % and 537 % for sample 1 & 4, sample 4 & 7 and sample 1 & 7, respectively. Though % increase between sample 1 & 7 is very large, the % phenol removal is only about 82 % for the 10 nm particles as against 52 % for the 70 nm particles. This is because the 10 nm particle is only the average size and so there may be particle loss for those whose size is smaller than 10 nm. This is also reported by Sarkar and Acarya [25]. They observed that phenol removal was maximum at lower particle size. Furthermore, Roostaei and Tezel [26], noted that adsorption capacity decreased by increasing the particle size. It can however, be concluded that the particle loss through the membrane impacts the phenol removal even though small size particle yields high TSA.

Effect of nanoparticles mixtures

In the individual particle study discussed in the previous section, either there was a particle loss effect for small particles or there was a low total surface area (TSA) effect for large particles which were particularly noticed at 60 ppm concentration. Hence we studied the effect of particle mixtures in various proportions for three concentrations, 60 ppm, 120 ppm and 180 ppm, for phenol removal. Results of the studies are shown in Fig. 6. Figure 6a depicts the effect of the three different particles i.e. 10 nm, 40 nm and 70 nm mixtures in equal proportions (1:1:1) for the three different concentrations. Accordingly 20 ppm of 10 nm particles, 20 ppm of 40 nm particles and 20 ppm of 70 nm particles were taken for forming a silver nanoparticles mixture solution. Similarly this was done for the other two concentrations i.e. for 120 ppm 40 ppm of each particle solutions and for 180 ppm 60 ppm of each particle solutions were taken. The effect of equal proportions (1:1:1) of the three nanoparticles (triparticle mixture) for three concentrations on various membrane are shown in Fig. 6a. Similarly, the other possible proportions of particles i.e. diparticle mixtures such as 1:1 (10 nm, 40 nm), 1:1 (10 nm, 70 nm,) and 1:1 (40 nm, 70 nm) were carried out. The results of such diparticle mixtures are shown in Fig. 6b, c and d, respectively.

It can be observed from Fig. 6a, b, c, d that the triparticle mixture and all the diparticle mixtures at 60 ppm result in less than 100 % phenol removal ranging from 70 % for triparticle mixture (Fig. 6a) to 58 % for diparticle (40 nm, 70 nm) (Fig. 6d) for the 10 nm pore membrane. The phenol removal % decreases drastically on higher pore size membranes (M2-M4) for the triparticle and all the diparticle combinations for 60 ppm concentrations. This clearly brings out the effect of particle loss, particle size and the TSA effect. The effect of particle size can be observed from Fig. 6a in that the phenol removal of the triparticle mixture for the 10 nm pore membrane is 70 % however the same is about 80 % for pure 10 nm particles for 60 ppm (Fig. 4a). This indicates mixing the 40 nm & 70 nm particles results in a net decrease in TSA and so a decrease in phenol removal. The particle size and the mixture effects can be perceived from Figs. 6b to c, in that for the diparticle mixture of 10 nm and 40 nm the phenol removal is again 80 % (Fig. 6b). This result can be explained that without 40 nm particles, there would be some loss of 10 nm particles and in this case that would be minimal as there would be 40 nm particles which would prevent the loss of particles. Hence, the decrease in TSA of pure 60 ppm 10 nm particle is countered by the 40 nm particles to prevent their loss and, additionally, the TSA contribution from the 40 nm particles together might contribut for the equivalent phenol removal % to that of pure 10 nm at 60 ppm silver nanoparticles. The particle loss factor and particle size can also be perceived at high concentrations of 180 ppm from Fig. 6b and c. It can be observed from these Figures that for the diparticle mixture of 10 nm and 40 nm the phenol removal is only about 90 % whereas in the other diparticle mixture of 10 nm and 70 nm it is 100 % for the 80 nm pore membrane. In the former case there would be notable loss of particles, though there might be some agglomeration effect, resulting in less than 100 % phenol removal. However, in the latter case, though there might be loss of 10 nm particles, the contribution of TSA by 70 nm particles coupled with the TSA of 10 nm particles (which are retained by the 70 nm particles and by agglomeration effects) is sufficiently high to attain 100 % phenol retention. Likewise, the combined particle based adsorption effect would be techno-economically and ecologically better for efficient removal of the toxic contents rather than using the individual small and large particles for a given pore size membrane. For example, it can be seen in Figs. 4 and 5 that 100 % phenol removal was not observed even at 180 ppm concentration for 70 nm particles. However, in the case of diparticle mixtures of 10 nm and 70 nm and 40 nm and 70 nm for the same 180 ppm concentration, 100 % phenol removal has been obtained for all pore size membranes including the 80 nm pore size. Though the phenol removal % could be increased with individual particles just by increasing the concentration above 180 ppm, it would not be economical. Similarly, low particle size would be preferred because, generally, it is difficult to synthesize smaller particles and for smaller the particles there are chances that these might interfere with the ecological system. Further, although small particles lead to higher surface areas, it also creates operational issues as small particles would not be stable in a treatment system. They could be easily swept away by the feed, thus posing difficulties with the maintenance of the system. Hence, for a given set of membranes, a combined particle mixture would yield a better toxic content removal due to the control of the loss of small particles by larger particles by a blocking effect and also by a particle agglomeration effect. This allows for exploiting the large total surface area (TSA) of small particles.

COD removal & toxicity study

As a further examination to the phenol removal % study, the filtrates were subjected to chemical oxygen demand (COD) studies as described in the experimental section. For this study, only four selected combinations were chosen from inference of the detailed phenol removal study done in the previous sections. The effect on COD removal by low and high particle concentrations i.e. 60 ppm and 180 ppm were studied for both individual particle (Fig. 7a & b) and for particle mixtures (Fig. 7c & d). It can be observed from these Figures that the COD removal % almost reflects the phenol removal %. It can be observed from Fig. 7a and b that the 10 nm particle is effective even at 60 ppm particle concentration, but only for the 10 nm & 30 nm membrane pore sizes. Beyond the 30 nm membrane pore size, the COD removal % decreases notable due to particle loss through the larger size pores. With high particle size, i.e. 70 nm, drastic

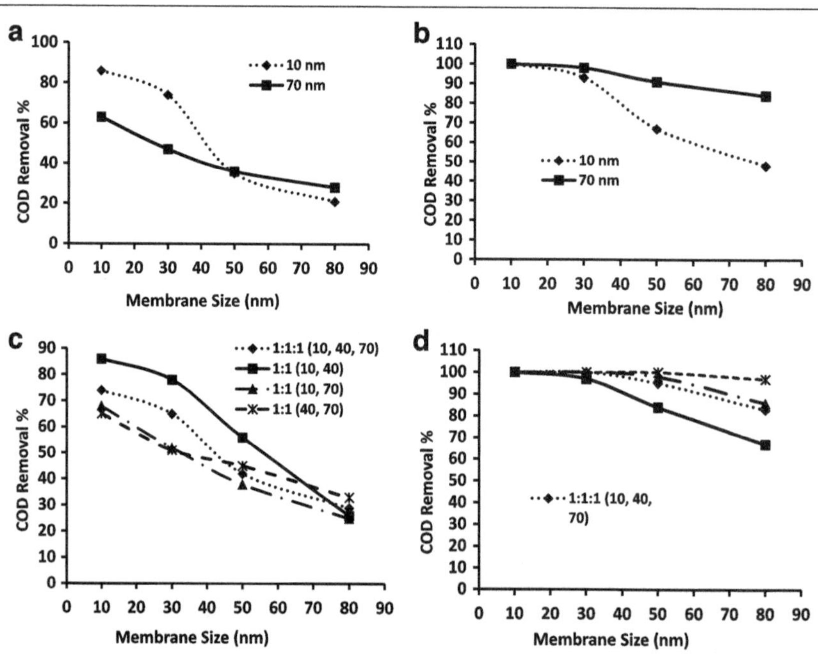

Fig. 7 COD removal % through membranes having various average pore sizes for individual particles **a**) 60 ppm, **b**) 180 ppm and for particle mixtures **c**) 60 ppm and **d**) 120 ppm

change is not noticed with membrane pore size, there is only a steady decrease. However, with the 180 ppm concentration, such effects are unnoticed due to the high concentration, which results in 100 % COD removal with only a slight decrease for the 50 nm and 80 nm membrane pore size due to phenol and particle loss effects. For the same study particle mixtures, it is interesting to note that the mixture yields a better effect than the results for individual particle sizes. It can be observed from Fig. 6c that the best result is achieved for a 10 nm and 40 nm diparticle mixture. The particle loss effect becomes dominant above the 30 nm membrane pore size. However, it can be observed from the same Figure that the 40 nm and 70 nm diparticle mixture performs better than the other particle mixtures for the 80 nm membrane pore size due to low particle loss and agglomeration effects. With high concentration, i.e. 180 ppm, almost all the diparticle mixtures give close to 100 % COD removal for most membrane pore sizes whereas the triparticle mixture has shows good results only up to the 30 nm membrane pore size and then shows a notable decrease due to the particle loss effects.

To further elucidate the mineralization potential of NANF, toxicity studies with resazurin were conducted for the four combinations for COD removal: the effect of low and high concentrations i.e. 60 ppm and 180 ppm were studied for both the individual particle study (Fig. 7a & b) and also for the particle mixture study (Fig. 7c & d). The results of this study are similar to those observed for the COD reductions and hence the results are not separately presented.

Thus it can be inferred that the small particles perform well even at low concentration for small membrane pore size while the large particles perform well only at high particle concentration for all membrane pore sizes. The diparticle mixture is most effective in obtaining good filtration results at low particle concentrations for all the membrane pore sizes.

Conclusion

This study explores the possibilities of improving toxic chemicals/particles removal through novel nanoparticles assisted nanofiltration (NANF). For this purpose, silver nanoparticles were prepared with three distinct average particle sizes of 10 nm, 40 nm and 70 nm and they were characterized in size and morphology by transmission electron microscopy (TEM). A model phenol compound was prepared and was used for filtration studies through nanoporous membranes with four distinct average pore sizes of 10 nm, 30 nm, 50 nm and 80 nm. This innovative approach yielded as high as 675 % improvement with respect to membrane filtration without nanoparticles. Further the effects of particle size and particle concentrations on filtration were studied. Tripartic mixtures in equal proportion (1:1:1) showed 100 % phenol removal at 180 ppm up to the 50 nm pore size membranes and was reduced to 85 % phenol removal for the 80 nm pore membrane. At 120 ppm, the triparticle mixture showed 100 % phenol removal up to the 30 nm pore size membrane and then decreased notably. The 10 nm and 40 nm diparticle mixture showed > 95 % phenol removal for both 120 ppm and 180 ppm up to the 50 nm pore size membrane and then decreased, whereas 40 nm with 120 ppm particles showed only 80 % phenol removal for the 50 pore membrane. Both the 10 nm & 70 nm and 40 nm & 70 nm diparticle mixtures at 180 ppm showed 100 % phenol removal on all the membranes. The triparticle and diparticle studies reveal that the particle mixtures result in more phenol removal % than that of the individual particles particularly for small particles for large pore size membranes and larger particles at low particle concentrations. Similar trends were observed both with the COD and toxicity reduction which shows the effectiveness of using particle mixtures. Overall, from this study it can be inferred that NANF has very good potential for treating waste water and the removal of the toxic contents.

Abbreviations

NANF: nanoparticle assisted nano filtration; M1: membrane with 10 nm pore size; M2: membrane with 30 nm pore size; M3: membrane with 50 nm pore size; M4: membrane with 80 nm pore size.

Acknowledgement

The corresponding author thanks MHRD (Ministry of Human Resource Development), Government of India, for providing financial support through their institute, National Institute of Technology, Trichy (in the form of seed money for research scholars).

Authors' contributions

LDN designed the entire experimental set-up and conducted the experiments. SS coordinated along with MG to interpret the data. MG also assisted LDN in analysis and nanoparticle preparation. SPS unique contribution is in nanoparticle preparation and nanoparticle analysis. PS exceptional input is in nanomembrane preparation and thorough verification of the results. All authors read and approved the manuscript.

Competing interests

The authors declare that they have no competing interests.

Author details

[1]Department of Chemical Engineering, National Institute of Technology, Tiruchirapalli 15, India. [2]Center for Environmental Engineering, PRIST University, Thanjavur, India. [3]Department of Textile Technology, PSG College of Technology, Coimbatore, India. [4]Department of Chemical Engineering, University of California Davis, Davis, CA 95616, USA.

References

1. Goel M, Chovelon JM, Ferronato C, Bayard R, Sreekrishnan TR. The remediation of wastewater containing 4-chlorophenol using integrated photocatalytic and biological treatment. Journal of Photochemistry & Photobiology, B: Biology. 2010;98:1–6.
2. Jin XW, Li EC, Lu SG, Qiu ZF, Sui Q. Coking wastewater treatment for industrial reuse purpose: combining biological processes with ultrafiltration, nanofiltration and reverse osmosis. J Environ Sci. 2013;25:1565–74.

3. Kamani H, Nasseri S, Khoobi M, Nodehi RN, Mahvi AH. Sonocatalytic degradation of humic acid by N-doped TiO2 nano-particle in aqueous solution. J Env Health Sc Engg. 2016;14:3–10.
4. Khazaei M, Nasseri S, Ganjali MR, Khoobi M, Nabizadeh R, Mahvi AH, Nazmara S, Gholibegloo E. Response surface modeling of lead (II) removal by graphene oxide-Fe3O4nanocomposite using central composite design. J Env Health Sc Engg. 2016;14:2.
5. Naghizadeh A, Nasseri S, Mahvi AH, Nabizadeh R, Kalantary RR, Rashidi A. Continuous adsorption of natural organic matters in a column packed with carbon nanotubes. J Env Health Sc Engg. 2013;11:14–9.
6. Serrano D, Suárez S, Lema JM, Omil F. Removal of persistent pharmaceutical micropollutants from sewage by addition of PAC in a sequential membrane bioreactor. Water Res. 2011;45:5323–33.
7. Sahar E, David I, Gelman Y, Chikurel H, Aharoni A, Messalem R, Brenner A. The use of RO to remove emerging micropollutants following CAS/UF or MBR treatment of municipal wastewater. Desalination. 2011;273:142–7.
8. Chon K, Kyongshon H, Cho J. Membrane bioreactor and nanofiltration hybrid system for reclamation of municipal wastewater: removal of nutrients, organic matter and micropollutants. Bioresour Technol. 2012;122:181–8.
9. Yang X, Flowers RC, Weinberg HS, Singer PC. Occurrence and removal of pharmaceuticals and personal care products PPCPs in an advanced wastewater reclamation plant. Water Res. 2011;45:5218–28.
10. Kuncoro EP, Roussy J, Guibal E. Mercury recovery by polymer-enhanced ultrafiltration: comparison of chitosan and polyethyleneimine used as macroligand. Sep Sci Technol. 2005;40:659–84.
11. Korus L, Bodzek M, Loska K. Removal of zinc and nickel ions from aqueous solutions by means ofthe hybrid complexation-ultrafiltration process. Separation and Purification Technology. 1999;172:111–6.
12. Cañizares P. Effect of polymer nature and hydrodynamic conditions on a process of polymer enhanced ultrafiltration. Journal of Membrane Science. 2005;2531–2:149–63.
13. Llorens J, Pujóla M, Sabate J. Separation of cadmium from aqueous streams bypolymer enhanced ultrafiltration: a two-phase model for complexation binding. Journal of Membrane Science. 2004;2392:173–81.
14. Cordova RSM, Dams RI, Cordova REV, Radetski MR, Corrêa AXR, Radetski CM. Remediation of phenol-contaminated soil by a bacterial consortium and Acinetobacteria calcoaceticus isolated from an industrial wastewater Treatment plant. Journal of Hazardous Materials. 2009;164:61–6.
15. Arutchelvan V, Kanakasabai V, Elangovan R, Nagarajan S, Muralikrishnan V. Kinetics of high strength phenol degradation using Bacillus brevis. Journal of Hazardous Materials. 2005;B17:234–40.
16. Gulrajani ML Gupta D, Periyasamy S, Muthu SG. Preparation and application of silver nanoparticles on silk for imparting antimicrobial properties. Journal of Applied Polymer Science. 2008;108(1):614–23.
17. Dannis M. Determination of Phenols by the Amino-Antipyrine Method. Sewage and Industrial wastes. 1951;23:1516–22.
18. Strotmann UJ, Butz B, Bias WR. The dehydrogenase assay with resazurin – practical performance as a monitoring system and pH-dependent toxicity of phenolic compounds. Ecotoxicologyand Environmental Safety. 1993;25:79–89.
19. Stroeve P, Rahman M, Naidu LD, Gilbert B, Morteza M, Ramireze P, Salvador M. Protein diffusion through charged nanopores with different radii at low ionic strength. PhysChemChemPhys. 2014;16:21570.
20. Chun KY, Stroeve P. Protein Transport in Nanoporous Membranes Modified with Self-Assembled Monolayers of Functionalized Thiols. Langmuir. 2002; 18:4653–8.
21. Periaysamy S. Studies on Modification of Silk using UV Radiation, PhD thesis, Department of Textile Technology, Indian Institute of Technology Delhi, New Delhi, India. 2007.
22. Geckeier KE. Polymer-metal complexes for environmental protection. Chemoremediation in the aqueous homogeneous phase. Pure and Applied Chemistry. 2001;73(1):129–36.
23. Li CW. Polyelectrolyte enhanced ultrafiltration (PEUF)for the removal ofCd(II): Effects oforganic ligands and solution pH. Chemosphere. 2008;72(4):630–5.
24. Theodore L. Nanotechnology: Basic Calculations for Engineers and Scientists, Chapter 5: Particle Size, Surface Area and Volume. New York: John Wiley & Sons; 2006.
25. Sarkar M, Acharya KP. Use of fly ash for the removal of phenol and its analogues from contaminated water. Waste Manage. 2006;26:559–70.
26. Roostaei N, Tezel FH. Removal of phenol from aqueous solutions by adsorption. J Environ Manage. 2004;70(2):157–64.

Human nail usage as a Bio-indicator in contamination monitoring of heavy metals in Dizajabaad, Zanjan province-Iran

Abdolhossein Parizanganeh[1*], Abbasali Zamani[1], Vahid Bijnavand[2] and Behzad Taghilou[2]

Abstract

Background: Due to pedogeochemical background and anthropogenic sources, heavy metal contamination of soil is a widespread problem in some parts of Zanjan province located in North West Iran. In this study an affected area located near National Iranian Lead and Zinc Company (NILZ) was selected for detailed study.

Methodology: Thirty soil samples and eighteen leaf samples were collected and analyzed for heavy metal contamination by Inductively Coupled Plasma-Optical Emission Spectroscopy (ICP-OES). Both soil and plant samples were found to be basically polluted by Pb (72.60 and 97.11), Zn (546.47 and 166.61), and Cd (1.80 and 1.55) mg. kg^{-1} respectively. High concentrations of these elements in soil and plant species signifies possible health risks to humans. The distribution maps drawn using Arc GIS (10) show high concentrations of these toxic metals around Dizajabaad village. To assess vulnerability and health risks of metal concentrations in human bodies' twenty nine fingernail samples were also collected from people living in this village.

Results and discussion: Analysis for different heavy metal contents of nail samples reveals very high levels of the same toxic elements (Pb = 15.15, Cd = 1.18, As = 15.47, Zn = 68.46 and Ni = 18.22 mg. kg^{-1}) compared with samples collected from an unaffected area and available values indicated in the literature. Further, high concentration of heavy metals in the nail samples suggests long term exposure of inhabitants to these toxic metals.

Keywords: Biomonitoring, Bio-indicator, Dizajabaad village, Finger nail, Heavy metals

Introduction

Heavy metals such as lead (Pb), cadmium (Cd), zinc (Zn) and nickel (Ni) occur naturally in water, soil and biota. Their concentrations depend on local geology, local addition from mining and industry and/or globally distributed pollution [1-5]. Elevated levels of these metals in the environment may arise from anthropogenic sources, including consumption of food from contaminated environments [6-12]. The increasing demand for environmental and food safety has stimulated research regarding the risk associated with environmental exposure and consumption of food contaminated with heavy metals [13]. Assessment of chronic exposure to essential/nonessential elements is an area of emerging interest in environmental and nutritional epidemiology.

Given the presence of multiple exposure pathways, the use of biomarkers lends promise to reduce measurement error in traditional exposure assessments, which often rely on recall or aggregate exposure measurements [14,15]. Biomonitoring of metal concentrations in human bodies have significant importance, as their amounts in biological samples (hair, nails, blood, saliva, and urine) can be utilized as a medium metal contamination index [16]. The simplicity with which hair and nails can be sampled, transported and handled, and generally higher elemental concentrations compared to other biological media, such as blood and urine makes hair and nails to be more suitable and convenient tissues for monitoring localized exposure to toxic metals [17,18].

This paper focuses on the utility of nail clippings in biomonitoring of metal concentrations in human bodies. Nail clippings have advantages over the other biological materials frequently analyzed for heavy metals content as they can be easily collected as against bone samples.

* Correspondence: h_zanganeh@znu.ac.ir
[1]Environmental Science Research Laboratory, Department of Environmental Sciences, Faculty of Science, University of Zanjan, Zanjan 45371-38791, Iran
Full list of author information is available at the end of the article

Human nails are composed of high-sulfur and hard keratins, the fibrous proteins [19]. Many elements bind to keratin present in fingernails and toenails [20]. The element content of nails tends to vary from one geographical region to another, depending on the natural background conditions [21-24]. The time window of exposure reflected by the nail clipping is dependent on the distance of the clipping and the rate of nail growth. Several known biological factors such as age, gender, and other individual-level characteristics can influence nail growth [25,26]. Although growth rates vary among individuals, fingernails grow at an average of 0.1 mm/day while a normal fingernail takes about 6 months to grow out completely [25]. Since concentration of metals in nails reflects their mean level in human body during a period of 12–18 months [27-31], its use is far from being the universal tool for monitoring long exposures to environmental pollutants [32].

Presence and high concentrations of some heavy metals; notably Pb, Zn, Cd and Ni, in soil & plant samples collected from NILZ Company and neighboring area (Dizajabaad village) were determined with ICP-OES. Previous studies have also reported high levels of these heavy metals in soil, which are likely to accumulate in food grown locally and cause heavy metal risk to consumers [33,34]. Extended data are variability of metals across a contaminated area can predict metal variation in nails. To this purpose, the concentration of five heavy metals (Pb, Cd, Zn, As, and Ni) in the nail samples of the inhabitants in the studied area were analyzed as biomarker of long term heavy metal exposures from contaminated environment. This study was carried out in summer 2012 in the Environmental Science Research Laboratory, University of Zanjan, Zanjan-Iran.

Materials and methods

The study area

Valuable reserves of lead and zinc are available in Angoran area in Zanjan province located in northwest Iran. Mining, transportation of concentrated ore by trucks and smelting units within the province presents a risk of contamination of soils, plants, surface/ground water resources and its entrance into the food chain. Many related industries are developed in Zanjan province, notably, NILZ Company which is one of the largest of its kind in the Middle East with a high market capitalization. The Company which plays an important role in the local economy was established in 1994 with a current consumption of about 300,000 tons of raw ore and an annual production of 55000 tons of lead and zinc. The tailing from NILZ,

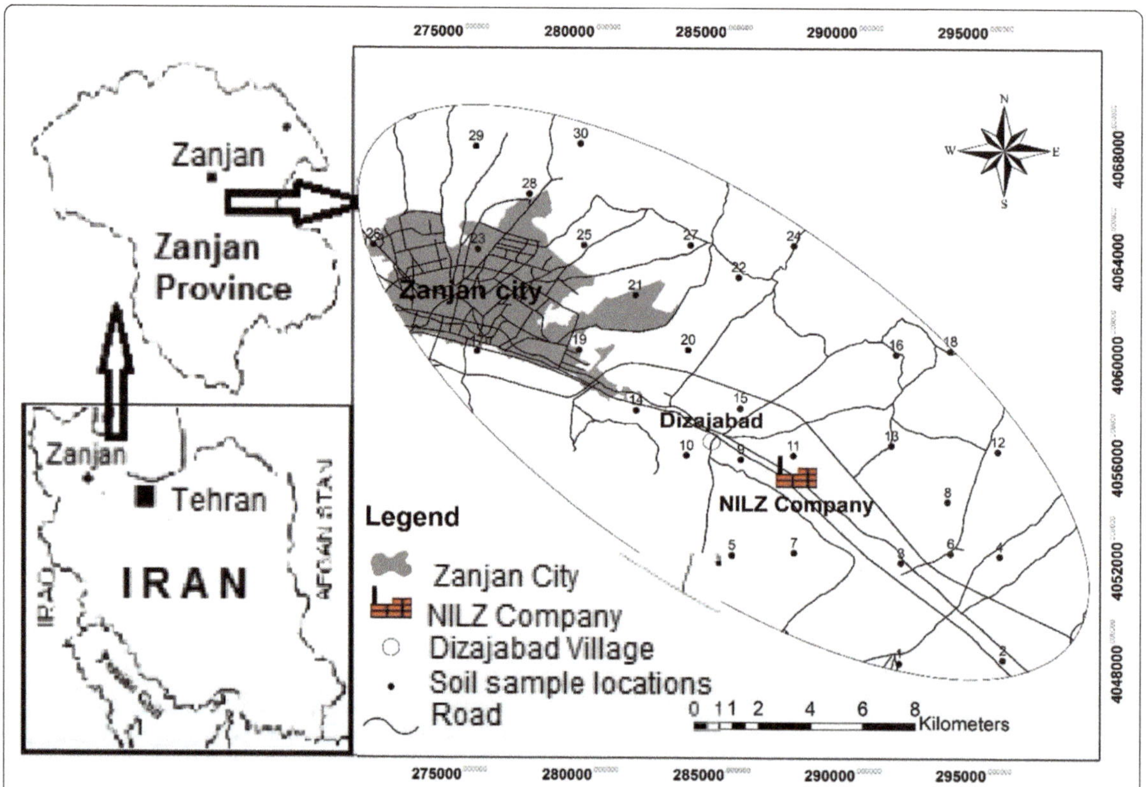

Figure 1 Location map of the studied area indicating sampling points.

Table 1 Statistical description of metal concentrations in top soil (mg. kg^{-1})

	Pb	Zn	Cd	Ni
N	30	30	30	30
Min.	16.00	26.66	0.87	1.07
Max.	615.33	13461.70	5.60	65.03
Variance	17339	5953797	0.81	241.91
Std. dev.	131.68	2440.04	0.90	15.55
Mean	72.60	546.47	1.80	32.13
USEPA	10	NR	-	40

estimated to be about 2.5 million tons, contains a variety of toxic elements, notably Pb, Zn, and Cd [35]. They are damped in the vicinity of the company and exposed to wind and rain contributing to soil, surface and groundwater contaminations. This research was focused on the environmental impacts of NILZ Company on its nearby inhabited areas notably people residing in Dizajabaad village (36° 66′ N, 48°48′ E). The investigated area has a cold and dry climate characterized by cool and dry summers and cold, snowy winters. Figure 1 shows location map of the studied area.

Soil sample collection, preparation and analysis

30 Soil samples were collected from the surface of the soil (0–30 cm) through systematic random sampling (4 × 4 km grid) and preserved by using the methods of soil analysis. Based on the information gathered through field trips and taking into account the type of land use and population concentrations, the number of samples extended by subdividing the grids in selected locations (2.2 km grid) (Figure 1). For each sampling point, four soil samples were collected and mixed properly to obtain a composite sample mixture. The soil sampling spots were cleared of debris prior to sampling. Each composite soil samples were placed in paper bags labeled then taken to the laboratory for pretreatment and analysis. The sampling tools, were washed with soap and rinsed with distilled water after each sampling [36]. In the laboratory, bulk soil samples were spread on trays and air dried at ambient conditions for two weeks. The samples were then grounded by mortar and pestle, sieved through a 2 mm mesh, and oven-dried at 50°C for about 48 hours. The samples were then stored in polyethylene bags and rehomogenized before being used. For analysis the soil samples were digested using the 11466 ISO standard methods (the aqua regia digestion method) [37] as 3 g of

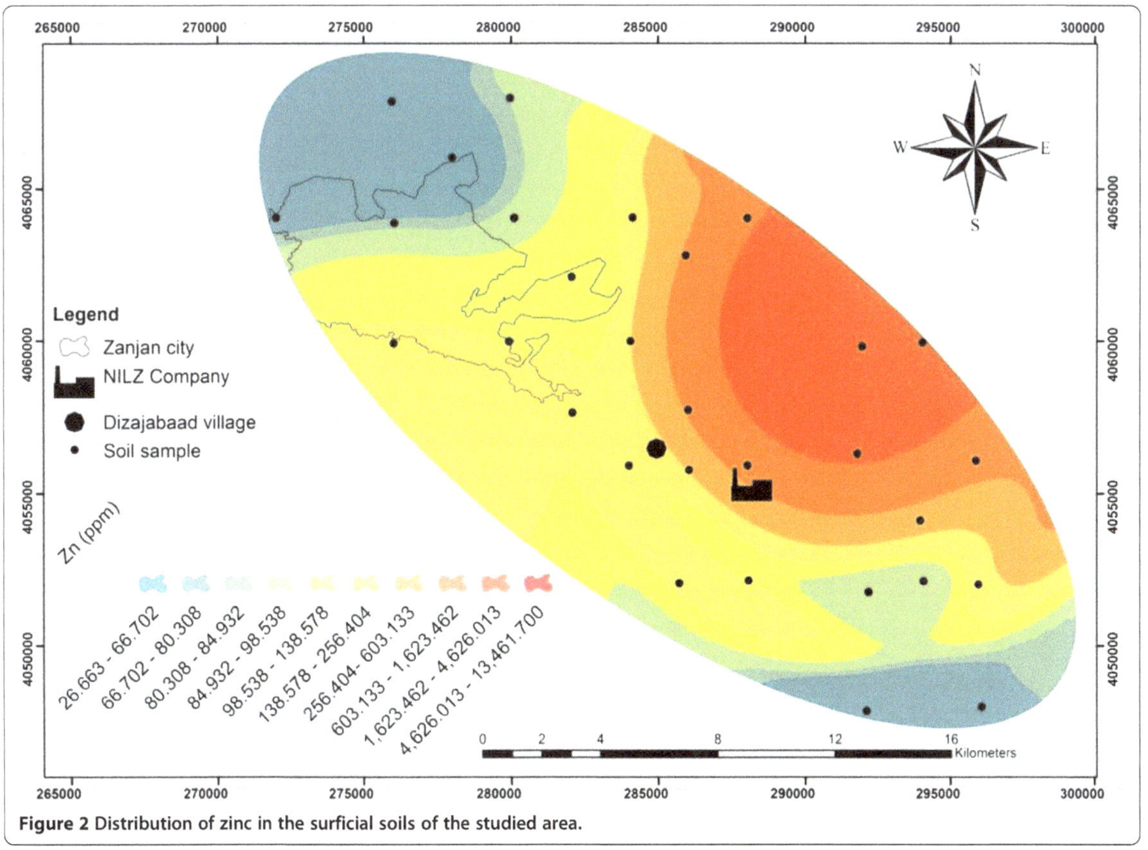

Figure 2 Distribution of zinc in the surficial soils of the studied area.

soil was placed in a 100 ml round bottom flask with 21 ml of concentrated HCl (35%) and 7 ml concentrated HNO_3 (65%). The obtained solution was kept at room temperature overnight before a water condenser was attached and the solution heated to boiling point for 2 hours. Then 25 ml of water was added down the condenser before filtration of the mixture through using a Whatman (No. 42) filter. The filtered residue was rinsed twice with 5 ml of water and the solution was made up to 100 ml. The metals (Zn, Pb, Ni and Cd) in the soil extracts were analyzed by Inductively Coupled Plasma-Optical Emission Spectroscopy (ICP-OES, Spectro Gensis). All chemicals were high purity grade reagents from Merck or Fluka chemical companies.

Leaf sample collection, preparation and analysis

Leaves of three native plant species namely; *Populus, Molus domestica* and *Salix alba* were collected from the studied village and compared with the leaves of the same trees from the vicinity of the NILZ company as well as from an unaffected location about 7 km away from the company (Mellat park east of Zanjan city). This was to see if the soil pollution by the studied heavy metals had positively affected the concentration of the metals in the studied plants.

Plant leaf (n = 18) samples were collected and analysed for their heavy metal content in August 2012. The leaf samples were stored separately in labelled polyethylene bags. Following cleaning and drying, leaf samples were accurately weighed into acid cleaned polypropylene tubes. 10 ml of aqua regia was added to each sample and the tubes were capped and left at room temperature. After 24 h, the digestive was heated to evaporate acids and diluted with 10 ml of deionised water, then filtered (0.4 μm) and diluted with HNO_3 (1.0 molar). The filtered samples were analysed by ICP-OES for different heavy metals.

Nail sample collection, preparation and analysis

Fingernail (n = 29) samples were collected in September 2012 from people living in the Dizajabaad village. For comparison, five fingernails were also collected from an unaffected area outside the province. Finger nail samples were collected using stainless steel nail clipper. The nail samples were stored separately in labelled polyethylene bags. The age range of the participants spanned between 9–65 years (median 35 years), and was skewed towards

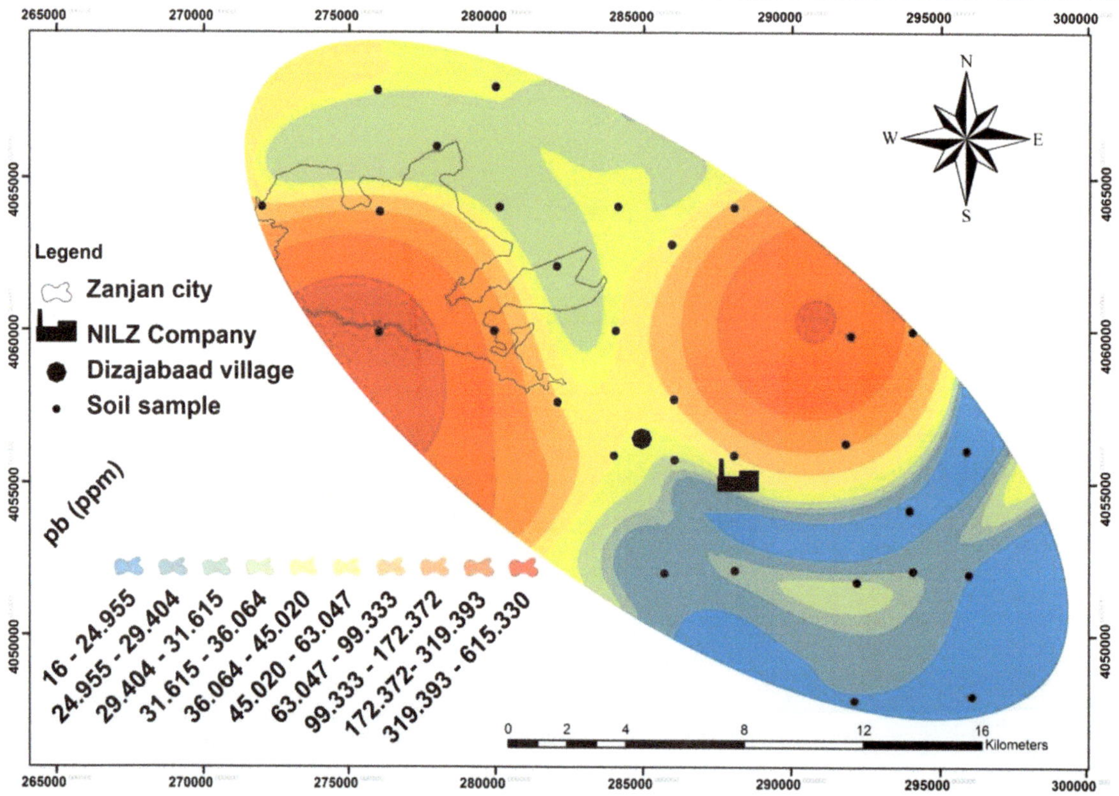

Figure 3 Distribution of lead in the surficial soils of the studied area.

men (nails: 6 women, 23 men). Following cleaning, nail sample was accurately weighed into acid cleaned polypropylene tubes. Nails were then cleaned 5 times with water and acetone, and were dried at room temperature. 1 ml concentrated HNO_3 (sub-distilled from Analytical grade 69% HNO_3, BDH) was added to each sample and the tubes were capped and left at room temperature. After 48 h, the digestive was diluted with 9 ml of deionised water, then filtered (0.4 μm) into a fresh acid cleaned polypropylene tube. The filtered samples were analyzed by ICP-OES.

Statistical analysis

SPSS (20.0 for Windows Evaluation Version) was used to perform statistical analyses of the data. Analysis of the experimental data was carried out by using Mann Whitney U and Kolmogorov-Smirnov. All significance statements reported in this study are at $P < 0.05$ level-of-significance. T-test and the Mann Whitney method, as parametric and nonparametric tests, respectively, allow testing the significant difference of the means between the two groups. The one-way analysis of variance (ANOVA method) and Kruskal-Wallis test allows testing the significant difference of the means. The latter is a non-parametric test without requirements announced for the ANOVA test. In this work both methods were tested for a comparison.

Parametric methods are known for their populations being approximately normal. In contrast, nonparametric methods are defined as the statistical techniques in which no assumption of normality is needed for the populations being studied. Normal distribution assumptions were checked by using Kolmogorov-Smirnov (K–S) test.

Principal Component Analysis (PCA), Factor Analysis (FA), Regression and Correlation analysis were used as statistical processes, for estimating the relationships among variables [33]. It helps to understand how the typical value of the dependent variable changes when any one of the independent variables varies, while others are held constant.

Results and discussions

Table 1 and Figures 2, 3, 4 and 5 show the concentration and distribution of each metal in soil samples within the studied area. Zinc concentration (Figure 2) is very high in Dizajabaad village located near NILZ Company

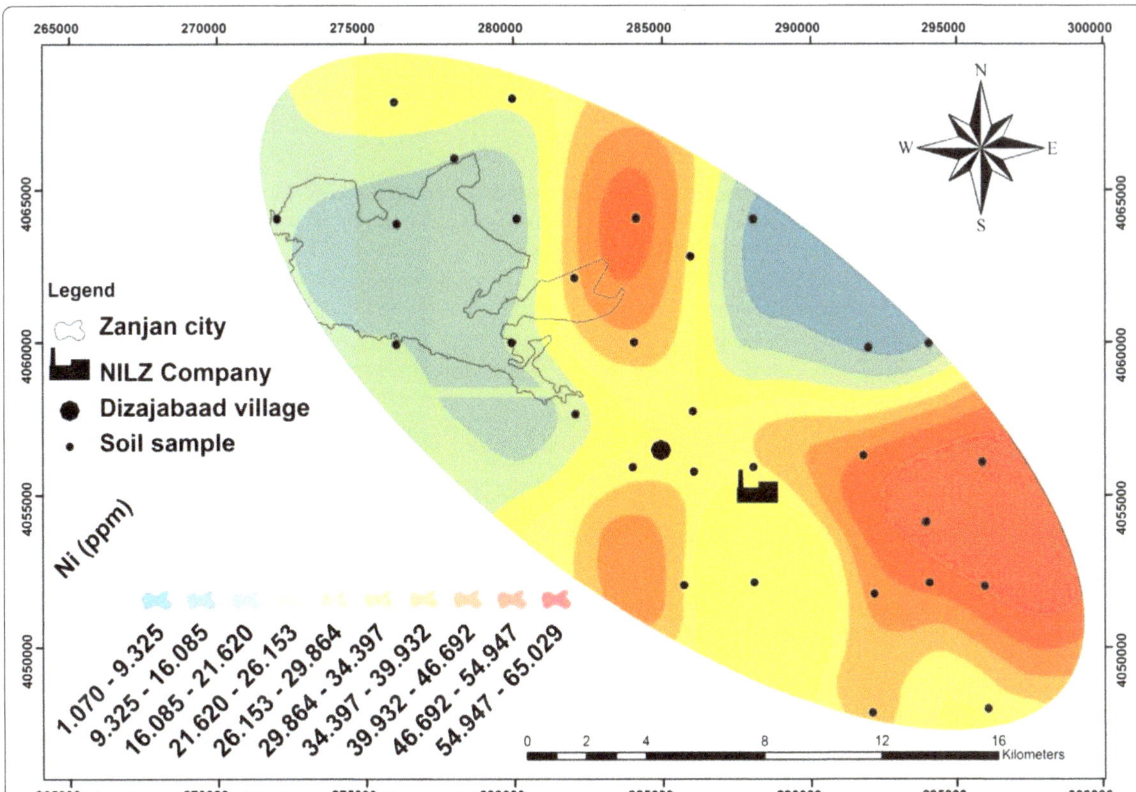

Figure 4 Distribution of nickel in the surficial soils of the studied area.

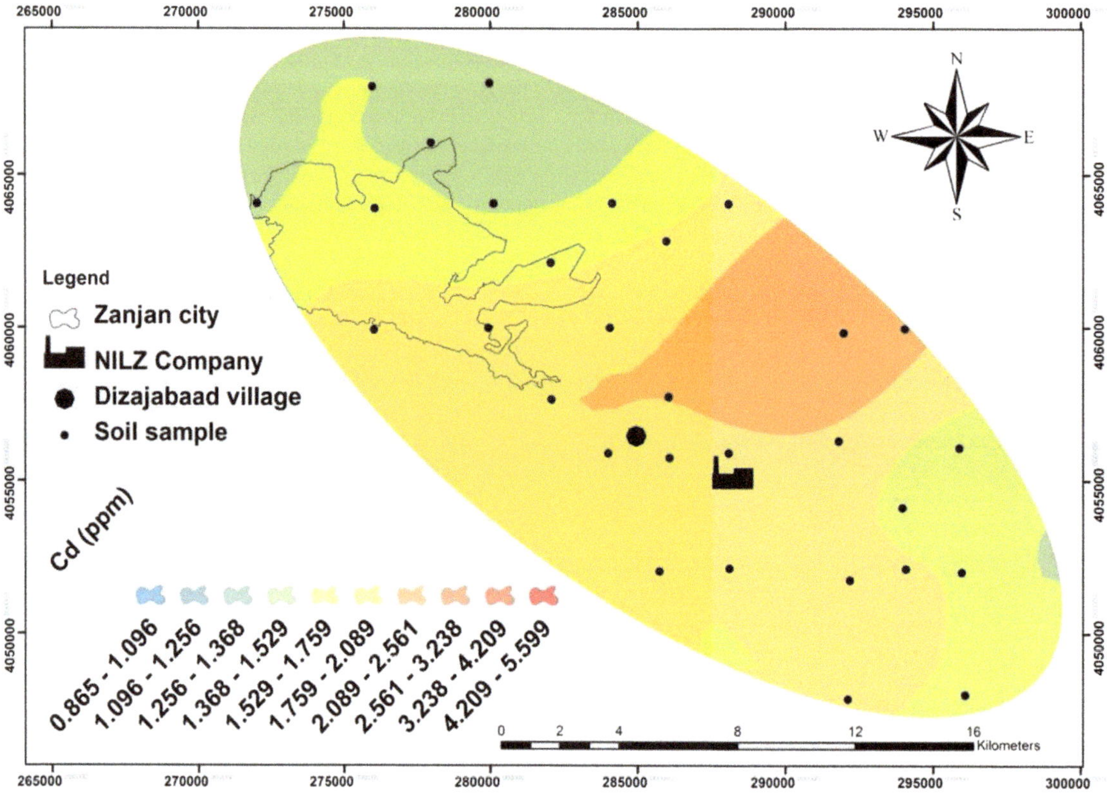

Figure 5 Distribution of cadmium in the surficial soils of the studied area.

and its damping site, indicating its anthropogenic source. Lead is also found in some localized areas with extremely high concentrations among samples of Dizajabaad village and NILZ damping area and can be related to this plant. High concentration of Pb in Zanjan city area may be attributed to the traffic and the use of leaded gasoline and also may be air borne (Figure 3). Nickel is also found in high concentration in surficial soils in south and south east of the area and its high concentration may be related to both agricultural fertilizer used in the cultivated lands in this component of the studied area as well as the NILZ company (Figure 4). For Cd the maximum concentration can be observed near the company (Figure 5).

The maximum concentration of all the studied metals were observed near the NILZ Company and the industrial areas within the area investigated. The average concentrations of Pb, Zn and Cd in sequence in top soils were 72.60, 546.47, 1.80 mg. kg^{-1} as against the average of the same metals according to the WHO standards 20, 50, 0.3 mg. kg^{-1} respectively. This means that the study area is polluted by these heavy metals. For Ni, the average concentration 32.13 mg. kg^{-1} is lower than the standard suggested by WHO (68 mg. kg^{-1}).

Evaluation of the soil's contamination levels were made by the assessment of Enrichment Factors (EF) and Geo-accumulation Index (I_{geo}) [38,39]. These factors and indexes depend on the background metal levels of the samples. The average metal content in the world shale (World Average Shale, W.A.S.) and in the world soils are often used to provide background metal levels [38,39]. As noted before, the iron content of the soil samples are not related to anthropogenic activities, thus the mean Earth's crust's Fe concentration ratio was chosen as reference for determining EF and I_{geo} index. EF was calculated using equations 1.

$$EF = \frac{\left(\frac{[M]}{[Fe]}\right)_{soil}}{\left(\frac{[M]}{[Fe]}\right)_{W.A.S.}} \tag{1}$$

I_{geo} was obtained by equation 2:

$$Igeo = \log_2\left(\frac{[M]_{soil}}{1.5\,[M]_{W.A.S.}}\right) \tag{2}$$

where M is the investigated element, and the subscripts "soil" and "W.A.S" indicates the medium the concentration refers to.

Evaluated Enrichment Factors also show moderate to severe enrichment of the soil samples by Zn and Pb, and very strong enrichment by Cd.

The major objective of factor analysis (FA) is to reduce the contribution of less significant variables to simplify even more the data structure given by PCA. This can be achieved by rotating the axis defined by PCA. This procedure allows constructing new variables (called factors). Factor loadings obtained by PCA with varimax for various metals are presented in Figure 6. The heavy metal grouping is shown in Figure 6 of the first two principal components generated from the main factors, Factors 1 and 2. Comparison of the results in nail and soil show that Pb, Zn and Ni have same source but entrance pathway of Cd in soil and nail is different. Recent studies on ground water samples using the same statistical methods also confirmed the anthropogenic source of the pollutants (Pb, Zn and Ni) in the studied area [33].

Average concentrations of heavy metals (Pb, Zn, Ni, Cd and As) in three different plant species (Populus, Molus domestica and Salix alba species) in Dizajabaad village, near NILZ Co, and in an unaffected area about 7 km away from the studied village are given in Table 2.

Monitoring of metal concentrations in leaf samples show a positive correlation with the concentration of metals in soil samples collected from the same area in which the highest concentrations are found near the NILZ Company and then decrease by increasing distance from it. High correlation coefficients were found for Zn (0.573) and Pb (0.584) with $p < 0.05$ and it was insignificant for Ni and Cd.

Linear relationship between the studied heavy metals in plant and soil show significant correlation between all

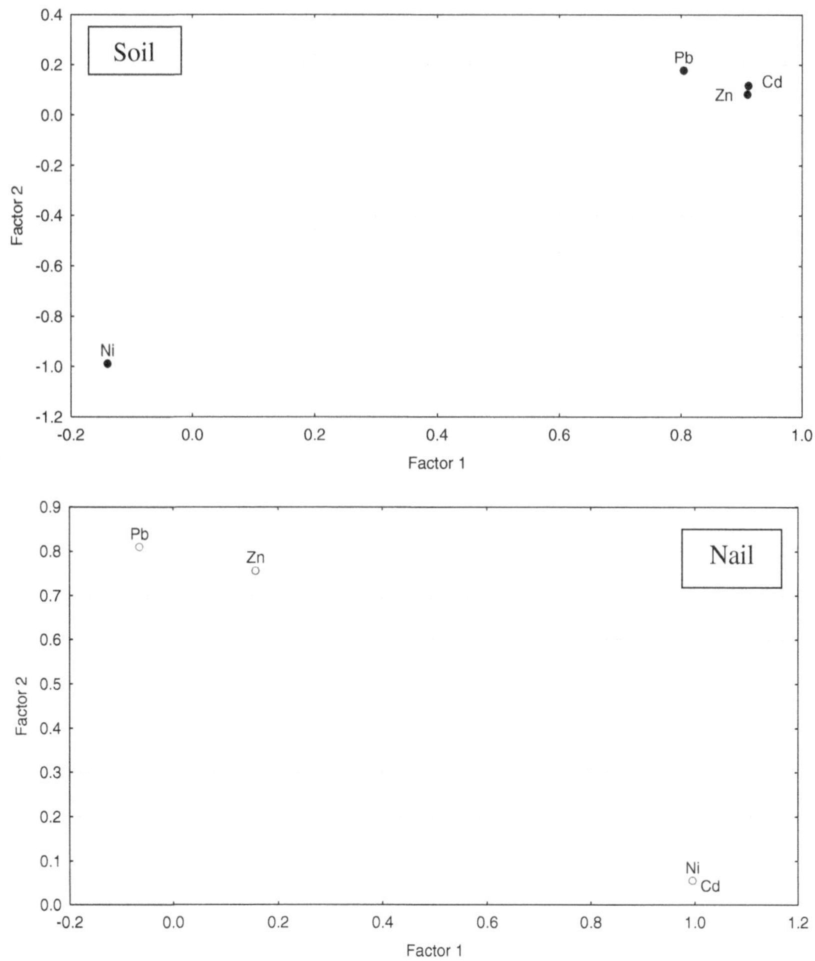

Figure 6 Factor loading for studied heavy metals (Rotation: Varimax normalized; Extraction: Principal components).

Table 2 Average concentrations of some heavy metals in three different plant species in Dizajabaad village, NILZ Co, and in an unaffected area (mg. kg^{-1})

Sample locations	Distance from NILZ Co. (km)	As	Pb	Cd	Zn	Ni
NILZ Company	0	15.92	135.55	2.642	260.83	1.10
Dizajabaad village	1	5.74	58.67	0.456	72.38	1.12
Avg. study area	0.5	10.83	97.11	1.55	166.61	1.11
An unaffected area*	7	0.74	21.49	0.001	11.83	0.88

*Mellat Park located in eastern periphery of Zanjan.

the studied heavy metals except for Ni. Linear correlation coefficients (R^2) were 0.452, 0.660 and 0.584 for Zn, Cd and Pb respectively.

In order to assess the relationship between heavy metal contents in soil and leaf samples, the sampled stations were divided into four distinct areas (Table 3). The results show a statistically significant difference ($p < 0.05$) depending on sampling locations. This also confirmed that by increasing heavy metal content in soil samples, the amount of its uptake by studied plants is raising.

Metal concentrations in fingernail samples collected from Dizajabaad village and from an unaffected area are given in Table 4 in mg. kg^{-1}. It was observed that the average Pb, Zn, Cd, Ni and As in sequence in nail samples from Dizajabaad village are: 15.15, 68.46, 1.18, 18.23, 15.48 mg. kg^{-1} as against the average of these metals in finger nail samples from an unaffected area 2.80, 29.20, 0.00, 4.60, 0.74 mg. kg^{-1} respectively. This means that the people living in the Dizajabad village are severely affected by these heavy metals. To the best of our knowledge, no human nail certified reference material is available for comparison. The absence of well-defined reference concentration ranges is also mentioned in the literature (as for example; [40-42]).

Conclusion

Soil samples collected from the studied area showed that the Dizajabaad village and its surrounding areas are polluted by heavy metals notably with lead, zinc and cadmium. Evaluated Enrichment Factors shows moderate to severe enrichment of the soil samples by Zn and Pb, and very strong enrichment by Cd. Principal Component Analysis and Factor Analysis also elucidate the anthropogenic source of the pollutants. Leaves of three native plant species namely; *Populus, Molus domestica and Salix alba* collected from the studied village compared with the leaves of the same trees from an unaffected location demonstrate that the soil pollution by the studied heavy metals had positively affected the concentration of the metals in the studied plant species. Analysis for different heavy metal contents of nail samples collected from people living in the studied area, compared with samples from an unaffected area, reveals very high levels of the same toxic elements found in soil/plant samples as well as the available values indicated in the literature. Furthermore, there were close associations established between the specific metals in soil, plants and nails. This means that the people living in the Dizajabad village are severely affected by these heavy metals and anthropogenic sources; notably NILZ Company is mostly responsible for the high concentration of these metals in soil, plants and biological tissue samples. The present study demonstrated that determination of metals in finger nails of the residence has potential utility as a biomarker to heavy metals from the environmental exposure. An overall view of the results obtained for toxic heavy metals establish the use of nails as a biomarker and diagnostic tool in medical sciences. More number of researches on a larger number of populations will enhance the merits of such investigations and their application in medical sciences.

Table 3 Classification of studied area considering the concentrations of some heavy metals in soil samples (mg heavy metals per kg soil)

Sample locations	Pb	Zn	Cd	Ni
Area 1	16.00- 30.00	26.00-80.00	0.86-1.26	1.07-16.08
Area 2	30.00-45.00	80.00-138.00	1.26-1.76	16.08-29.86
Area 3	45.00-99.00	138.00-603.00	1.76-2.56	29.86-39.93
Area 4	99.00-615.00	603.00-13461.00	2.56-5.60	39.93-65.03

Table 4 Metal concentrations in fingernail samples collected from people living in Dizajabaad village and from an unaffected area (mg. kg^{-1})

Location		Pb	Zn	Cd	Ni	As
Dizajabaad Village	N	29	29	29	29	29
	Minimum	0.0	32.0	0.0	0.0	0.0
	Maximum	123.0	176.0	11.0	79.0	50.0
	Variance	547.14	886.63	6.97	211.88	237.28
	Mean	15.16	68.46	1.18	18.23	15.48
	Std. Dev.	23.39	29.78	2.64	14.55	15.40
Outside Zanjan province (Unaffected area)	N	5	5	5	5	5
	Minimum	0.0	3.0	0.0	3.0	0.0
	Maximum	8.0	46.0	0.0	7.0	3.0
	Variance	15.20	262.20	0.00	2.80	1.68
	Mean	2.80	29.20	0.00	4.60	0.74
	Std. Dev.	3.89	16.19	0.00	1.67	1.29

Competing interests
The authors declare that they have no competing interests.

Authors' contributions
This work is parts of two MSc thesis of VB and BT where AHP and AAZ, supervised these thesis, suggested the problem, participated in determination of sample points, sample preparation procedure, and wrote and edited the manuscript. All authors read and approved the final manuscript.

Acknowledgements
Sincere gratitude to Dr F. Ghasemian, Head department of Biology, University of Zanjan and Dr A. Delavar, Department of Soil Science, University of Zanjan for their critical revision of the text and partial financial support.

Author details
[1]Environmental Science Research Laboratory, Department of Environmental Sciences, Faculty of Science, University of Zanjan, Zanjan 45371-38791, Iran. [2]Department of Environmental Science, Faculty of Science, University of Zanjan, Zanjan 45371-38791, Iran.

References

1. Cui YJ, Zhu YG, Zhai RH, Chen DY, Huang YZ, Qiu Y, Liang JZ: **Transfer of metals from soil to vegetables in an area near a smelter in Nanning.** *China Environ Int* 2004, **30**:785–791.
2. Cui YJ, Zhu YG, Zhai RH, Huang YZ, Qiu Y, Liang JZ: **Exposure to metal mixtures and human health impacts in a contaminated area in Nanning.** *China Environ Int* 2005, **31**:784–790.
3. Zheng N, Wang QC, Zhang XW, Zheng DM, Zhang ZS, Zhang SQ: **Population health risk due to dietary intake of heavy metals in the industrial area of Huludao City, China.** *Sci Total Environ* 2007, **387**:96–104.
4. Khan S, Cao Q, Zheng YM, Huang YZ, Zhu YG: **Health risks of heavy metals in contaminated soil and food crops irrigated with waste water in Beijing, China.** *Environ Pollut* 2008, **152**:686–692.
5. Hang X, Wang H, Zhou J, Ma C, Du C, Chen X: **Risk assessment of potentially toxic element pollution in soils and rice (Oryza sativa) in a typical area of the Yangtze River Delta.** *Environ Pollut* 2009, **157**:2542–2549.
6. Airey D: **Mercury in human hair due to environment and diet, are review.** *Environ Health Perspect* 1983, **52**:303–316.
7. Wang X, Sato T, Baoshan X, Tao S: **Health risk of heavy metals to the general public in Tianjin, China via consumption of vegetables and fish.** *Sci Total Environ* 2005, **350**:28–37.
8. Zheng N, Wang QC, Zheng DM: **Health risk of Hg, Pb, Cd, Zn, and Cu to the inhabitants around Huludao Zinc Plant in China via consumption of vegetables.** *Sci Total Environ* 2007, **383**:81–89.
9. Sridhara CN, Kamala CT, Samuel SRD: **Assessing risk of heavy metals from consuming food grown on sewage irrigated soils and food chain transfer.** *Ecotoxicol Environ Saf* 2008, **69**:513–524.
10. Whyte ALH, Hook GR, Greening GE, Gibbs-Smith E, Gardner JPA: **Human dietary exposure to heavy metals via the consumption of green shell mussels (Pernacanaliculus Gmelin 1791) from the Bay of Islands, northern New Zealand.** *Sci Total Environ* 2009, **407**:4348–4355.
11. Zhuang P, Mc Bride MB, Xia H, Li N, Li Z: **Health risk from heavy metals via consumption of food crops in the vicinity of Dabaoshanmine, South China.** *Sci Total Environ* 2009, **407**:1551–1561.
12. Metian M, Charbonnier L, Oberhaensli F, Bustamante P, Jeffree R, Amiard JC, Warnau M: **Assessment of metal, metalloid, and radionuclide bioaccessibility from mussels to human consumers, using centrifugation and simulated digestion methods coupled with radio tracer techniques.** *Ecotoxicol Environ Saf* 2009, **72**:1499–1502.
13. D'Mello JPF: *Food Safety, Contaminants and Toxins.* Wallingford, Oxon, UK, Cambridge: CABI Publishing; 2003.
14. Rothman KJ: **Methodologies frontiers in environmental epidemiology.** *Environ Health Perspect* 1993, **101**(S4):19–21.
15. Decaprio AP: **Biomarkers: coming of age for environmental health and risk assessment.** *Environ Sci Technol* 1997, **31**:1837–1848.
16. Zielhuis RL, Henderson PTH: **Definitions of monitoring activities and their relevance for the practice of occupational health, Int. Arch. Occup.** *Environ Health Perspect* 1986, **57**:249–257.
17. Rodushkin I, Axelsson D, Rodushkin I, Axelsson D: **Application of double focusing sector field inductively coupled plasma-mass spectrometry multi-elemental characterization of human hair and nail Part II.** *A study of inhabitants of Northern Sweden Sci Total Environ* 2000, **262**:21–36.
18. Sukumar A, Subramanian R: **Relative element levels in paired samples of scalp hair and fingernails of patients from New Delhi.** *Sci Total Environ* 2007, **372**:474–478.
19. McLean WH, Epithelial Genetics Group: **Genetic disorders of palm skin and nail.** *J Anat* 2003, **202**:133–142.
20. Hopps HC: **The biological bases for using hair and nail for analyses of trace elements.** *Sci Total Environ* 1977, **7**:71–89.
21. Eads EA, Lambdin CE: **A survey of trace metals in human hair.** *Environ Res* 1973, **6**:247–252.
22. Chattopadhyay A, Jervis RE: **Hair as an indicator of multi element exposure of population groups.** In *Proceedings of the Trace Substances in Environmental Health, VIII. A Symposium.* Edited by Hemphill DD. Columbia: University of Missouri; 1974:31–38.
23. Hefferre JJ: **Usefulness of chemical analysis of head hair for exposure to mercury.** *J Am Dent Assoc* 1976, **92**:1213–1216.
24. Teraoka H: **Chemical investigation of inorganic elements in human hair.** *Jpn J Hyg* 1981, **36**:712–717.
25. Fleckman P: **Basic science of nail unit.** In *Basic Nail Surgery.* 3rd edition. Edited by Scher RK, Daniel CD, Fleckman P. Philadelphia: WB Saunders Co; 1997:37–54.
26. Dawber R, Bahrain R: **Nail growth.** *Curtis* 1987, **39**:99–103.
27. Suzuki Y, Mikasa H, Fujil N, Nishiyama K: **Adsorption and elution of metals on hair.** *Biol Trace Elem Res* 1988, **16**:59–66.
28. Wilhelm M, Hafner D: **Monitoring of cadmium, copper, lead and zinc status in young children using toenails, comparison with scalp hair.** *Sci Total Environ* 1991, **103**:199–207.
29. Hayashi M, Yamamoto K, Yoshimura M, Hayashi H, Shitara A: **Cadmium, lead, and zinc concentrations in human fingernails.** *Bull Environ Contam Toxicol* 1993, **50**:547–553.
30. Chen KLB, Amarasiriwardena CJ, Christiani DC: **Determination of total arsenic concentration in nails by inductively coupled plasma mass spectrometry.** *Biol Trace Elem Res* 1999, **67**:109–125.
31. Were FH, Njue W, Murungi J, Wanjau R: **Use of human nails as bioindicator of heavy metals environmental exposure among school age children in Kenya.** *Sci Total Environ* 2008, **393**:376–384.
32. Oyoo-Okoth E, Admiraal W, Osano O, Ngure V, Kraak M, Omutange E: **Monitoring exposure to heavy metals among children in Lake Victoria, Kenya.** *Environ Saf* 2010, **73**:1797–1803.
33. Zamani AA, Yaftian MR, Parizanganeh AH: **Multivariate statistical assessment of heavy metal pollution sources of groundwater around a lead and zinc plant.** *Iran J Environ Health Sci Eng* 2012, **9**:29.
34. Parizanganeh AH, Bijnavand V, Zamani AA, Hajabolfath A: **Concentration, distribution and comparison of total and bioavailable heavy metals in top soils of Bonab District in Zanjan province.** *Open J Soil Sci* 2012, **2**:123–132.
35. Quantity and Quality of Special Wastes in Zanjan Province: *Research Report.* Zanjan-Iran: Zanjan Department of the Environment; 2007.
36. Whitten MG, Ritchie GSP: **Calcium chloride extractable cadmium as an estimate of cadmium uptake by subterranean clover.** *Aust J Soil Res* 1991, **29**:215–221.
37. **Soil quality-Extraction of trace elements soluble in aqua regia,** *Int Stand* 1995, [www.iso.org/iso/catalogue_detail.htm?csnumber=19418]
38. Vega FA, Covelo EF, Andrade ML: **Impact of industrial and urban waste on the heavy metal content of salt marsh soils in the southwest of the province of Pontevedra (Galicia, Spain).** *J Geochem Exploration* 2008, **96**:148–160.
39. Rafiei B, Bakhtiari Nejad M, Hashemi M, Khodaei AS: **Distribution of heavy metals around the Dashkasan Au Mine.** *Int J Environ Res* 2010, **4**:647–654.
40. Bencze K: **What contributions can be made to biological monitoring of hair analysis?** *Fresenius J Anal Chem* 1990, **337**:867–876.
41. Gulson BL: **Nails, concern over their use in lead exposure assessment.** *Sci Total Environ* 1996, **177**:323–327.
42. Kelepertsis A, Alexakis D, Kita I: **The Environmental geochemistry of soils and waters of Susaki area, Korinthos, Greece.** *Environ Geochem Health* 2001, **23**:117–135.

21

Assessment of different organic supplements for degradation of *Parthenium hysterophorus* by vermitechnology

R Hiranmai Yadav

Abstract

Globally, there are a number of treatments indicated for the control of invasive alien plant species like parthenium. The production and use of vermicomposts from weeds or other wastes in agriculture is economical and eco-friendly. The unique advantage of using vermicomposting is that it helps to build and sustain soil condition and fertility for sustainable agricultural activities. The present study was an attempt to produce the vermicompost from Parthenium, farm and animal wastes and to analyze its nutrient content and suitability to be used as manure. The raw materials Parthenium, farm and animal wastes were collected and decomposed by tank method using *Eisenia foetida*. There were four different treatments in three replications of parthenium mixed with farm wastes and animal manures @10:1:1 ratio. The pH, organic carbon, organic matter, macro and micro nutrients and exchangeable bases were analyzed by standard methods. Addition of different farm and animal wastes helped to degrade the Parthenium and improve the nutrient value. Different treatments have shown improvements in the degraded product in terms of pH, organic carbon, organic matter, macro and micro nutrients and exchangeable bases. The vermicompost was found to have a good quality comparable to any organic manure. The results revealed the economic feasibility of the vermicompost (the organic manure) production as it uses the locally available materials and eco friendly nature of its technology. This methodology can be adopted by farmers to improve the crop productivity and maintain the soil fertility using the locally available organic waste materials.

Keywords: *Eisenia fetida*, Field wastes, Nutrient content, Vermicompost, Weeds, Parthenium

Introduction

Organic farming involves the use of eco friendly manures in agriculture. The increased use of chemical fertilizers improved the production but at same time the soil fertility is getting reduced due to inadequate organic matter. To combat this, the use of organic materials are recommended. The vermitechnology is a promising technique for recycling the weeds and wastes and the resultant product improves the soil fertility and crop production without harming the nature. It is easy to practice, ecologically safe and used to reduce the pollution problems. Different experiments are being carried out to emphasize the importance and utility of this technology. The practice of vermiculture is at least a century old and now it is receiving worldwide attention as a waste management technique in terms of weeds utilization and reduction in quantity of accumulated wastes. The resultant product of vermicomposting is a nutrient rich input for crop production [1].

Reduction of wastes by co composting using chicken slurry and pineapple wastes by [2] have been found to be a digestibility of solid waste and improved nutrient balance. The co composting by improved nutrient release reduces the cost and use of fertilizers. The cattle manure, food waste and paper vermicomposts was found to be rich in nutrients and vermicomposts contains plant growth regulating materials like plant growth hormones and humic acids that can increase the germination, growth and yield of plants [3, 4].

Studies by [5] have also reported that the macronutrients were improved by the decomposition of Lantana with animal manures and earthworms. Earlier studies by [6] showed that the decomposition of Lantana residue

Correspondence: mayahiranbt@gmail.com
School of Natural Resources Management and Environmental Sciences, College of Agriculture and Environmental Sciences, Haramaya University, P.O. Box #337, Dire Dawa, Ethiopia

with oak pine leaf litter observed that there is high rate of nutrient release from the decomposing materials. Mixing of Lantana with other mulch materials positively influenced the decomposition and release of nutrients. Banta and Dev [7] have also reported low C/N ratio of the composts prepared from Lantana.

Earthworm excreta are excellent soil conditioning material with a high water holding capacity and a natural time release for releasing Nitrogen (N) into the soil [8]. The higher concentration of N in vermicompost coupled with its greater availability would be expected to make it a superior soil amendment for plant growth [9]. Total porosity and non-capillary porosity were improved in soil due to municipal compost application [10]. The highest reduction in percentage of phenols was observed in *Parthenium* compost. *Parthenium* recorded higher content of phenols in fresh forms [11]. Suthar and Sharma [12] have vermicomposted Lantana with cowdung using *Eisenia foetida* and recorded a decrease in pH, total organic carbon and C/N ratio while there was an increase in ash content, total nitrogen, available phosphorus, exchangeable potassium and calcium and nitrate nitrogen.

Parthenium hysterophorus commonly called as congress grass or carrot grass is a poisonous, allergic and obnoxious weed of fields, barren land and forest areas [13]. Since its introduction the weed is reported as relentlessly spreading throughout the agricultural lands, forests, orchards, poorly managed arable crop lands and rangelands almost throughout the country [14]. The Parthenium weed infestation increases the total stand density and biomass. It has the ability to utilize the drought prone area that leads to the deterioration of indigenous species and leave much bare ground cover [15].

The invasion of Parthenium to agricultural land is a serious problem and there is no practically effective weeding method for its eradication or control. It is increasing every year and the removal or elimination of the plant is becoming more complicating. The studies by have also shown that *Parthenium hysterophorus* can be a raw material for vermicomposting when mixed with cow dung and pressmud in equal proportion [16]. Integrated efforts should be made to suppress and get rid of the weed. The farm wastes and animal wastes of diverse types are available in field after harvest. These nitrogen rich materials could be used as an additive to enhance the decomposition of weeds like Parthenium. The wastes those are available in the farm when added to the vermicomposting process that will enhance the degradation and enrich the nutrient value of the resultant product. The present work was an attempt to decompose Parthenium with farm wastes and different animal manures and assess its manurial value.

Methods

A study was carried out to assess the impact of application of farm wastes and animal manures in decomposing the parthenium.

Substrates and combinations used in compost preparation

The materials were collected within the campus field and animal farms. There were four different treatments of parthenium mixed with farm wastes and animal manures @10:1:1 ratio. The nutrient content of fresh Parthenium is expressed in Table 1.

T_1- *Parthenium hysterophorus* plants + farm wastes + Cow dung
T_2- *Parthenium hysterophorus* plants + farm wastes + Goat manure
T_3- *Parthenium hysterophorus* plants + farm wastes + Poultry manure
T_4- *Parthenium hysterophorus* plants + farm wastes + Swine manure

Vermicomposting process

The process was carried out in cement tanks of 1 m × 1 m × 1 m size. The cement tanks were constructed under shade. The compost unit is constructed with permanent materials for continuous process. The Parthenium plants were collected before flowering stage and chopped into smaller pieces of 1–2 in. mechanically and decomposed with farm wastes and animal wastes for about 20 days. This was to reduce the size of the particles that can facilitate the feeding of earthworms. The earthworm *Eisenia fetida* was used for the decomposition process. The climatic conditions were favorable and temperature and moisture were maintained by sprinkling water regularly. After 20 days the materials were mixed and temperature was checked to release the

Table 1 Nutrient status of fresh Parthenium

S No	Nutrient content	Quantity
1.	Organic carbon (%)	2.02
2.	Organic matter (%)	3.49
3.	Nitrogen (%)	0.62
4.	Phosphorus (%)	0.10
5.	Potassium (%)	0.32
6.	Zinc (ppm)	178.00
7.	Copper (ppm)	76.00
8.	Iron (ppm)	84.00
9.	Manganese (ppm)	125.00
10.	C:N ratio	3.26
11.	C: P ratio	20.20

Source: PhD thesis, R. Hiranmai Yadav

Table 2 Nutrient status of Parthenium vermicomposted with farm and animal wastes

S No	Parameters	T1	T2	T3	T4
1.	pH	9.08	8.72	8.63	8.52
2.	OC(%)	24.56	25.90	24.48	21.83
3.	OM(%)	37.61	42.96	44.73	43.12
4.	N(%)	1.83	2.16	2.39	2.23
5.	P(ppm)	112.83	111.94	112.27	112.87
6.	K(ppm)	619.88	555.58	581.63	522.46
7.	Na(ppm)	154.58	145.40	133.92	137.25
8.	Ca(ppm)	955.94	1050.70	1204.98	1127.52
9.	Mg(ppm)	146.25	144.23	143.40	145.30
10.	Cu(ppm)	0.70	0.56	0.60	0.59
11.	Fe(ppm)	7.31	3.57	3.10	4.20
12.	Zn(ppm)	4.56	4.57	4.6	4.67
13.	Mn(ppm)	3.39	2.85	3.22	3.51
14.	C/N	13.42	11.99	10.24	9.79
15.	C/P	0.22	0.23	0.22	0.19

earthworms. About 1Kg earthworms were released into each tank. After release of worms to the tank, precautions were taken to protect the worms from predators. The vermicomposting process was carried out for duration of 60 days.

Analysis of physicochemical parameters

The nutrient contents were analysed by standard procedures. The vermicompost was collected and analysed for different parameters to assess the nutrient status.

pH

pH was measured using digital pH meter.

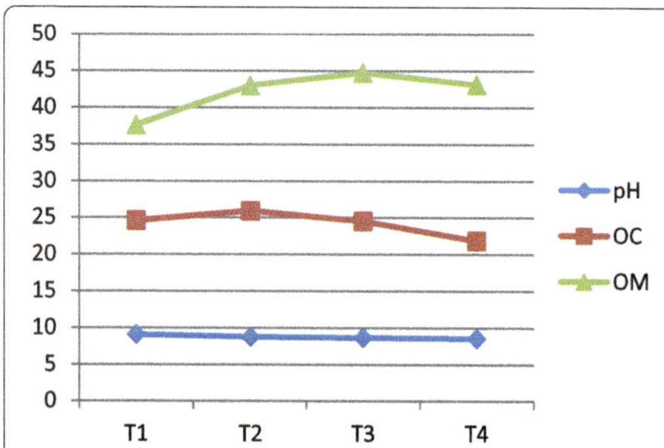

Fig. 1 pH, Organic carbon (%) and Organic matter content (%) of different vermicomposts

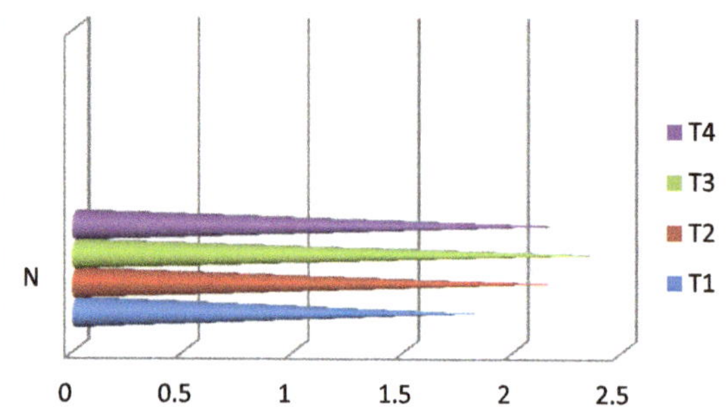

Fig. 2 Nitrogen (%) content of different vermicomposts

Estimation of organic carbon and organic matter

The Organic carbon (C) content was estimated following [17] and organic matter was calculated from it. For the analysis 0.5 g of soil was weighed (passed through 0.2 mm sieve) and transferred it to a 250 ml/500 ml conical flask. 10 ml of 1 $NK_2Cr_2O_7$ was added and mixed well by swirling the flask. Then 20 ml of con. H_2SO_4 was added and mixed by gentle rotation for one minute to ensure complete contact of the reagent with the soil. Allowed the contents to stand for 20–30 min. The flask was kept preferably on an asbestos sheet to avoid burning of table due to intense heat. The reaction mixture was diluted with 200 ml distilled water. 10 ml of phosphoric acid and 1 ml of diphenylamine indicator were added. Titrated the solution with 0.5 N ferrous ammonium sulphate. The colour was dull green at the beginning and then shifted to a turbid blue as the titration proceeded. The end point was very sharp. At the end point the colour sharply shifted to a brilliant bright green. The organic carbon content in the sample was determined using the formula,

$$\text{Per cent organic carbon} = \frac{(\text{B.V.}-\text{S.V.}) \times 10 \times 0.003 \times 100}{\text{B.V.} \times \text{W}}$$

Organic matter was calculated by multiplying the value of organic carbon by a standard value of 1.724.

Estimation of total Nitrogen (N)

The macro nutrients namely total Nitrogen (N) was calculated by Micro kjeldahl method [18]. One g of sample was transferred into a 100 ml conical flask. 15 ml of diacid mixture (sulphuric acid and perchloric acid in the ratio of

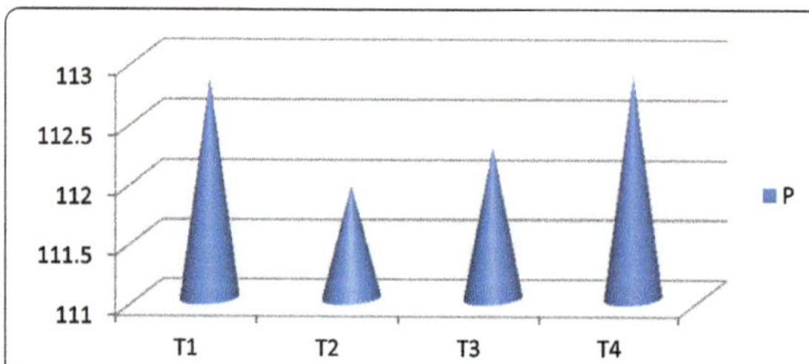

Fig. 3 Phosphorus content of different vermicomposts

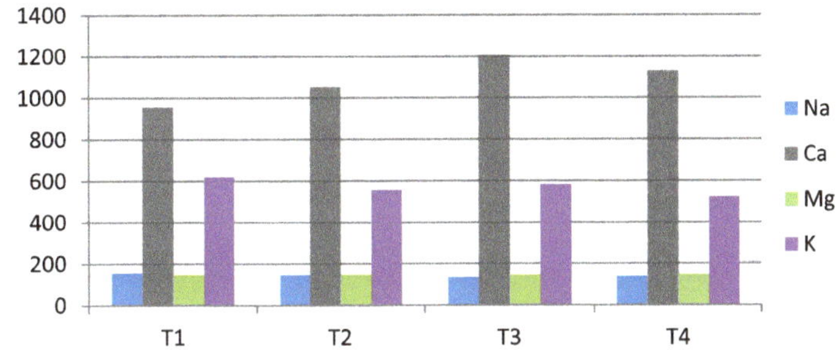

Fig. 4 Exchangeable bases of different vermicomposts

5: 2) was added and covered the mouth of the flask with a funnel. Digested the contents of the flask over a sand bath till a clear solution was obtained. The contents of the flask were transferred into a 250 ml volumetric flask using minimum quantity of water and made up the volume to 250 ml.10 ml of the diacid extract was pipetted out into a distillation flask. 25 ml of 2 % boric acid was pipetted out into an ice tumbler/beaker and two drops of double indicator (bromocresol green methyl red) was added and kept it at the delivery end of the distillation set.10 ml of 40 % sodium hydroxide was added to the distillation flask and distilled the contents and the distillate was collected in 2 % boric acid. Tested the completion of distillation with a moistened red litmus paper. Absence of blue colour indicated that all ammonia had been distilled. The delivery tube was washed with distilled water and the washings were collected in ice tumbler. Titrated the contents with N/10 sulphuric acid. The end point was the change of colour from green to pinkish.

Calculation

Weight of the sample taken = 1 g

Volume of diacid extract prepared = 100 ml

Volume of diacid extract pipetted out for analysis = 10 ml

Volume of N/10 H_2SO_4 consumed = X ml

1 ml of N/10 H_2SO_4 = 0.0014 g of N

This is present in 10 ml of the diacid extract

Therefore in 250 ml $= \frac{X \times 0.0014 \times 100}{10} \times$ g of N

This is present in 1.0 g of the sample in 100 g $= \frac{X \times 0.0014 \times 100 \times 100}{10} \times$ g of N

% of nitrogen on moisture free basis $= \frac{X \times 0.0014 \times 100 \times 100}{10} \times \frac{100}{(100-M)}$ (M = Moisture content of the sample)

Estimation of Phosphorus (P) and micronutrients

The amount of phosphorus and micronutrients in the sample was estimated by the method of [19].

Estimation of phosphorus (P)

5 ml of the triple acid extract was pipetted out into a 25 ml volumetric flask. 5 ml of Barton's reagent was added, shaken well and made up the volume to 25 ml and made it a homogenous solution. Allowed to stand for about 30 min. for the colour to develop. The intensity of yellow colour developed was read in a spectrophotometer at 470 nm. By referring to the standard curve prepared calculated the phosphorus content of the sample.

Calculation

Weight of sample taken = w g/g

Volume of triple acid extract prepared = V ml

Aliquot taken for colour development = 5 ml

Concentration of the solution as read from the spectrophotometer = x ppm

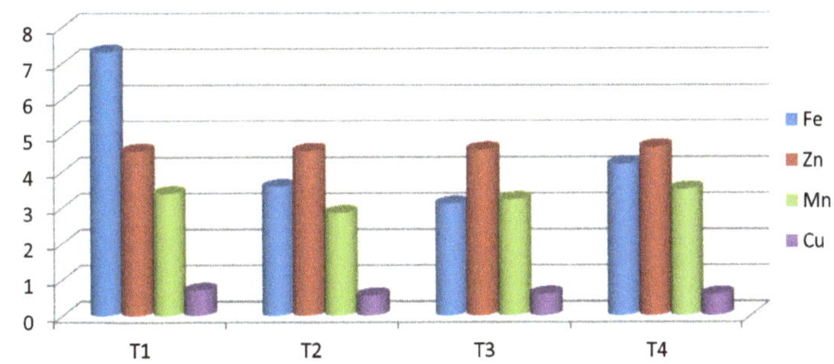

Fig. 5 Micro nutrients contents of different vermicomposts

Therefore P content on moisture free basis = $\frac{x}{10^6} \times \frac{25}{5} \times V \times \frac{100}{1w} \times \frac{100}{(100-M)}$ (M – Moisture content of the sample)

Estimation of micronutrients

The micronutrients like Zinc (Zn), Copper (Cu), Iron (Fe) and Manganese (Mn) were also estimated. From the working standard solutions, the standard curve was prepared for each nutrient. In between two standard solutions, introduced the blank and ensured that there is no change in the zero point. Prepared a graph by plotting the absorbance values against concentrations. Transferred 10 g of air dried soil sample to 150 ml conical flask/polythene bottle and added 20 ml of the DTPA extractant solution. Closed the bottle and shaked for 2 h in a horizontal shaker. Filtered through whatman No. 42 filter paper. Take a reading of the filtrate in the atomic absorption spectrophotometer and by referring to the standard curve calculate the concentration of each micronutrient in the sample. Shaking time, concentration and pH of the DTPA extractant and temperature will influence the quantity of nutrients extracted. The exchangeable bases Sodium (Na) by 1 M ammonium acetate method, Calcium (Ca), Magnesium (Mg) [20] and Potassium (K) [21] contents were also analysed. C: N and C: P ratios were calculated.

Results and discussion

The vermicompost harvested after sixty days of decomposition was dark brown in color and homogeneous in nature. The product had an earthy aroma characteristic of the matured compost. The biomass of earthworms was also increased considerably from the initial added number. It indicates that the substrates were suitable for the multiplication of the organisms. Growth and reproduction of worms require sufficient nutrient from the substrates. The palatability of the substrate materials are of significance for multiplication and thereby conversion into a nutrient rich organic manure by the earthworm activity. The decomposition of the organic wastes improves the nutrient value of the end products (Table 2). The pH was alkaline in all the vermicomposted samples that ranged from 9.09 in Parthenium vermicomposted with farm wastes and cow dung (T_1) to 8.52 in T_4. The organic carbon and organic matter varied considerably among the different treatments (Fig. 1). The nitrogen and phosphorus contents were also found to be significant in different treatments (Figs. 2 and 3). The exchangeable bases are depicted in Fig. 4 and micro nutrients in Fig. 5. The exchangeable bases potassium, sodium and magnesium and micro nutrients copper and iron were also found to be high in T_1. The calcium content was higher in T_3 whereas the zinc and manganese were in T_4. The C/N and C/P ratios are represented in Figs. 6 and 7. They were within the recommended range for organic manures. All the nutrient contents were improved by the treatments with different organic supplements.

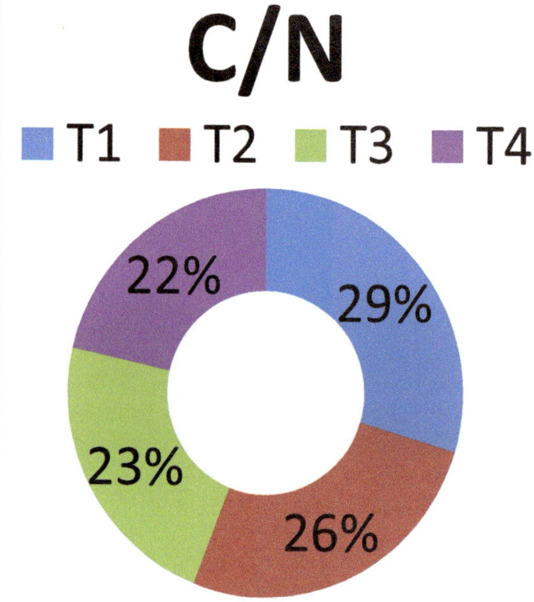

Fig. 6 C/N ratio of different vermicomposts

Similar observations are made by [22] in decomposition of Parthenium with *Eisenia foetida*. They reported that utilization of organic supplements for the degradation of Parthenium helps to improve the nutrient value in addition to the reduction of weed biomass. The resultant product could be a good source of organic manure.

The earlier studies also showed that the pH of the fresh Parthenium was slightly acidic (6.84) whereas it

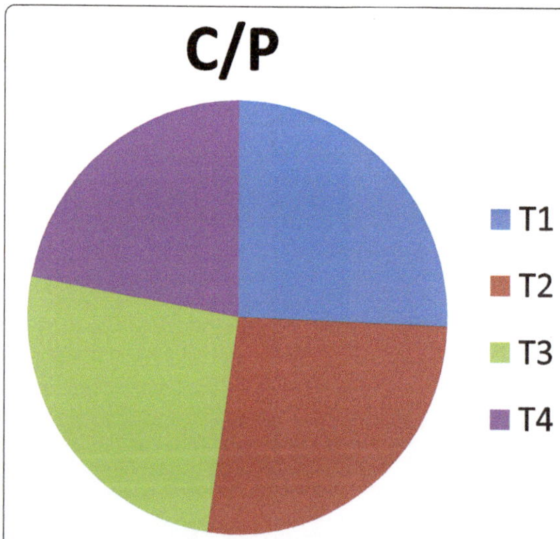

Fig. 7 C/P ratio of different vermicomposts

turned alkaline in composted (8.89) and vermicomposted (8.53) samples. The content of organic carbon (C) and organic matter varied among the fresh and decomposed samples. It was reduced in composted and vermicomposted Parthenium. The fresh Parthenium recorded the lowest per cent of total Nitrogen (N), Phosphorus (P) and Potassium (K) compared to decomposed materials. Zinc (Zn) and Copper (Cu) contents were lower in fresh Parthenium compared to the composted and vermicomposted Parthenium. The Iron (Fe) content was lower in composted Parthenium compared to both fresh and vermicomposted samples. Manganese (Mn) content was observed to be low in fresh, than composted and vermicomposted Parthenium [23].

A tremendous increase in organic carbon and potassium content of vermicompost prepared from farm and weed wastes (papaya leaf litter, glyricidia loppings and pigeon pea stalks) was recorded [24]. The application of this compost increased the nitrogen, potassium and organic carbon content of the soil and thereby improving the congenital release and utilization of the nutrients by the crops. The application of vermicompost prepared from celotia and Parthenium was also found beneficial in reducing the weed infestation in the groundnut crop. The results of [25] indicated that the Parthenium compost at low amendments with cowdung help its eradication with better utilization. The improved nitrogen, phosphorus, potassium and calcium and decrease in pH and organic carbon was observed by [26] in Parthenium vermicomposted with *Eisenia fetida*. Their results indicated that the Parthenium with cowdung in appropriate quantity can be a raw material for vermicomposting. The vermicompost prepared with Parthenium and cow dung in 3:2 ratio using *Eudrillus eugeneae* and *Eisenia fetida* and found that Parthenium plants without flowers had high pH, N and K than with flowers whereas the phosphorus content was high in parthenium plants composted with flowers [27].

Conclusions

The adoption of vermicomposting to eradicate the weeds and reduce the problems arising due the weeds in agricultural fields is gaining momentum among farmers. The technology should be utilized effectively in this regard. The present work indicates that the process can be carried out in simple methods and using materials that are available on farm. The weeds can be collected and mixed with organic farm wastes that will add the nitrogen content and increase the nutritive value of the resultant product. The nutrient status of the resultant products promises the utilization of animal wastes and farm wastes based on the availability of farmer's field for decomposition of Parthenium weed. The vermitechnology in addition to weed reduction and waste utilization also gives an opportunity to entrepreneurship.

Competing interests

The author declares that they have no competing interests.

Author's contributions

The author have carried out the research work and prepared the present article solely completely.

References

1. Joshi PC, Chauhan A. Composting of some organic materials using *Eisenia foetida* and conventional microbial methods: a comparative study. Uttar Pardesh J Zool. 2006;6:123–5.
2. Angelidaki I, Ahring BK. Co digestion of olive mill waste water with manure, house hold waste or sewage sludge. Biodegradation. 1997;8:221–6.
3. Atiyeh RM, Lee S, Edwards CA, Arancon NQ, Metzger JD. The influence of humic acids derived from earthworms- processed organic wastes on plant growth. Bioresour Technol. 2002;84:7–14.
4. Arancon NQ, Edwards CA, Lee S, Byrne R. Effects of humic acids from vermicomposts on plant growth. Eur J Soil Biol. 2006;46:65–9.
5. Singh KP, Angiras NN. Allelopathic effect of wild sage (*Lantana camara* L.) compost on wheat and associated weeds under north western Himalayas. J Env Bio-Sci. 2011;25:91–2.
6. Kumar P, Pant M, Negi GCS. Soil Physico-Chemical Properties and Crop Yield Improvement Following *Lantana* Mulching and Reduced Tillage in Rainfed Croplands in the Indian Himalayan Mountains. J Sustain Agric. 2009;33:636–57.
7. Banta G, Dev SP. Field evaluation of nitrogen enriched with phosphocompost prepared from green biomass of *Lantana camara* in wheat. Indian J Ecol. 2009;36:39–44.
8. Harris GD, Platt WL, Price BC. Vermicomposting in a rural community. Biocycle. 1990;31:48–51.
9. Christopher MSM. Recycling of plantation agro-wastes. Planters' Chron. 1996;91:53–61.
10. Mathan KK. Impact of biological wastes on soil physical properties and yield of maize and finger millet. Madras Agric J. 2000;87:618–20.
11. Ramaswami PP. 1997. In: Proc. First International Conference on Parthenium Management, 1: 77–80.
12. Suthar S, Sharma P. Vermicomposting of toxic weed- *Lantana camara* biomass: chemical and microbial properties changes and assessment of toxicity of ebnd product using seed bioassay. Ecotoxicol Environ Saf. 2013;95:179–87.
13. Bakthavalsalam R, Geetha T. Reproductive influence of earthworm, *Lambito mauriti* cultured in different media of paddy chaff and weed plants materials. Environ Ecol. 2004;22:574–8.
14. Tefera T. Allelopathic effects of Parthenium hysterophorus extracts on seed germination and seedling growth of Eragrotis tef. Blackwell Verlag, Berlin. J Agronomy Crop Science. 2002;188:306–10.
15. Etana A, Kelbessa E, Soromessa T. Impact of *Parthenium hysterophorus* L.(Asteraceae) on herbaceous plant biodiversity of Awash National Park(ANP), Ethiopia. Manag Biolog Invasions. 2011;2:69–80.
16. Anbalagan M, Manivannan S. Assessment of impact of invasive weed *Parthenium hysterophorus* L. mixed with organic supplements on growth and reproduction preference of *Lampito mauritii* (Kinberg) through vermitechnology. Int J Environ Biol. 2012;2:88–91.
17. Walkley A, Black CA. An examination of the Degtiareff method for determining soil organic matter and proposed modification of chromic acid titration method. Soil Sci. 1936;37:29–38.
18. Humphries EC. Modern methods of plant analysis. Springer-Verlag. 1956;1:468–502.
19. Jackson ML. Soil chemical analysis. New Delhi: Prentice Hall of India Pvt. Ltd; 1973. p. 10–50. 461–64.
20. Hesse PR. A text book of soil chemical analysis. London, U.K: John Murry (Publishers) Ltd; 1971. p. 528.
21. Piper CS. Soil and plant analysis. Bombay: Hans Publishers; 1966. p. 368.
22. Anbalagan M, Manivannan S, Arul Prakasm B. Biomanagement of *Parthenium hysterophorus*(Asteraceae) using an earthworm, *Eisenia fetida*(Savigny) for recycling the nutrients. Adv Appl Sci Res. 2012;3:3025–31.

23. Hiranmai Yadav R, Eyasu M, Vijayakumari B. Bioconversion of *Parthenium hysterophorus* into valuable organic manure. Bahir Dar, Ethiopia, East Africa: Proceedings of the National Conference on Science, Technology, and Innovations for Prosperity of Ethiopia; 2012. p. 54–8.
24. Babu M, Vijaya S, Adinarayana G, Subbaiah KR, Balaguravaiah D, Yellamanda RT. Vermicompost with different wastes and problematic weeds. Indian J Agr Res. 2008;42:52–6.
25. Sivakumar S, Kasthuri H, Senthilkumar P, Subbhuraam CV, Song YC. Efficiency of composting Parthenium plant and neem leaves in the presence and absence of an oligochaete, *Eisenia fetida*. Iranian J Environ Health Sci Eng. 2009;6:201–8.
26. Yadav A, Garg VK. Vermicomposting an effective tool for management of invasive weed *Parthenium hysterophorus*. Bioresour Technol. 2011;102:5891–5.
27. Indira S, Koireng RJ, Singh NR. Vermicomposting techniques and nutrient contents of Parthenium based organic manure involving manure worm (*Eisenia fetida and Eudrillus eugeniae*). Green Farming. 2009;2:769–70.

Removal of inorganic mercury from aquatic environments by multi-walled carbon nanotubes

Kamyar Yaghmaeian[1], Reza Khosravi Mashizi[1*], Simin Nasseri[1,2], Amir Hossein Mahvi[1], Mahmood Alimohammadi[1] and Shahrokh Nazmara[1]

Abstract

Background: Mercury is considered as a toxic heavy metal in aquatic environments due to accumulation in bodies of living organisms. Exposure to mercury may lead to different toxic effects in humans including damages to kidneys and nervous system.

Materials and methods: Multi-walled carbon nanotubes (MWCNTs) were selected as sorbent to remove mercury from aqueous solution using batch technique. ICP instrument was used to determine the amount of mercury in solution. Moreover, pH, contact time and initial concentration of mercury were studied to determine the influence of these parameters on the adsorption conditions.

Results: Results indicate that the adsorption strongly depended on pH and the best pH for adsorption is about 7. The rate of adsorption process initially was rapid but it was gradually reduced with increasing of contact time and reached the equilibrium after 120 min. In addition, more than 85 % of initial concentration of 0.1 mg/l was removed at 0.5 g/l concentration of sorbent and contact time of 120 min. Meanwhile, the adsorption process followed the pseudo second-order model and the adsorption isotherms could be described by both the Freundlich and the Langmuir models.

Conclusion: This study showed that MWCNTs can effectively remove inorganic mercury from aqueous solutions as adsorbent.

Keywords: Adsorption, Aqueous solutions, Heavy metals, Nano material

Background

Water resource pollution with industrial effluents is known as a serious environmental problem, nowadays [1].Of these, scientists have focused mostly on the presence of mercury due to its bioaccumulation in organisms, toxic effects and persistence in environment [2, 3]. In addition, it should be noted that mercury has been widely used in various industrial fields as an element including chlor-alkali, pharmaceutical, producing barometer and thermometer, mining, dental practices. It is proved that high concentrations of mercury are released constantly from the mentioned industries to the environment [4, 5].

From the toxicological point of view, the level of toxicity of mercury is highly related to its chemical form [6]. To put it another way, mercury transforms biologically, physically, and chemically through its cycle in nature, which results in the formation of various forms of mercury. Organic mercury is the most toxic among these forms [7]. Mercury is mostly in its inorganic (Hg^{+2}) or methylmercury forms in aquatic environments [6]. However, at the presence of specific kind of bacteria the inorganic form of mercury transforms to methylmercury which is highly toxic for human and other organisms at food chain [8]. Exposing to mercury results in neurological disorders, damage to central nervous systems, and also negatively affects the kidney and liver [9]. Considering these facts, there should be a proper way to handle this element and remove it from the environment.

Various methods have been used to remove mercury from water and wastewater, including chemical precipitation, ion exchange and membrane methods [10].

* Correspondence: khosravireza60@yahoo.com
[1]Department of Environmental Health Engineering, School of Public Health, Tehran University of Medical Sciences, Tehran, Iran
Full list of author information is available at the end of the article

Nevertheless, these methods have their own weaknesses including high level of either energy or chemical compound is needed, and most importantly these methods are not able to remove low concentration of mercury from the environment. However, adsorption due mostly to its high performance, recoverability, and reactive ability of adsorbent can be considered as a suitable method in terms of economy [11, 12].

In adsorption process, it is needed to have an adsorbent with wide specific surface. The surface results from the existence of tiny pores at it which the chemical property, area, size and distribution of these pores influence the level of an adsorbent's specific surface. Different materials, such as fruit shell [13], chitosan [14], marine macroalga [15], bagasse pith [16], furfural [17], and rubber [18] have been applied as adsorbents to remove mercury from aqueous environments.

After the discovery of carbon nanotubes, scientists paid specific attention to them because of their particular efficiency in construction, electricity, chemistry, and physics. Also, these materials have widely been used to produce nano-structured materials, nanocomposites, sensors, and gas adsorption. In 2004, when EPA proposed a research into the environmental application of these materials [19], wide ranges of studies were conducted on these nanotubes which have tiny pores with uniform size and also wide specific surface [20]. Usage of carbon nanotubes was studied to remove pollutants, such as fluoride [21], dichlorobenzene [22], trihalomethanes [23], zinc [24], chromium [25], nickel [26, 27], and cadmium [28] from water and waste water. In this paper, Multi-walled carbon nanotubes) $MWCNT_S$ (were used to remove inorganic mercury from aqueous solutions.

Materials and methods

Mercury solution was prepared by using $HgCl_2$ (Merk) and deionized water. Carbon nanotubes were obtained from research division of Iranian Petroleum Industry. The characteristics of applied nanotubes in this study were as follow: the BET surface area of 270 m^2/g; diameters of 10–30 nm, respectively; length of 10 μm, and over 95 % purity. Furthermore, the morphology and size of carbon nanotubes were characterized by transmission electron microscope (TEM) and scanning electron microscope (SEM). The surface functional groups of multi-walled carbon nanotubes were detected by Fourier transform infrared spectroscopy (FTIR). Figure 1 shows the TEM, SEM and FTIR images of carbon nanotubes.

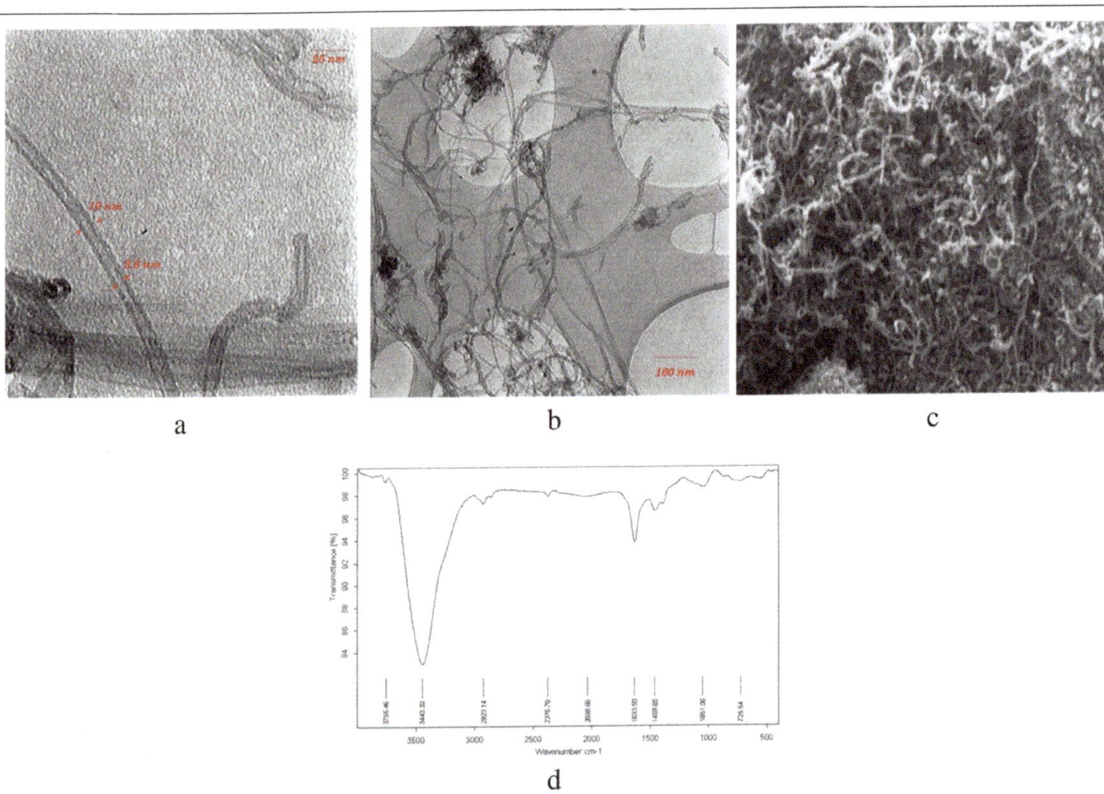

Fig. 1 TEM (**a** and **b**), SEM (**c**) and FTIR spectra (**d**) of MWCNTs

Table. 1 considered variables in previous studies the removal of metal ions from aquatic environments by carbon nanotubes

Metal ion	Variables	Ref	Metal ion	Variables	Ref
zinc	pH, contact time, initial metal ion concentration. Isotherm models	[24]	Zinc	adsorption kinetic, Isotherm models	[30]
Chromium	pH, contact time, agitation speed	[25]	copper	pH, ionic concentration, Isotherm models	[31]
nickel	pH, contact time, initial metal ion concentration, adsorbent's concentration, Isotherm models	[26]	Lead	pH, contact time, agitation speed, adsorbent's concentration, adsorption kinetic, Isotherm models	[32]
cadmium	pH, contact time, initial metal ion concentration, temperature, adsorption kinetic, Isotherm models	[28]	Lead	contact time, pH, ionic strength, foreign ions	[33]

Batch reactors with the same volumes of 250 ml were used. The reactors were filled with 100 ml of mercury solution with the concentrations of 0.1, 1 and 10 mg/l. The pH of the solutions were adjusted by nitric acid 0.1 N and NaOH 0.1 N (Merk). After adjusting the pH, the solutions were agitated under the temperature of 25 °C and 150 rpm on an Incubator Shaker (Innova 4340, USA). Thereafter, the solution was passed through 0.2 μm Millipore filter in order to separate the adsorbents from the aqueous solutions. Then, the pH was adjusted and lowered to <2 by using nitric acid. It should be noted that the prepared solutions were kept in glass containers at 4 °C. Besides, Cold vapour inductively coupled plasma optical emission spectrometry (Spectro, Germany) was applied to measure the concentration of mercury. This method has high sensitivity, excellent detection limits, rapid analysis and Easy to use. Also chemical and spectral interference is less than other methods [29]. For the reliable determination of mercury all variables were measured at least twice.

Some variables were considered in previous studies the removal of metal ions from aquatic environments by carbon nano tubes (Table 1). The effects of contact time, primary concentration of the solution, adsorbent's concentration, ionic strength of the solution, and pH on the process of adsorption were assessed in this research. Freundlich and Langmuir adsorption isotherms were used to study the adsorption isotherm. Also pseudo-first- order and pseudo-second-order kinetics models were applied to determine the adsorption kinetic model.

The amount of adsorbed Hg^{2+} was calculated through the following equation [24–30]:

$$q = \frac{(C_o - C_t) \times V}{m} \quad (1)$$

where q denotes the amount of Hg^{2+} adsorbed on adsorbent at any time (mg/g), C_0 denotes the initial Hg^{2+} concentration (mg/l), C_t denotes the concentration of Hg^{2+}at any time (mg/l), V denotes the volume of the solution (l) and m denotes the adsorbent mass (g).

Results and discussion

The surface functional groups of MWCNTs

Figure 1(d) depicts the FTIR spectra of multi-walled carbon nanotubes. This spectrum displays major peaks at 3755, 3443, 2923, 1633, 1459 and 1051 cm^{-1}. The peak at 3755 cm^{-1} is associated with free hydroxyl groups. The peak at 3443 cm^{-1} is attributed to the O–H stretch from carboxyl groups (O = C - OH and C - OH). The peak at 2923 cm^{-1} is associated with (CH2). The peak at 1633 cm^{-1} is related to carbonyl groups. The peak at

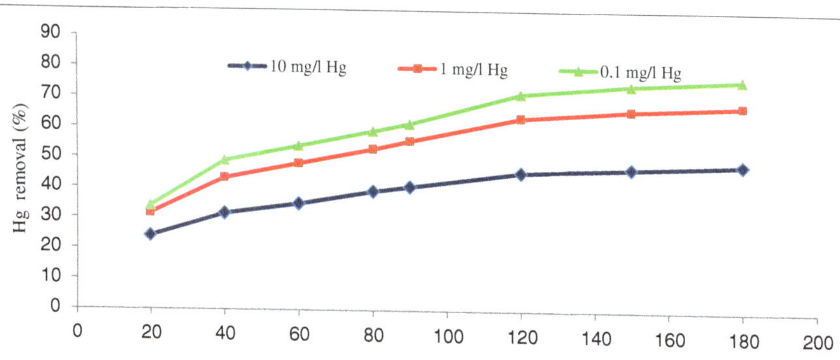

Fig. 2 Effect of contact time on the removal of mercury by $MWCNT_S$ at different mercury concentrations (adsorbent dose = 0.5 g/l, T = 25 °C, pH =5, agitation speed = 150 rpm)

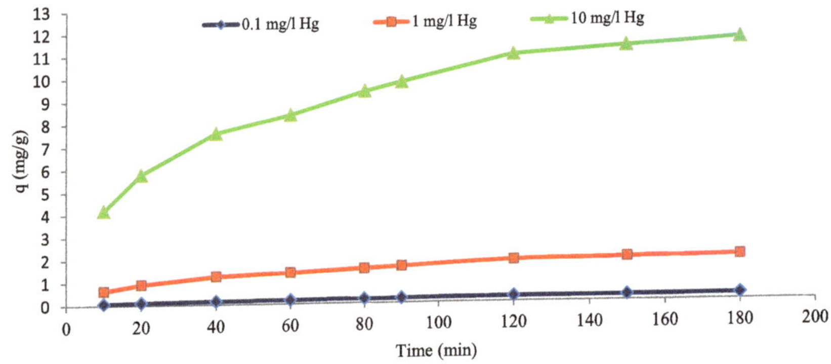

Fig. 3 Effect of contact time on the q at different mercury concentrations adsorbed on to $MWCNT_S$ (adsorbent dose = 0.5 g/l, T = 25 °C, pH =5, agitation speed = 150 rpm)

1459 cm^{-1} is associated with carboxylic acids and phenolic groups (O–H). The peak at 1051 cm^{-1} is attributed to the (C–O). These functional groups can prepare numerous chemical sorption sites on surface of the MWCNTs [24, 28, 30, 32, 33].

Effect of contact time

As it is indicated in the Fig. 2, the effect of contact time on the level of adsorption of inorganic mercury was evaluated on the surface of multi-walled carbon nanotubes. In this test, the primary concentrations of inorganic mercury were 0.1, 1, and 10 mg/l and the other factors including temperature, pH, concentration of adsorbent, and agitation speed were kept constant. The adsorption of mercury increased with time until it reached equilibrium. The contact time for attainment to equilibrium was 120 min. Also Fig. 2 shows that the increase in the initial concentrations of mercury ions did not affect the equilibrium time. Similar findings have been reported for sorption of Ni^{+2} on carbon nanotubes [26] In addition, the rate of adsorption was reduced with the increase of contact time, which this reduction can be due to the saturation of adsorption points on the carbon nanotubes. '*q*' which is the amount of adsorbed mercury per the weight of carbon nanotubes raised by increase of contact time (Fig. 3).

Effect of primary concentration

The level of mercury removal in the pH of 5, contact time of 120 min, temperature of 25 °C, agitation speed of 150 rpm, and primary concentration of 0.1, 1, and 10 were 71.4, 63.6 and 45.6 % respectively (Fig. 4). the percentage of mercury removal was declined when the concentration of mercury increased from 0.1 to 10 mg/l. Primary concentration also affects the removal of chrome from aqueous solutions by carbon nanotubes [34]. When the concentration of adsorbent in the solution is constant, the number of adsorbent places is also constant for metallic ions. In this condition, the increase in the number of such ions results in the increase of competence among these ions to be adsorbed and it leads to the decrease of metallic ions' removal due mainly to electrical repulsive force [34]. It is noteworthy that the increase of primary concentration do not

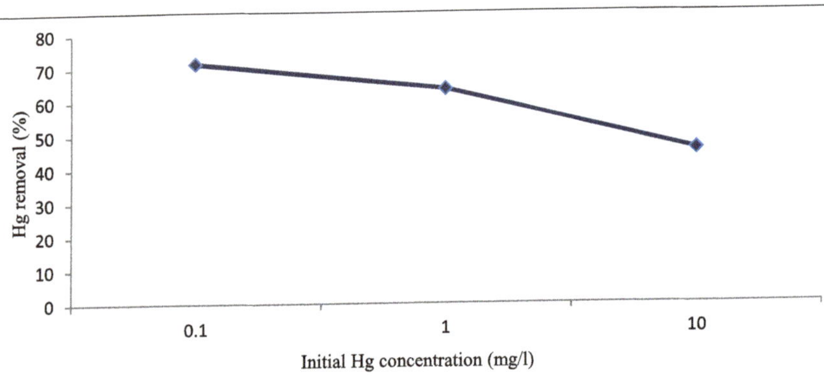

Fig. 4 Effect of primary Hg concentration on the removal of mercury by $MWCNT_S$ (adsorbent dose = 0.5 g/L, T = 25 °C, pH =5, agitation speed = 150 rpm, contact time = 120 min)

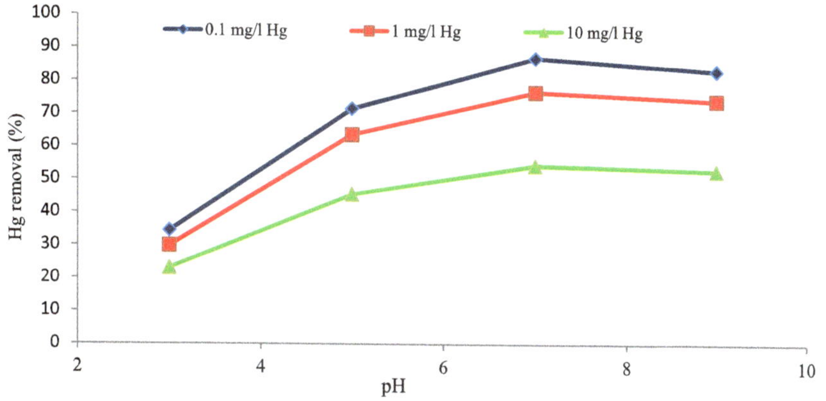

Fig. 5 Effect of pH on the removal of mercury by $MWCNT_S$ at different mercury concentrations (adsorbent dose = 0.5 g/l, T = 25 °C, contact time = 120 min, agitation speed = 150 rpm)

influence the balance time but it significantly affects 'q' (Fig. 3). '*q*' raised with increasing of zinc primary concentration in the removal of zinc from aqueous solutions by carbon nanotubes [24].

Effect of solution's pH

pH is known as an important and effective parameter on adsorption process. The pH effects not only include the type of ion but also include the properties of adsorbent surface such as surface active groups [35]. As it is shown in the Fig. 5, the effects of pH on the level of mercury's adsorption on carbon nanotubes in the *conditions* of mercury's concentrations of 0.1, 1, and 10 mg/l and temperature of 25 °C, contact time of 120 min, adsorbent dosage of 0.5 g/l and agitation speed of 150 rpm were assessed. Based on the results, pH has a significant effect on the process of adsorption and the increase of it raises the capacity of mercury's adsorption. Besides, the appropriate pH achieved in this study is 7, and below the 5, mercury transforms to Hg^{+2} which competes with H^+ for the adsorbent places. Hence, at the low pH, the level of adsorption decreases. It is mainly due to this fact that the H^+ level declines when the pH increases. The results of the present study are in the line with the studies conducted by Gao et al. [36] on the removal of Cu, Cd, Ni, and Zn by carbon nanotubes, and Atieh et al. [32] on the removal of Pb by the same adsorbent.

After the test, the pH was measured again and it was cleared that in the primary pH of upper than 5, the final pH decreases. The release of H^{2+} from the surface of carbon nanotubes into the solution could be the reason of this decrease. In the same contact time, the decrease of final pH was higher for the high concentration of mercury then the low concentration of mercury. When

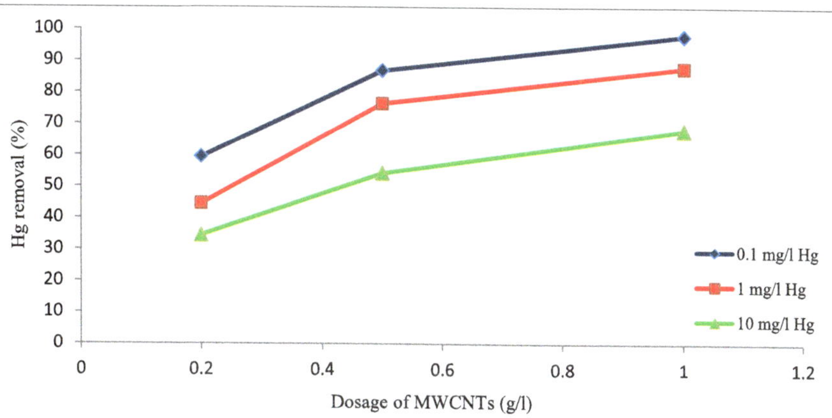

Fig. 6 Effect of adsorbent dosage on the removal of mercury by $MWCNT_S$ at different mercury concentrations (T = 25 °C, contact time = 120 min, agitation speed = 150 rpm, pH = 7)

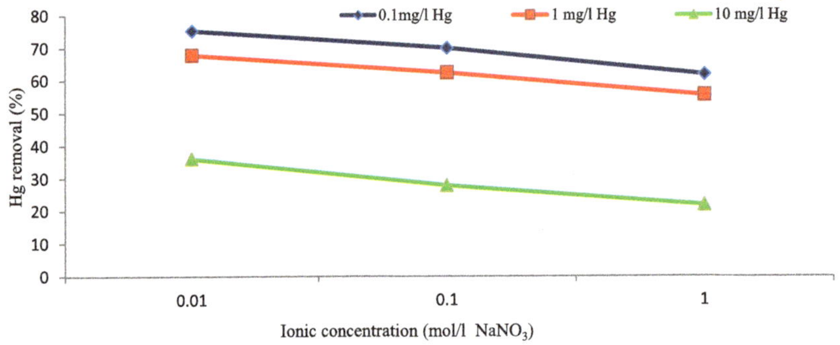

Fig. 7 Effect of Ionic concentration on the removal of mercury by $MWCNT_S$ at different mercury concentrations (adsorbent dose = 0.5 g/l, T = 25 °C, contact time = 120 min, agitation speed = 150 rpm, pH = 7)

the primary concentration of mercury increases, the amount of adsorption of Hg^{+2} also gets higher by which the rate of H^{2+} release from the adsorbent surface increases and causes to the drop of the pH level of the solution [37]. This process can be the indicator of chemical adsorption of inorganic mercury on the surface of the carbon nanotubes in pH of over 5.

Effect of adsorbent's concentration

The results indicated that the percentage of mercury's removal increased when the concentration of carbon nanotubes increased from 0.2 to 0.5 and finally to 1 mg/l (Fig. 6). This is mostly due to this reason that the increased concentration of adsorbent makes more active surface [33]. Similar findings have been reported for sorption of Ni^{+2} on carbon nanotubes [27].

Effect of ionic concentration

Based on the results, high ionic concentration has negative effect on the removal of mercury (Fig. 7). The reason is probably the impact of ionic concentration on the transfer of Hg^{+2} from the solution to the surface of adsorbent. In a study conducted by Lu and Liu [27], the authors found that the rate of Ni^{2+} adsorption on the surface of carbon nanotubes decreases when the ionic concentration increase.

Adsorption isotherm

Adsorption isotherms describe the distribution of metal ions between the liquid and solid phase at equilibrium. Adsorption balance of metallic ions is usually studied by the Freundlich and Langmuir adsorption isotherm. Langmuir isotherm is the indicator of active surface adsorption on the homogenous surface, while the Freundlich isotherm is used for heterogeneous surfaces [11]. The linear form of Langmuir and Freundlich equations are as below.

Langmuir

$$\frac{C_e}{q_e} = \frac{1}{kq_m} + \frac{C_e}{q_m} \tag{2}$$

Freundlich

$$\log q_e = \log K_F + 1/n \log C_e \tag{3}$$

Where C_e denotes the equilibrium concentration of Hg^{+2} (mg/l), q_e denotes the amount adsorbed (mg/g), q_m denotes the theoretical saturated adsorption capacity (mg/g), K denotes the Langmuir constant (l/mg). The values of K and q_m were calculated by plotting C_e/q_e versus C_e. K_F and n are the Freundlich constants related to adsorption capacity and adsorption intensity, The Freundlich constants n and K_F were calculated by plotting log q_e versus log C_e.

The information of the two adsorption isotherms of inorganic mercury on the surface of carbon nanotubes are indicated in Table 2. As can be inferred from the information, the process of adsorption follows both Freundlich and Langmuir isotherms. Li et al. [38] depicted that adsorption of lead on to carbon nanotubes follows both Langmuir and Freundlich equations. Besides, the maximum adsorption capacity obtained from the Langmuir isotherm is 25.641 mg/g. Numerous low cost organic and inorganic adsorbents (e.g. activated carbon) have been

Table. 2 The parameters of Langmuir and Freundlich isotherm models for the removal of Hg^{+2} by $MWCNT_S$

Isotherm models	Parameters	value
Langmuir	q_m (mg/g)	25.641
	k (l/mg)	0.565
	R^2	0.948
Freundlich	K_F (l/ g)	0.077
	N	0.7173
	R^2	0.999

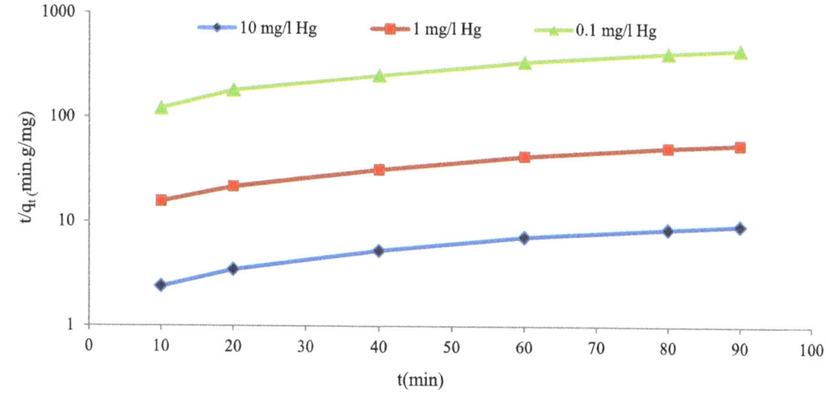

Fig. 8 Pseudo- second -order kinetics plot for the removal of Hg^{+2} by $MWCNT_S$ at different mercury concentrations (adsorbent dose = 0.5 g/l, T = 25 °C, agitation speed = 150 rpm, pH = 5)

explored for mercury removal [39, 40]. Zabihi et al. [41], Di Natale et al. [42], Asasian et al. [43] used different activated carbons to remove mercury that adsorption capacities of activated carbons were different. These amounts were strongly dependent on adsorption conditions such as solution's pH, adsorbent's concentration, temperature, ionic concentration and especially initial concentration of mercury [42]. However, activated carbons have disadvantages like weak physical stability, low selectivity for mercury, poor reactive ability of adsorbent and the release of mercury vapor into the atmosphere [39].

Adsorption kinetic

The adsorption kinetic model can provide suitable information for designing the removal of pollutants from water and wastewater. In order to assess the adsorption kinetic of inorganic mercury on the surface of carbon nanotubes, the pseudo-first and pseudo-second orders of kinetic equations were applied.

The pseudo first-order equation is written as:

$$\frac{dq_t}{dt} = k_1(q_e\text{-}q_t) \tag{4}$$

Where q_t is the amount of Hg^{+2} adsorbed at any time (mg/g), q_e is the amount of Hg^{+2} adsorbed at equilibrium (mg/g), K_1 is the adsorption rate constant (1/min).

The integrating for the boundary conditions $t = 0$ to $t = t$ and $q_t = 0$ to $q_t = q_e$, gives the linear relationship between the amount of Hg^{+2} adsorbed (q_t) and time (t).

$$\text{Log}(q_e\text{-}q_t) = \text{Log}\, q_e \text{-} \frac{k_1}{2.303} t \tag{5}$$

A straight line log (q_e -q_t) versus t indicates the applicability of the pseudo-first order kinetic model.

The pseudo second-order equation can be expressed as:

$$\frac{dq_t}{dt} = k_2(q_e\text{-}q_t)^2 \tag{6}$$

Where k_2 denotes the second-order sorption rate constant (g/mg min). Integrating for the boundary conditions, $t = 0$ to $t = t$ and $q_t = 0$ to $q_t = q_e$, gives the following equation:

$$\frac{t}{q_t} = \frac{1}{k_{2qe^2}} + \frac{1}{q_e} t \tag{7}$$

The information regarding these two equations is provided in Fig. 8 and Table 3. The achieved q_e from the pseudo-first order adsorption kinetic for the concentration of 0.1, 1, and 10 mg/l are 0.17, 1.40, and 8.1 mg/g, respectively. The corresponding scores for the pseudo-second order adsorption kinetic are 0.24, 1.99, and 11.76 mg/g, respectively, while the experimental q_e for the primary concentrations of mercury are 0.23, 1.84,

Table. 3 Kinetic parameters for the removal of Hg^{+2} by $MWCNT_S$ at different mercury concentrations

Kinetic model	primary Hg^{+2} concentration (mg/l)		
	0.1	1	10
Pseudo-first order			
q_e, exp (mg/g)	0.23	1.84	10.94
q_e, cal (mg/g)	0.17	1.40	8.1
K_1 (1/min)	0.0184	0.0184	0.0207
R^2	0.99	0.991	0.992
Pseudo-second-order			
q_e, cal (mg/g)	0.24	1.99	11.76
k_2 (g/mg min)	.0187	0.022	0.0042
R^2	0.996	0.997	0.995

and 10.94, respectively. The process of adsorption of inorganic mercury on the surface of multi-walled carbon nanotubes mostly fits with the second order adsorption kinetic and it is because of this that the q_e achieved from the pseudo-second order equations is closer to q_e of the base experiments, compared to the pseudo-first order equations. Also, R^2 conceived from the pseudo-second order adsorption kinetic graph is higher than that of pseudo-first order graph. Previous studies also showed that absorbed cadmium [28], zinc [30], lead [33] and chromium [34] on carbon nanotubes follow pseudo-second order adsorption kinetic.

Qu et al. [44] reported that nanomaterials have been widely used to remove heavy metals from water due to their large surface area, high reactivity, short intra particle diffusion distance and low temperature modification. In spite of that Tang et al. [45] suggested the reuse and management of the used nanomaterials is an important issue and has not been considered seriously. Only a few relevant studies are available in literature. It would be worthwhile to investigate the reusability of the used nanomaterials.

Conclusion

Multi-walled carbon nanotubes were assessed as adsorbent to remove inorganic mercury from aqueous solutions. The adsorption rate of Hg^{2+} on the surface of adsorbent is highly affected by pH, and the increase of pH from 3 to 7 increases the percentage of removal. The best contact time is 120 min. Also, the increases of primary concentration of inorganic mercury and ionic concentration solution have negative effect on adsorption process. Finally, the process of adsorption follows both Freundlich and Langmuir isotherms, and the pseudo-second order adsorption kinetic can well describe adsorption process. The present study indicated that Carbon nanotubes have high efficiency in adsorbing of mercury.

Abbreviations

ICP-OES-EOP: Inductively coupled plasma optical emission spectrometry (End-on-Plasma); MWCNTS: Multi-walled carbon nanotubes.

Competing interests

The authors declare that they have no competing interests.

Authors' contribution

Authors participated in this research including design, experiments and data analysis, and manuscript preparation. All authors read and approved the final manuscript.

Acknowledgements

This research was supported by Tehran University of Medical Sciences & health Services (Project No. 17314).
Our aim of this research is Removal of mercury from water and waste water that it plays an important role to maintain environmental and human health. Usage of nano materials as a new technology has been increased in different research. Hence, efficiency of carbon nanotubes to remove mercury from water and waste water was studied in this paper.

Author details

[1]Department of Environmental Health Engineering, School of Public Health, Tehran University of Medical Sciences, Tehran, Iran. [2]Center for Water Quality Research (CWQR), Institute for Environmental Research (IER), Tehran University of Medical Sciences, Tehran, Iran.

References

1. Jaafari J, Mesdaghinia AR, Nabizadeh R, Hoseini M, Kamani H, Mahvi AH. Influence of upflow velocity on performance and biofilm characteristics of Anaerobic Fluidized Bed Reactor (AFBR) in treating high-strength wastewater. J Environ Health Sci Engineer. 2014;12:1–10.
2. Li Y, Wu CY. Role of moisture in adsorption, photocatalytic oxidation, and reemission of elemental mercury on a SiO_2–TiO_2 nanocomposite. Environ Sci Technol. 2006;40:6444–8.
3. Girginova PI, Daniel-da-Silva AL, Lopes CB, Figueira P, Otero M, Amaral VS, et al. Silica coated magnetite particles for magnetic, removal of Hg^{2+} from water. J Colloid Interface Sci. 2010;345:234–40.
4. Bailey SE, Olin TJ, Bricka RM, Adrian DD. A review of potentially low-cost sorbents for heavy metals. Water Res. 1999;33:2469–79.
5. Gibicar D, Horvat M, Logar M, Fajon V, Falnoga I, Ferrara R, et al. Human exposure to mercury in the vicinity of chlor-alkali plant. Environ Res. 2009;109:355–67.
6. Balarama KMV, Karunasagar D, Rao SV, Arunachalam J. Preconcentration and speciation of inorganic and methyl mercury in waters using polyaniline and gold trap-CVAAS. Talanta. 2005;68:329–35.
7. Zhang L, Chang X, Hu Z, Zhang L, Shi J, Gao R. Selective solid phase extraction and preconcentration of mercury (II) from environmental and biological samples using nanometer silica functionalized by 2, 6-pyridine dicarboxylic acid. Microchim Act. 2010;168:79–85.
8. Wang X, Pehkonen SO, Ray AK. Photocatalytic reduction of Hg (II) on two commercial TiO_2 catalysts. Electrochim Acta. 2004;49:1435–44.
9. Morel FMM, Kraepiel AML, Amyot M. The chemical cycle and bioaccumulation of mercury. Annual Review of Ecology and Systematics. 1998;29:543-66. doi:10.1146/annurev.ecolsys.29.1.543.
10. Miretzky P, Cirelli AF. Hg(II) removal from water by chitosan and chitosan derivatives: a review. Hazard Mater. 2009;167:10–23.
11. Rao PD, Lu C, Su F. Sorption of divalent metal ions from aqueous solution by carbon nanotubes: a review. Sep Purif Technol. 2007;58:224–31.
12. Skodras G, Diamantopoulou IR, Zabaniotou A, Stavropoulos G, Sakellaropoulos GP. Enhanced mercury adsorption in activated carbons from biomass materials and waste tires. Fuel Process Technol. 2007;88:749–58.
13. Inbaraj SB, Sulochana N. Mercury adsorption on a carbon sorbent derived from fruit shell of terminalia catappa. Hazard Mater. 2006;133:283–90.
14. Jeon C, Holl WH. Chemical modification of chitosan and quilibrium study for mercury ion removal. Water Res. 2003;37:4770–80.
15. Herrero R, Lodeiro P, Castro CR, Vilarin T, Vicente MESD. Removal of inorganic mercury from aqueous solutions by biomass of the marine macroalga Cystoseira baccata. Water Res. 2005;39:3199–210.
16. Krishnan KA, Anirudhan TS. Removal of mercury (II) from aqueous solutions and chlor-alkali industry effluent by steam activated and sulphurised activated carbons prepared from bagasse pith: ineticsand equilibrium studies. Hazard Mater. 2002;92:161–83.
17. Budinova T, Savova D, Petrov N, Razvigorova M, Minkova V, Ciliz N, et al. Mercury adsorption by different modifications of furfural adsorbent. Ind Eng Chem Res. 2003;42:2223–9.
18. Vizuete ME, Garc AM, NGisbert A, Gonz CF, Serrano VG. Adsorption of mercury by carbonaceous adsorbents prepared from rubber of tyre wastes. Hazard Mater. 2005;119:231–8.
19. Feng X: Application of single walled carbon nanotubes in environmental engineering: adsorption and desorption of environmentally relevant species studied by infrared spectroscopy and temperature programmed desorption. Doctoral Dissertation 2006, University of Pittsburgh.
20. Mohan D, Pittman CU. Activated carbons and low cost adsorbents for the remediation of tri- and hexavalent chromium fromwater. Hazard Mater. 2006;137:762–811.

21. Li YH, Wang S, Zhang X, Wei J, Xu C, Luan Z, et al. Adsorption of fluoride from water by aligned carbon nanotubes. Mater Res Bull. 2003;38:469–76.
22. Peng X, Li Y, Luan Z, Di Z, Wang H, Tian B, et al. Adsorption of 1,2-dichlorobenzene from water to carbon nanotubes. Chem Phys Lett. 2003;376:154–8.
23. Lu C, Chung YL, Chang KF. Adsorption of trihalomethanes from water with carbon nanotubes. Water Res. 2005;39(6):1183–9.
24. Lu C, Chiu H. Adsorption of zinc (II) from water with purified carbon nanotubes. Chem Eng Sci. 2006;61:1138–45.
25. Gupta VK, Agarwal S, Saleh TA. Chromium removal by combining the magnetic properties of iron oxide with adsorption properties of carbon nanotubes. Water Res. 2011;45(6):2207–12.
26. Kandah MI, Meunier JL. Removal of nickel ions from water by multi-walled carbon nanotubes. Hazard Mater. 2007;146:283–8.
27. Lu C, Liu C. Removal of nickel (II) from aqueous solution by carbon nanotubes. Chem Technol Biotechnol. 2006;81:1932–40.
28. Vukovi GD, Marinkovi AD, Coli M, Risti MD, Aleksi R, Peri'c-Gruji AA, et al. Removal of cadmium from aqueous solutions by oxidized and ethylenediamine-functionalized multi-walled carbon nanotubes. Chem Eng J. 2010;157:238–48.
29. Hellings J, Adeloju S B, Verheyen T V: Rapid determination of ultra-trace concentrations of mercury in plants and soils by cold vapour inductively coupled plasma-optical emission spectrometry. Microchem Journal 2013, http://dx.doi.org/10.1016/j.microc.2013.02.007.
30. Lu C, Chiu H. Chemical modification of multiwalled carbon nanotubes for sorption of Zn^{2+} from aqueous solution. Chem Eng. 2008;2008(139):462–8.
31. Sheng G, Li J, Shao D, Hu J, Chen CH, Chen Y, et al. Adsorption of copper(II) on multiwalled carbon nanotubes in the absence and presence of humic or fulvic acids. J Hazard Mater. 2010;178:333–40.
32. Atieh MA, Bakather OY, Al-Tawbini B, Bukhari AA, Abuilaiwi FA, Fettouhi MB. Effect of carboxylic functional group functionalized on carbon nanotubes surface on the removal of lead from water. Bioinorg Chem Appl. 2010;2010:1–9.
33. Xu D, Tan X, Chen C, Wang X. Removal of Pb(II) from aqueous solution by oxidized ultiwalled carbon nanotubes. Hazardous Material. 2008;154:407–16.
34. Hu J, Chen C, Zhu X, Wang X. Removal of chromium from aqueous solution by using oxidized multiwalled carbon nanotubes. Hazard Mater. 2009;162:1542–50.
35. Qu R, Sun C, Ma F, Zhang Y, Ji Ch XQ, Wang C, et al. Removal and recovery of Hg (II) from aqueous solution using chitosan-coated cotton fibers. Hazard Mater. 2009;167:717–27.
36. Gao Z, Bandosz TJ, Zhao Z, Han M, Qiu J. Investigation of factors affecting adsorption of transition metals on oxidized carbon nanotubes. Hazard Mater. 2009;167:357–65.
37. Wang XK, Chen CL, Hu WP, Ding AP, Xu D, Zhou X. Sorption of 243Am(III) to multi-walled carbon nanotubes. Environ Sci Technol. 2005;39:2856–60.
38. Li YH, Wang S, Wei J, Zhang X, Xu C, Luan Z, et al. Lead adsorption on carbon nanotubes. Chem Phys Lett. 2002;357:263–6.
39. Presto AA, Granite EJ. Impact of sulphur oxides on mercury capture by activated carbon. Environ Sci Technol. 2007;41:6579–84.
40. Anirudhan TS, Divya L, Ramachandran M. Mercury (II) removal from aqueous solutions and wastewaters using a novel cation exchanger derived from coconut coir pith and its recovery. Hazardous Material. 2008;157:620–62.
41. Zabihi M, Haghighi Asl A, Ahmadpour A. Studies on adsorption of mercury from aqueous solution on activated carbons prepared from walnut shell. Hazard Mater. 2010;174:251–6.
42. Di Natale F, Erto A, Lancia A, Musmarra D. Mercury adsorption on granular activated carbon in aqueous solutions containing nitrates and chlorides. Hazard Mater. 2011;192:1842–50.
43. Asasian N, Kaghazchi T, Soleimani M. Elimination of mercury by adsorption onto activated carbon prepared from the biomass material. Ind Eng Chem. 2012;18:283–9.
44. Qu XL, Alvarez PJJ, Li QL. Applications of nanotechnology in water and wastewater treatment. Water Res. 2013;47(12):3931–46.
45. Tang WW, Zeng GM, Gong JL, Liang J, Xu P, Zhang C, et al. Impact of humic/fulvic acid on the removal of heavy metals fromaqueous solutions using nanomaterials: A review. Sci Total Environ. 2014;468–469:1014–27.

Assessment of tetracycline contamination in surface and groundwater resources proximal to animal farming houses in Tehran, Iran

Allahbakhsh Javid[1], Alireza Mesdaghinia[2*], Simin Nasseri[2,3], Amir Hossein Mahvi[2,4], Mahmood Alimohammadi[2] and Hamed Gharibi[1]

Abstract

Background: Antibiotics have been increasingly used for veterinary and medical purposes. The overuse of these compounds for these purposes can pollute the environment, water resources in particular. Tetracycline, among other forms of antibiotics, is one of the most applied antibiotic in aquaculture and veterinary medicine. The present study aimed to tack the traces of tetracycline in the effluents of municipal and hospital wastewater treatment plants, surface and groundwater resources and finally the drinking water provided from these water resources.

Methods: The samples were taken from Fasha-Foyeh Dam, wells located at Varamin Plain, and Yaftabad; and also, wastewater samples were collected from the wastewater treatment plant effluents of Emam Khomeini Hospital and a municipal wastewater treatment plant which its effluent is being released to the surface water of the area covered in this work. 24 samples were collected in total during July 2012 to December 2012. The prepared samples were analyzed using high-performance liquid chromatography.

Results: Based on the results, mean tetracycline levels in surface and ground water at nearby of animal farms was found to vary from 5.4 to 8.1 ng L^{-1}. Furthermore, the maximum TC concentration of 9.3 ng L^{-1} was found to be at Yaft-Abad sampling station. Although tetracycline traces could not be detected in any investigated Hospital WWTP effluents, it was tracked in MWWTP effluent samples, in the concentration range of 280 to 540 ng l^{-1}.

Conclusion: The results showed that the concentration of TC in water resource near the animal farms is higher than the other sampling stations. This is related to the usage of antibiotic for animals. In fact, it caused the contamination of water resources and could contribute to radical changes in the ecology of these regions.

Keywords: Tetracycline, Groundwater resources, Surface water resources, Drinking water, Animal farms

Background

Antibiotics have been increasingly used for veterinary and medical purposes. The increase in the use of these compounds affects both the environment and human health; in other words, the active forms of the antibiotics are being excreted from the body via urine and/or feces into the environment. Considering this, the overuse of these compounds can pollute the water resources [1]. It should be noted that there are various pathways in which these compounds enter into both surface and groundwater resources, including run-off, leakage from lagoons, leaching of manure applied to fields, and leaching from animal housing areas [Watanabe et al.]. The presence of antibiotics, TC in particular, in water and soil can cause some allergies and toxicity, since these compounds are still active [2]. For instance, excreted antibiotics in the environment affect almost all the bacterial species forcing them to develop a resistance toward these compounds [3]. Based on the reports of previously conducted studies, the residues of various

* Correspondence: mesdaghinia@sina.tums.ac.ir
[2]Department of Environmental Health Engineering, School of public Health, Tehran University of Medical Sciences, Tehran, Iran
Full list of author information is available at the end of the article

forms of antibiotics have been detected in the samples taken from surface and groundwater resources and also drinking water [4, 5]. Furthermore, the antibiotics have also been found in the samples taken from the effluents of both municipal and hospital wastewater treatment plant [6, 7].

Tetracycline (TC), among other forms of antibiotics, is one of the most applied antibiotics in aquaculture and veterinary medicine [8]. It should be noted that this antibiotic has been applied in livestock and poultry productions more than the aquaculture medicine. In addition, tetracycline is being discharged into the environment, water resources in particular, through wastewater effluent of drug manufacturing companies, disposal of non-consumable compounds and expired drugs containing tetracycline, and also from animal and agricultural wastes [9, 10]. TC has been classified among the antibiotics frequently detected in sewage, domestic wastewaters, surface and groundwater resources, drinking water, and sludge [11]. Considering the increase in the usage of TC and also the inefficiency of most conventional wastewater treatment processes in removing this antibiotic, the surface and ground water resources are now at more risk of being polluted with TC. Furthermore, it should be noted that there is not any regulation for routine sampling and analyzing TC level in the water resources. Previously conducted studies showed that one of the most widely used antibiotic in animals is tetracycline [8].

This study aimed to tack the traces of tetracycline in a specific route; in other words, tetracycline was strived to detect in the samples taken from the effluents of municipal and hospital wastewater treatment plants, surface and groundwater resources and finally the drinking water provided from these water resources. In this work, TC was selected due to its wide usage and persistence in the environment. Figure 1 shows the sampling locations. It should be noted that these locations are related to each other. In fact, the dam represents the surface water resource receiving the effluents of selected municipal and hospital wastewater treatment plants. The wells, in addition, are located in the vicinity of the main animal farming houses in Tehran, Iran and also the dam. Furthermore, drinking water prepared from using the water resources was also studied.

Methods

Sampling stations

In the present work, four samples over 6 months from July 2012 to December 2012 were taken; in total, 24 samples were collected and analyzed in this study. Water samples were taken from tap water, Fasha-Foyeh Dam and also specific wells located at Varamin Plain and Yaftabad. In addition, wastewater samples were collected from the effluents of Emam Khomeini Hospital wastewater treatment plant and a municipal wastewater treatment plant. Figure 1 shows Geographical position of places which the samples were taken.

In this study, two locations were selected to take sample from, namely, Varamin Plain and Yaftabad (i. e. wells nearby animal farming houses) to determine whether TC leached to groundwater resources. In addition, Fasha-Foyeh Dam was selected as surface water resource nearby animal farming houses to assess whether TC has been transported by run-off into the most nearby surface water resource; In other words, the sampling took place at nearby water resources of animal farms which their wastewater effluents are being discharged into the resources. Furthermore, the effluents of municipal and hospital wastewater plants which are being released into the same water resources were considered and analyzed in order to detect TC. The wastewater samples were also taken directly from the effluents of WWTPs to assess the amount of TC released from human sources and compare it with water samples mentioned above. Another spot was also selected (i. e. tap water) to determine whether the drinking water is polluted.

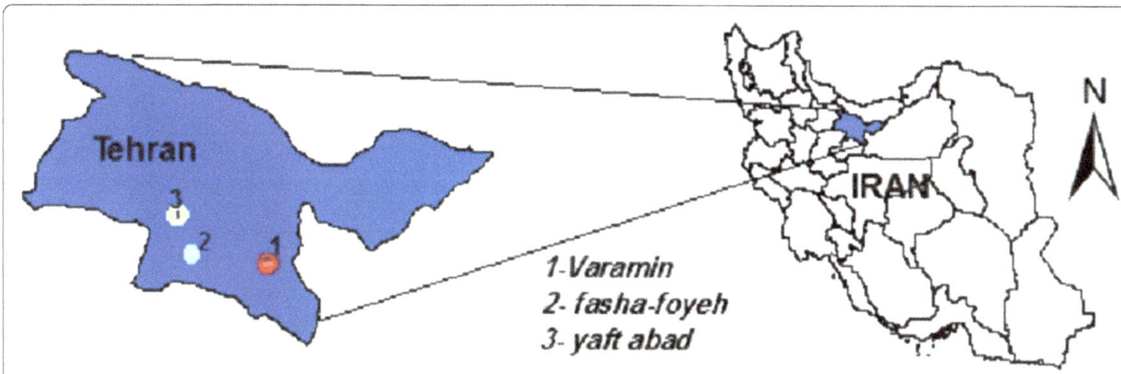

Fig. 1 Geographical position of study area and sampling points

Preparing and analyzing the samples

The method of analyzing the samples is based on a method applied and proposed in a study by A Pena et al., [3]. At first, samples were allowed to settle at 4 °C for 12–24 h in the dark. Careful handling prevented re-suspension of settled particles, and the supernatant of the samples was used for analysis instead of a filtered aliquot [12]. Since antibiotics are present in surface and ground waters and other resources at trace levels, pre-conditioning by suitable sorbent is a necessary step in sample preparation. The average recoveries 80 ± 3 were obtained using strata C18-E cartridge. Detection frequency of tetracycline in water sources and WWTP effluent is shown in Table 1.

According to a study conducted by Shalaby et al., the strata C18-E cartridge was applied to buffer extract the residue; after sample loading, the cartridge was washed by 10 ml of 5 % methanol mixed in water, and tetracycline was eluted with a mixture of 10 ml of methanol and 0.01 M oxalic acid [13].

The samples were taken by syringe and filtered through 0.45 μm membrane at pre-selected time intervals; then, the samples were measured using high-performance liquid chromatography (HPLC). HPLC consisted of a Knauer LPG pump, an EZ-chrom HPLC system manager program and a UV detector (k-2500). The UV–detector was set at the maximum absorption wavelength of 365 nm. Aliquots of 100 μL were injected manually using a model SGE injection valve (SGE. Australia). MZ-analysentechnik ODS-3 C18 (4.6 mm × 250 mm) packed with 5 μm spherical particles was used for separation. An Acetonitrile (A) aqueous oxalic acid 0.01 M (B) mixture was used as mobile phase at 30^0C temperature with a constant flow rate of 1.0 $mLmin^{-1}$. The used mobile phase are shown in Table 1 [3].

Results and discussion

This work is the first survey of the contamination of surface and ground water resources with TC in Tehran, Iran. The results showed that the mean concentration of TC in water resources, both surface and groundwater, varies from 5.4 to 8.1 ng L^{-1}among which Yaft-abad station with 9.3 ng L^{-1} level of TC was found to have the maximum TC concentration; The results are shown in Table 2. The concentration of TC in the samples taken

Table 1 HPLC gradient program used for determination of TC

Time (min)	Mobile phase A (%)	Mobile phase B (%)
0	10	90
7	40	60
7.5	50	50
12.95	10	90
13		End

Table 2 Mean concentrations of tetracycline in water resource and wastewater treatment plant effluent

Location	TC concentration (ngL^{-1})	
Yaft-abad – (well)	Mean	8.1
	Max	9.3
	Min	6.9
	n = 4	
Fasha-foyeh- (dam)	Mean	6.4
	Max	8.7
	Min	5.7
	n = 4	
Varamin- (well)	Mean	5.4
	Max	7.2
	Min	4.4
	n = 4	
Tap water	ND	
	n = 4	
HWWTP effluent	ND	
	n = 4	
MWWTP effluent	Mean	540
	Max	630
	Min	280
	n = 4	

ND non-detectable

from surface water resource (i. e. Fasha-foyeh- (dam)), was found in the range of 5.7 to 8.7 ng/L. Detected levels of this antibiotic in surface water resources could be explained by the fact that tetracycline of veterinary wastes finds its way into the surface and ground water resources. Run-off is one of the main ways in which antibiotics leach into particularly surface water resources [14, 15]. Considering the location of animal farming houses in vicinity of Fasha-foyeh- (dam), this could be a probable cause of finding trace amount of TC in the taken samples from this location. There is also another source of polluting the surface water resources, as taken into account in this study, and it is the release of wastewater effluents into the resources [15]. The concentration of TC in MWWTP effluent samples ranged from 280 to 540 ng l^{-1}. Considering the fact that most of the antibiotics have high affinity to soil compounds [16, 17], tetracycline in particular, the main source of TC in the surface water resource studied in this work could be the effluent of MWWTP. This result is in line with the results of previously conducted studies on the presence of antibiotics in the water resources [1, 13]. In addition, the concentration of TC was negligible in the samples taken from the effluents of Hospital wastewater treatment plant. The main reason is that TC is no longer being used among hospitalized patients. In a study conducted in the

United States of America, the concentration of tetracycline in the effluent of MWWTP was reported to be between 170 and 850 ng L^{-1} [18]. In a study conducted by Miao et al., 150 to 970 ngL^{-1} concentration of TC was reported in Canada [19]. Comparing with these findings, the concentrations of TC detected in the samples taken from WWTPs effluents in this study are at lower levels.

As mentioned above, the highest concentration of TC was found in the samples taken from ground water resources. Although TC has high affinity to soil compounds, the presence of this antibiotic in the samples even in this range is of high importance. It should be noted that the overuse of these drugs causes them to be found in the water resources due mainly to the saturation of soil capacity; in fact, this phenomenon is known as terracumulation [20, 21]. When the soil becomes saturated with TC, the infiltration of water originated from rain into groundwater resources can carry this antibiotic contributing to the pollution of these resources [22]. Considering the fact that TC is still an active antibiotic when it is bounded to soil particles, the presence of it in groundwater resource is indeed a major problem.

In addition, four samples were taken from tap water to determine whether any concentration of TC is in drinking water. The water treated and distributed among the people of the region we took samples from came from a water treatment plant which uses the water resources (i. e. surface and groundwater) mentioned above. As shown in Table 2, the concentration of TC in these samples was negligible.

Conclusion

This work was aimed to survey the contamination of surface and ground water resources with TC in Tehran, Iran. Based on the results, the water sources studied in this work are contaminated with TC and the presence of other types of antibiotics is also probable. The release of TC from the animal farming areas can be implied to be the main source of groundwater resources pollution with this antibiotic. In addition, the release wastewater effluent from WWTP can be considered as a potential source for the contamination of surface water resource with TC. In fact, the presence of this antibiotic in the environment can cause the mutation of the bacterial species and make them resistant to the antibiotics; and also, negatively affect those who use the water from this resource. In this regard, regular monitoring of the presence of antibiotics mostly used in these areas in order to prevent further damage to the environment. In conclusion, so we recommend more monitoring of the existence of antibiotics residues in water resource near the animal farms.

Abbreviations

TC: tetracycline; HPLC: high-performance liquid chromatography; WWTP: wastewater treatment plant; MWWTP: municipal wastewater treatment plant.

Competing interests

The authors declare that they have no competing interests.

Authors' contributions

AJ was the main investigator, collected the data, SN and AM supervised the study. AHM, and MA were advisors of the study. HGH performed the statistical analysis, and drafted the manuscript. All authors read and approved the final manuscript.

Acknowledgments

This work is part of a PhD dissertation and supported by Tehran University of Medical Sciences. The authors wish to thank the staff of laboratory of School of Public Health. The second part has been published as "Performance of photocatalytic oxidation of tetracycline in aqueous solution by TiO2 nanofibers".

Author details

[1]School of Public Health, Shahroud University of Medical Sciences, Shahroud, Iran. [2]Department of Environmental Health Engineering, School of public Health, Tehran University of Medical Sciences, Tehran, Iran. [3]Center for Water Quality Research, Institute for Environmental Research, Tehran University of Medical Sciences, Tehran, Iran. [4]Center for Solid Waste Research, Institute for Environmental Research, Tehran University of Medical Sciences, Tehran, Iran.

References

1. Xu WH, Zhang G, Zou SC, Li XD, Liu YC. Determination of selected antibiotics in the Victoria Harbour and the Pearl River, South China using ighperformance iquid chromatographyeelectrospray ionization tandem mass spectrometry. Environ Pollut. 2007;145:672–9.
2. Kummerer K. The presence of pharmaceuticals in the environment due to human use - present knowledge and future challenges. J Environ Manage. 2009;90:2354–66.
3. Pena A, Albert-Garcia JR, Silva LJG, Lino CM, Calatayud JM. Photo-induced fluorescence of magnesium derivatives of tetracycline antibiotics in wastewater samples. J Hazard Mater. 2010;179:409–14.
4. Yiruhan, Wang QJ, Mo CH, Li YW, Gao P, Tai YP, et al. Determination of four fluoroquinolone antibiotics in tap water in Guangzhou and Macao. Environ Pollut. 2010;158(7):2350–8.
5. Chee-Sanford J, Aminov R, Krapac I, Garrigues-Jeanjean N, Mackie R. Occurrence and diversity of tetracycline resistance genes in lagoons and groundwater underlying two swine production facilities. Appl Environ Microbiol. 2001;67:1494–1502.
6. Brown KD, Kulis J, Thomson B, Chapman TH, Mawhinney DB. Occurrence of antibiotics in hospital, residential, and dairy effluent, municipal wastewater, and the Rio Grande in New Mexico. Sci Total Environ. 2006;366(2-3):772–83.
7. Homem V, Santos L. Degradation and removal methods of antibiotics from aqueous matrices–a review. J Environ Manage. 2011;92(10):2304–47.
8. Sarmah AK, Meyer MT, Boxall ABA. A global perspective on the use, sales, exposure pathways, occurrence, fate and effects of veterinary antibiotics (VAs) in the environment. Chemosphere. 2006;65:725–59.
9. Boxall AB, Kolpin DW, Halling-Sørensen B, Tolls J. Are veterinary medicines causing environmental risks? J Environ Sci Technol. 2003;37:286A–94.
10. Mompelat S, LeBot B, Thomas O. Occurrence and fate of pharmaceutical products and by-products, from resource to drinking water. Environ Int. 2009;35:803–14.
11. Penghua W, Pow-Seng Y, Lim TT. C–N–S tridoped TiO2 for photocatalytic degradation of tetracycline under visible-light irradiation. Appl Catal Gen. 2011;399:252–61.
12. Nödler K, Licha T, Bester K, Sauter M. Development of a multi-residue analytical method, based on liquid chromatography–tandem mass spectrometry, for the simultaneous determination of 46 micro-contaminants in aqueous samples. J Chromatogr A. 2010;1217:6511–21.

13. Shalaby AR, Salama NA, Abou-Raya SH, Emam WH, Mehaya FM. Validation of HPLC method for determination of tetracycline residues in chicken meat and liver. Food Chem. 2011;124:1660–6.
14. Alder AC, McArdell CS, Golet EM, Ibric S, Molnar E, Nipales NS, et al. Occurrence and fate of flouroquinolone, macrolide, and sulfanamide antibiotics during wastewater treatment and in ambient waters in Switzerland. Symposium Series. Washington: American Chemical Society; 2001. 791, p. 56–69.
15. Pedersen J, Yeager M, Suffet I. Xenobiotic organic compounds in runoff from fields irrigated with treated wastewater. J Agric Food Chem. 2003;51:1360–72.
16. Hamscher G, Sczesny S, Ho¨per H, Nau H. Determination of persistent tetracycline residues in soil fertilized with liquid manure by high-performance liquid chromatography with lectrospray ionization tandem mass spectrometry. Anal Chem. 2002;74:1509–18.
17. Samuelsen OB, Torsvik V, Ervik A. Long-range changes in oxytetracycline concentration and bacterial resistance towards oxytetracycline in a fish farm sediment after medication. Sci Total Environ. 1992;114:25–36.
18. Batt AL, Snow DD, Aga DS. Occurrence of sulphonamide antimicrobials in private water wells in Washington County, Idaho, USA. Chemosphere. 2006;64:1963–71.
19. Karthikeyan KG, Meyer T. Occurrence of antibiotics in wastewater treatment facilities in Wisconsin, USA. Sci Total Environ. 2006;361:196–207.
20. Miao XS, Bishay F, Chen M, Metcalfe CD. Occurrence of antimicrobials in the final effluents of wastewater treatment plants in Canada. Environ Sci Technol. 2004;38:3533–41.
21. Rooklidge SJ. Environmental antimicrobial contamination from terraccumulation and diffuse pollution pathways. Sci Total Environ. 2004;325:1–13.
22. Tolls J. Sorption of veterinary pharmaceuticals in soils: a review. Environ Sci Technol. 2001;35:3397–406.

Sonocatalytic degradation of humic acid by N-doped TiO_2 nano-particle in aqueous solution

Hossein Kamani[1], Simin Nasseri[1,2], Mehdi Khoobi[3], Ramin Nabizadeh Nodehi[1] and Amir Hossein Mahvi[1,4,5*]

Abstract

Background: Un-doped and *N*-doped TiO_2 nano-particles with different nitrogen contents were successfully synthesized by a simple sol–gel method, and were characterized by X-ray diffraction, field emission scanning electron microscopy, Energy dispersive X-ray analysis and UV–visible diffuse reflectance spectra techniques. Then enhancement of sonocatalytic degradation of humic acid by un-doped and N-doped TiO_2 nano-particles in aqueous environment was investigated. The effects of various parameters such as initial concentration of humic acid, N-doping, and the degradation kinetics were investigated.

Results: The results of characterization techniques affirmed that the synthesis of un-doped and *N*-doped TiO_2 nano-particles was successful. Degradation of humic acid by using different nano-particles obeyed the first-order kinetic. Among various nano-particles, *N*-doped TiO_2 with molar doping ratio of 6 % and band gap of 2.92 eV, exhibited the highest sonocatalytic degradation with an apparent-first-order rate constant of 1.56×10^{-2} min^{-1}.

Conclusions: The high degradation rate was associated with the lower band gap energy and well-formed anatase phase. The addition of nano-catalysts could enhance the degradation efficiency of humic acid as well as *N*-doped TiO_2 with a molar ratio of 6 %N/Ti was found the best nano-catalyst among the investigated catalysts. The sonocatalytic degradation with nitrogen doped semiconductors could be a suitable oxidation process for removal of refractory pollutants such as humic acid from aqueous solution.

Keywords: Humic acid, *N*-doped TiO_2, Sonocatalytic degradation

Background

Humic substances, as part of natural organic matters, have been a major issue in water treatment plants due to their non-biodegradability and their water-soluble formation [1, 2]. These substances can affect the water quality such as odor, taste and color. It has been also confirmed that these substances act as precursors to form disinfection by-products when water treated with chlorine [1, 3, 4]. Hence, removal of humic substances has been widely investigated for the protection of public health. In water treatment plants, portion of these substances are removed from raw water by conventional methods such as; coagulation, precipitation, filtration and adsorption [5–7]. Wang et al. reported that the removal of humic substances by using conventional processes is only 5-50 % [8].

In addition, application of high coagulant dosage isn't reasonable due to high cost operation and problem in sludge disposal. Besides, the presence of humic substances in water may reduce the efficiency of water treatment processes when membranes or microporous adsorbents are applied.

Chemical degradation is one of the best technologies that have been widely accepted for removal of humic substances [3, 9, 10]. Recently, sonolysis process attracted considerable attention as an advanced oxidation process (AOP) for degradation of pollutants in water [11–14]. However, this method consumes considerable energy and its efficiency is low compared to other methods. In order

* Correspondence: ahmahvi@yahoo.com
[1]Department of Environmental Health Engineering, School of Public Health, Tehran University of Medical Sciences, Tehran, Iran
[4]Center for Solid Waste Research, Institute for Environmental Research, Tehran University of Medical Sciences, Tehran, Iran
Full list of author information is available at the end of the article

to increase the degradation efficiency semiconductors have been added to the sonolysis processes [15, 16].

In recent years, application of heterogeneous sonocatalysis using TiO2 has become an environmentally sustainable treatment and cost-effective option for degradation of pollutants. Moreover, TiO2 is the most suitable photocatalyst for water treatment due to its high photocatalytic activity, long-time stability, relative low cost and non-toxicity [17–19]. It is well known that mechanism of sonocatalysis is similar to the photocatalysis [20, 21]. Thus, various techniques, including dye sensitization, semiconductor coupling and doping with metal and non-metal elements may enhance the sonoactivity of TiO2. According to previous studies, the doping of TiO2 with non-metal has been verified to be the most feasible method to improve photocatalytic activity of this catalyst [22]. It is also important to mention that the doping with nitrogen may be more effective than other non-metals because of its comparable atomic size with oxygen and small ionizing energy [23].

In the present study, un-doped and *N*-doped TiO_2 nano-particles with different nitrogen contents were successfully synthesized by a simple sol–gel method and were characterized by X-ray diffraction (XRD), field emission scanning electron microscopy (FE-SEM), energy dispersive X-ray analysis (EDX) and UV–visible diffuse reflectance spectra (UV-vis DRS) techniques.

The sonocatalytic activity of the as-synthesized TiO_2 for degradation of humic acid was investigated under ultrasonic irradiation with respect to the effects of nitrogen doping content, the initial concentration of humic acid and the addition of doped nanocatalyst into sonolysis process. Furthermore, the possible mechanism of sonocatalysis of N-doped TiO_2 was proposed.

Methods

Materials

Titanium tetraisopropoxide (TTIP, $Ti(OC_3H_7)_4$), Ethanol (EtOH), triethylamine, nitric acid (HNO_3), Hydrochloric acid (HCl) and sodium hydroxide (NaOH) were purchased from Merck Company, Germany, as analytical grade and were used without further purifying. Humic acid was purchased from Aldrich Company as sodium salt, and it was used after preparation. The stock solution of humic acid was prepared according to the methods [24]. The humic acid solution was prepared by addition of humic acid powder into deionized water and was heated up to 60 in order to accelerate the dissolution of humic acid. Then, the humic acid suspension cooled down to room temperature and was filtered through a 0.45-μm Milipore syringe filter. The residue of humic acid on the filter was dried in an oven at 105 until stable weight. The humic acid in filtered solution was calculated by gravimetric method and stored as a stock solution for experimental use.

Synthesis of N-doped TiO_2

All catalyst samples were synthesized using a sol–gel method. To synthesize *N*-doped TiO_2 with a nominal molar doping of the dopant, 3 % "TN1", 6 % "TN2" and 12 % "TN3", 3 mL Titanium tetraisopropoxide and a certain amount of triethylamine was dissolved in 20 mL of ethanol, and the solution was stirred for 15 min (solution A). 2 mL deionized water was added into 10 mL of ethanol that contained nitric acid, this solution was also stirred for 15 min (solution B). Solution B was added drop wise to the solution A under magnetic stirring. After constantly stirring for 30 min, the semitransparent sol was obtained. Subsequently, the obtained semitransparent sol was set for 5 h at room temperature and then dried at 80 °C for 24 h in an oven. The dried powder was ground and calcinated under air at 500 °C for 1 h with a heating rate of 16 °C min^{-1}. For comparison, un-doped TiO_2 was also synthesized without the addition of dopant under the same conditions.

Characterization of N-doped TiO_2

In order to determine the effect of *N*-doping on the nano-particle structure, the analysis by X-ray diffraction (XRD), surface morphology, elemental analysis and photophysical properties were carried out. A Philips X'Pert X-ray Diffractometer with a diffraction angle range 2θ = 10–70° using Cu Kα radiation (λ = 1.5418A) was used to collect XRD diffractograms. The accelerating voltage and emission current were 40 kV and 30 mA, respectively. The average crystallite size was determined according to the Scherrer equation using the full-width at half-maximum (FWHM) of the (1 0 1) peak. The UV–visible diffuse reflectance spectra (UV-DRS) were recorded using a UV–vis spectrophotometer (Avaspec-2048-TEC, Avantes, Netherland) with $BaSO_4$ as the reflectance standard. Then, the recorded data were converted to the absorbance units by using the Kubelka–Munk theory. The surface morphology and shape of the as-synthesized *N*-doped TiO_2 was observed through a field emission scanning electron microscope (FE-SEM, TESCAN) by gold-coated samples. Energy dispersive X-ray analysis (EDX) in the FE-SEM was also taken for the elemental analysis of the doped samples.

Sonocatalytic activity

Each suspension was prepared by adding 20 mg of each synthesized catalyst into a 100 mL of humic acid solution at concentrations 5, 10, and 20 mg L^{-1} in a reaction vessel. Prior to ultrasonic irradiation, the suspension was stirred using magnetic stirrer for 30 min in darkness to ensure a good dispersion and also to complete adsorption/desorption equilibrium of humic acid on the catalyst surface. All experiments were carried out in laboratory scale and in batch system. The ultrasonic irradiation was generated by an Elma ultrasonic bath (model TI-H5) which was operated at a frequency of 130 kHz and a maximum output power of 100 W. During the sonocatalytic processes, the solution temperature was maintained at 25 ± 2 °C using a water cooling system in ultrasonic bath. After the desired reaction time, 5 mL aliquots were withdrawn at certain interval and centrifuged at 6000 rpm for 20 min to separate the catalysts by a centrifuge (Hettich, Germany, model D-78532). The residual humic acid concentration in supernatant solution was determined by UV-vis spectrophotometer (Perkin Elmer, USA) at 254 nm. For comparison of reaction rate among different condition, the kinetic model was used.

Results and discussion

X-ray diffraction pattern

An X-ray diffraction pattern was used to investigate the type of crystalline in material and also to know if any change was occurred after doping of TiO_2. Figure 1a shows the XRD patterns of the un-doped and *N*-doped TiO_2 samples. As shown in the XRD pattern, all synthesized samples had a sharp diffraction peak indicating a good characteristic crystal. The distinctive peaks at 2θ = 25.49°, 37.14°, 37.99°, 38.76°, 48.35°, 54.12°, 55.33°, 62, 90° and 68.95°; correspond to the anatase (JCPDF Card No. 20-0387) were observed. The patterns also showed that the anatase was the main phase in un-doped and *N*-doped TiO_2 under all synthesis conditions.

These results revealed that the peak positions were nearly the same and no detectable dopant-related peaks were observed, implying that the structure of TiO_2 has not been changed and also suggesting that nitrogen dopants do not react with TiO_2 to form new crystalline [25, 26]. It is noteworthy, that many documents have also reported that doping with the nitrogen ions have not exhibited additional phase except anatase [22, 27]. The pure anatase phase in *N*-doped TiO_2 could be due to the fact that the nitrogen dopants are so low and they have also moved into either the interstitial positions or into the substitution sites of the TiO_2 crystal structure [25, 28]. Compared to the un-doped $TiO_{2,}$ the peak of *N*-doped TiO_2 samples exhibited a slight shift toward the lower angle corresponding to (1 0 1) plane of anatase (Fig. 1b), indicating a lattice distortion of the *N*-doped TiO_2. These defects and disorderly state in the particles caused by nitrogen dopants are reported as key factor for absorption edge shift towards the visible-light region [25, 27].

The average crystallite size of un-doped TiO_2 and *N*-doped TiO_2 were calculated according to the Debye-Scherrer formula as the following:

$$D = \frac{k\lambda}{\beta cos\theta} \tag{1}$$

where:
D = the average crystallite size,
k = a dimensionless shape factor (usually = 0.9),
λ = the wave length of the X-ray radiation (0.15418 nm for Cu Ka),
β = the full width at half-maximum of the diffraction, and
θ = the corresponding diffraction angle in degree [21].

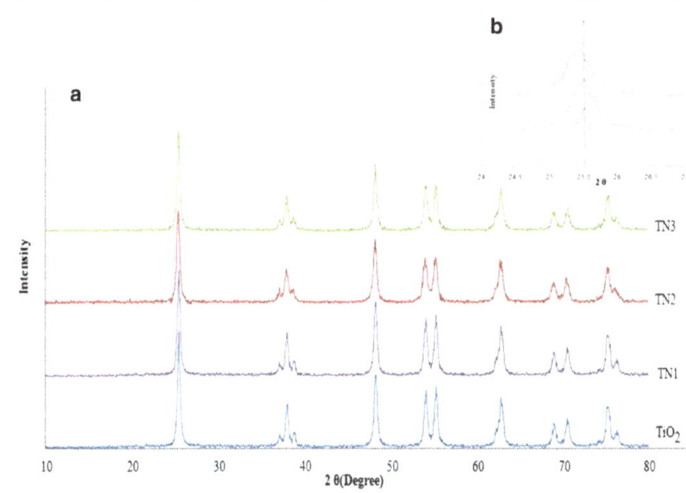

Fig. 1 XRD patterns (**a**) and lattice distortion (**b**) of un-doped and *N*-doped nano-particles

The calculated results were 30, 30, 26 and 34 nm for un-doped TiO_2, NT1, NT2 and NT3 nano-particles, respectively.

FE-SEM and EDX

The FE-SEM was used to show the shape and morphology of un-doped and *N*-doped TiO_2 particles (Fig. 2). The prepared nano-particles were found to be fine, irregular shape, slightly smooth surface and tend to agglomerate to form larger irregular grains. The diameter of particles was found to be 30-40 nm, which is in a good agreement with the crystal size obtained by XRD indicating that both un-doped and *N*-doped particle is nano-sized particles (Additional file 1: Figure S1). The energy dispersive X-ray Spectroscopy (EDX) of *N*-doped TiO_2 for different points of sample shows the appearance peaks of N, O and Ti atoms, which indicating that *N*-doped TiO_2 are mainly composed of these elements and confirm the N-doping process [29, 30]. The EDX spectra and the EDX elemental mapping (Additional file 1: Figure S2) also indicate no impurities in the samples and a good uniform distribution of N ions.

UV-vis diffuse reflectance spectra (UV-vis-DRS)

UV-visible diffuse reflectance spectra are the easiest and most convenient method to have a rough measure of the influence of doping [31]. As shown in Fig. 3a, doping of TiO_2 with nitrogen ion is clearly indicated by the reflectance spectra in the range of 300–700 nm. It is confirmed by various studies that N-doping has positive effect on the activity of the TiO_2 photocatalyst [31, 32]. As expected, N-doping caused a red shift from UV to the visible-light region. This red shift led to a better light absorption and consequently high radical generation and degradation efficiency.

Changing toward higher light absorption and red shift of absorption edge, which is in consistent with the yellowish color of nano-particles, can be attributed to narrowing of the band gap of synthesized nano-sized particles [26].

The band gap energies (E_g) of nano-sized particles can be determined according to the following equation [33]:

$$(\alpha h\nu) = A\left(h\nu - E_g\right)^{r} \tag{2}$$

where α is the absorption coefficient, h is Planck's constant, ν is the frequency of light, A is the absorption constant, E_g is the optical energy gap of the nano-sized particle and r is a number for characterizing the transition process, which is equal to 2 for indirect transition and 0.5 for direct transition. Therefore, the band gap energy of un-doped and *N*-doped TiO_2 can be determined from plots of the square root of $(\alpha h\nu)^{0.5}$ versus photon energy (Fig. 3 b).

The calculated optical band gaps were 3.02, 2.92, 2.91 and 3.09 eV for the TN1, TN2, TN3 and un-doped TiO_2 nano-particles, respectively. In all synthesized nano-particles the optical band gaps were lower than the band gap of commercial TiO_2 (3.2 eV) that is reported in various literatures [34]. This narrower band gap enhances transition of electrons from the valence band to the conduction band in the doped TiO_2 under ultrasonic irradiation and therefore it can increase sonocatalytic activities [34].

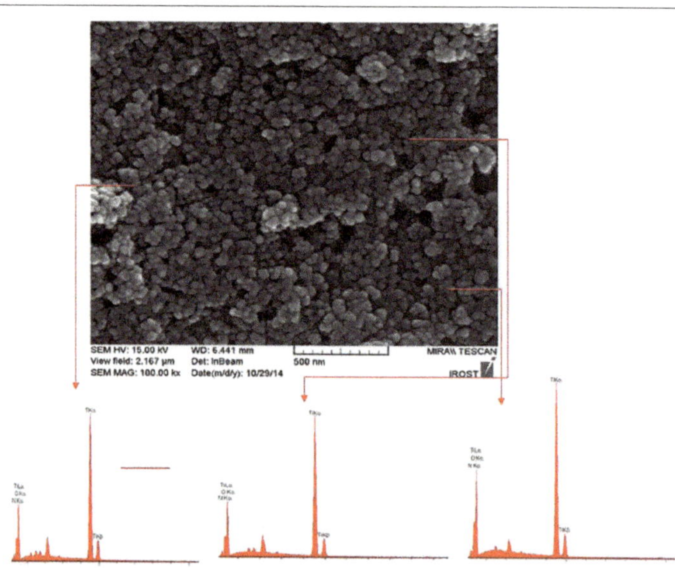

Fig. 2 FE-SEM image and EDX spectra of *N*-doped TiO_2 (sample TN2)

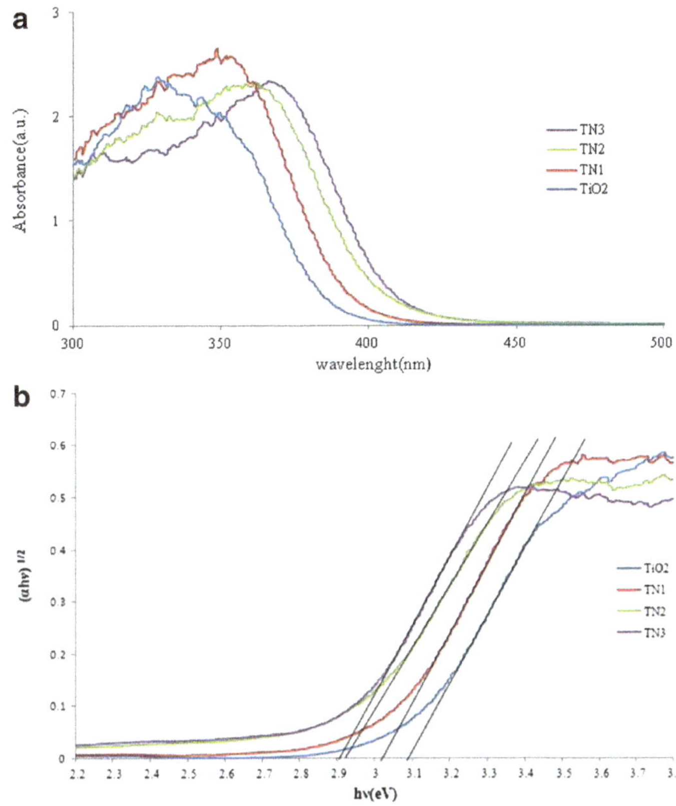

Fig. 3 a UV-Vis Diffuse Reflectance Spectra and (**b**) energy band gap of un-doped and *N*-TiO_2

The decrease in the band gap of *N*-doped TiO_2 can be attributed to the localized *N* 2p states in the structure of TiO_2 lattice in the form of substitutional and/or interstitial N states. It has been reported that substitutional N doping decreases the band gap by mixing of the O 2p and N 2p orbitals, while interstitial doping creates an additional state between the valence band and conduction band [22, 34].

Sonocatalytic performance of various sonocatalysts

The degradation of humic acid was studied using sonolysis, sonocatalysis with TiO_2, and sonocatalysis with different nitrogen contents doped in TiO_2. Figure 4 shows the degradation of humic acid under using different sonocatalysts at the neutral pH. The amount of adsorption for humic acid on the surface of the nano-particles was less than 3 % in darkness without ultrasonic irradiation; therefore it was negligible for un-doped and N-doped TiO_2.

As shown in Fig. 4, only 32 % of humic acid was degraded under ultrasonic irradiation after 90 min (without sonocatalyst), while the degradation efficiency of TiO_2, TN1, TN2 and TN3 sonocatalysts were 49.0, 55.0, 72.0 and 60.0.%, respectively. These results indicate that presence of sonocatalyst increases the degradation efficiency. This improvement could be due to this fact that the added sonocatalysts act as nuclei for bubble formation in aqueous solution as well as formation of oxygen vacancies in N-doped TiO_2 crystallite [15, 21]. These oxygen vacancies act as electron-trapping sites and prevent the recombination of hole-electron pairs, while, the additional amount of surface defect such as oxygen vacancies could increase the recombination of hole-electron pairs [21, 23].

As shown in Fig. 4, the highest sonocatalytic activity was achieved by TN2 with 72.0 % for humic acid degradation after 90 min of ultrasonic irradiation. According to the reported studies, the sonocatalytic activity of doped TiO_2 under ultrasonic irradiation is affected by different parameters such as surface area, phase of catalyst, oxygen vacancies, crystalinity of nano-particles and band gap energy [21, 23]. Therefore, the high sonocatalytic activity of TN2 could be attributed to the band gap narrowing resulting from doping of nitrogen and well-formed anatase phase. Figure 4 also indicates that the

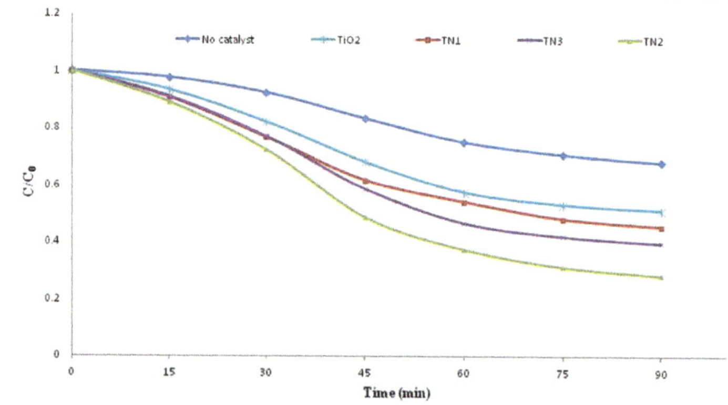

Fig. 4 Sonocatalytic degradation of humic acid for TiO_2 and different *N*-doped TiO_2 nano-particles

sonocatalytic activity of *N*-doped TiO_2 initially increased with the increase of N dopant but further increasing of dopant decreased the activity. Therefore for improvement of sonocatalytic activity of TiO_2, optimum amount of dopant is essential.

Kinetic study

The sonocatalytic degradation of humic acid can be well explained by a pseudo-first-order reaction and its kinetics can be expressed with the following equation:

$$\ln\left(\frac{C_0}{C}\right) = k_{app}t \qquad (3)$$

where k_{obs} is the apparent reaction rate constant, C_0 and C are the humic acid concentrations at initial and at time t, respectively. The k_{obs} were determined from the slopes of straight lines obtained by plotting $ln(C_0/C)$ versus irradiation time.

The values of apparent reaction rate constants (k_{app}) related to the various synthesized nano-sized particles are presented in Table 1. The correlation coefficients above 0.98 indicated the sonodegradation of humic acid by un-doped and *N*-doped TiO_2 suspensions obey the first-order kinetic model in solution. These results also indicated that reaction rate of humic acid can be improved by doping of nitrogen into the TiO_2 structure. The apparent reaction rate constant of un-doped TiO_2 was 0.84×10^{-2} min^{-1}, while the apparent reaction rate constant of TN2 was 1.56×10^{-2} min^{-1}. In addition, enhancement of the reaction rate constants of TN1, TN2 and TN3 were 1.98, 3.25 and 2.40 times higher than the reaction rate constant of sonolysis without catalyst, respectively. These results are in accordance with the those reported by Huang et al. [35] and Wu et al. [33] who studied the degradation of organic pollutants by un-doped TiO_2 and ion- doped TiO_2.

Effect of initial humic acid concentration

The initial concentration of solute in aqueous environment is a key factor on sonocatalytic degradation. As shown in Fig. 5, the degradation efficiency of humic acid increased with decrease in its initial concentration. Sonocatalytic degradation of humic acid with the initial concentrations of 5, 10, and 20 mg L^{-1} for 90 min lead to the conversion of 82.0, 76.0 and 68.0 % of humic acid, respectively. This result indicates that the high degradation efficiency could be obtained at lower humic acid concentration. Our results are in good agreement with the results reported in literature [36]. This result can be due to this fact that under the same conditions, the amount of formed radicals during the sonocatalytic reaction was equal in the entire volume of the solution; therefore, the reaction of humic acid molecules with radicals becomes more likely at lower humic acid concentrations [15].

Langmuir–Hinshelwood model is widely used for analysis of heterogeneous sonocatalytic degradation kinetics as well as to realize the dependence of observed initial reaction rate on the initial concentration of solute in the aqueous environment [9, 29, 37, 38]. The L-H kinetic model is defined as the following equation:

Table 1 Results of kinetic constant, k_{app}, relative increase and removal efficiency of different *N*-doped TiO_2

Catalyst	$k_{app}.10^{-2}(min^{-1})$	Relative increase	R^2	Removal efficiency after 90 min
Absent of catalyst	0.48	1.00	0.9868	32.0
TiO_2	0.84	1.75	0.9851	49.0
TN1	0.95	1.98	0.9895	55.0
TN2	1. 56	3.25	0.9869	72.0
TN3	1. 15	2.40	0.9846	60.0

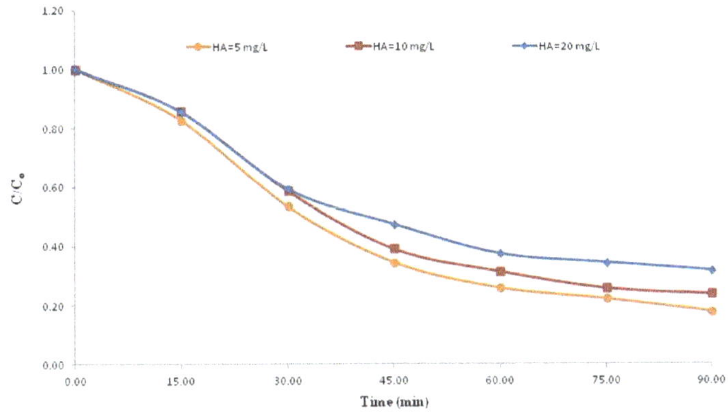

Fig. 5 Effect of initial humic acid concentration on sonocatalytic degradation of humic acid by *N*-doped TiO_2 (TN2) (catalyst concentration: 100 mg L^{-1})

$$r = -\frac{dc}{dt} = k_r\theta_x = \frac{k_rKC}{1+KC} \tag{4}$$

where r is the reaction rate (mg L^{-1} min^{-1}), C is the concentration of solute at any time (mg L^{-1}), t is the reaction time (min), k_r is the Langmuir-Hinshelwood reaction rate constant, related to the limiting rate of reaction at maximum coverage for the experimental condition (mg L^{-1} min^{-1}) and K is the Langmuir adsorption constant reflecting the proportion of solute molecules which adhere to the catalyst surface (L mg^{-1}) and θ is the fraction of the surface of TiO_2 covered by solute. A linear expression of L-H model can be obtained by linearzing the Eq. (4) as follows:

$$\frac{1}{r_0} = \frac{1}{k_r} + \frac{1}{k_rKC_0} \tag{5}$$

The parameters k_r and K which were calculated by plotting the reciprocal initial rate against the reciprocal initial concentration were 0.62 mg L^{-1} min^{-1} and 0.04 L mg^{-1}, respectively (Fig. 6). As shown in Fig. 6, from the correlation coefficient above 0.98 it could be observed that the experimental data are in good agreement with L-H model. According to the L-H model, the reaction is first order at low concentration and zero order at high concentration.

Possible mechanism

In sonolysis process, the sono-luminescence and localized hot-spots with high temperatures up to 5000 K and high pressures (approximately1800 atm) caused by acoustic cavitation and collapse of micro-scale bubbles will occur [11, 12, 39]. These hot spots can pyrolysis water molecules to $OH^{'}$ and $H^{'}$ radicals as below Eq. (6):

$$H_2O +)))\rightarrow OH^{'} + H^{'} \tag{6}$$

In addition, the sono-luminescene could induce the formation of flash light/energy which equals or exceeds the band gap energy of TiO_2 to excite the all synthesized nano-sized particles. The electron excitation from the

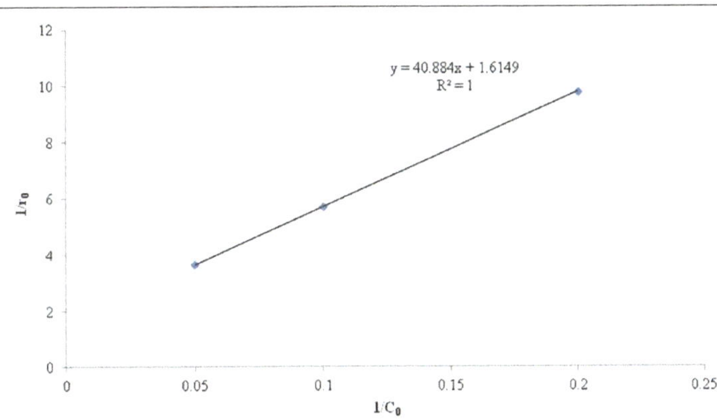

Fig. 6 Variation of reciprocal initial rate versus the reciprocal initial concentration of humic acid

local state of N 2p result in the generation of conduction band electrons (e−) and valence band holes (h^{+}) as shown by Eqs. (7) and (8):

$$))) \rightarrow \text{light or energy} \quad (7)$$

$$\text{N-doped-}TiO_2 +))) \rightarrow h^+ + e^- \quad (8)$$

These charges migrate to the surface and finally react with a suitable electron donor and acceptor. The electrons are captured by Ti^{4+} to form Ti^{3+} states. Subsequently, the 3d orbital of Ti^{3+} ions are localized at 0.75–1.18 eV below the bottom of the conduction band. Ti^{3+} is known to be the most reactive site for oxidation process because it may cause more oxygen vacancy sites, as well as oxygen molecule is more easily adsorbed on TiO_2 surface. Besides, the electrons will react with these surface adsorbed oxygen molecules (O_2) to form superoxide radical anion ($O_2^{'}$) (Eq. 3) and is transformed further to hydroxyl radical ($OH^{'}$) as shown in Eqs. (9) – (14).

$$e^- + Ti^{4+} \rightarrow Ti^{3+} \quad (9)$$

$$e^- + O_{2(ads)} \rightarrow O_2^{'-} \quad (10)$$

$$2O_2^{'-} + 2H_2O \rightarrow 2H_2O_2 + O_2 \quad (11)$$

$$O_2^{'-} + H^+ \rightarrow HOO^{'} \quad (12)$$

$$HOO^{'} + H_2O \rightarrow H_2O_2 + OH^{'} \quad (13)$$

$$H_2O_2 +))) \rightarrow 2OH^{'} \quad (14)$$

The holes migrate to the surface and react with water molecules or chemisorbed OH^- on the surface of N-doped TiO_2 and result in formation of $OH^{'}$ radicals (Eqs. (15) and (16)). Besides, the holes can directly oxidize organic substances adsorbed on the surface of catalyst (Eq. (17))

$$h^+ + Ti\text{-}OH^- \rightarrow Ti\text{-}OH^{'} \quad (15)$$

$$H_2O + h^+ \rightarrow OH^{'} + H^+ \quad (16)$$

$$\textit{organic substances} + h^+ \rightarrow \textit{degraded products} \quad (17)$$

where ")))" denotes to the ultrasonic irradiation. It is widely accepted that $O_2^{'}$ and $OH^{'}$ have strong oxidative degradation potential. Wu et al. found that the amounts of the produced $OH^{'}$ radicals increase with doping of TiO_2 [33] . In this study, from degradation efficiency it can be understand that the highest amount of radicals is generated on the surface of TN2 because narrower band gap of TN2 facilitates the transition of electron from the valence band to the conduction band and eventually increases sonocatalytic activity. Thus, optimum amount of nitrogen dopant play an important role in improving sonocatalytic activity.

Conclusions

In this study, a simple sol-gel method was used to synthesize of un-doped and *N*-dope TiO_2 for activity enhancement of sonolysis and sonocatalysis processes. The characterization of synthesized nano-particles was carried out by XRD, FE-SEM, EDX and UV-vis spectra. The characterization experiments confirmed that nitrogen doping has been successfully done in the TiO_2 structure.

The degradation of humic acid was used to evaluate the sonocatalytic activity of synthesized nano-particles. On the basis of the above results and discussion, addition of nano-catalysts could enhance the degradation efficiency of humic acid as well as *N*-doped TiO_2 with a molar ratio of N/Ti as 0.06 was found the best nano-catalyst among the investigated catalysts. The synthesized *N*-doped TiO_2 showed about 1.86 times higher sonocatalytic activity for humic acid degradation compared to the un-doped TiO_2.

The sonocatalytic degradation of humic acid with different catalysts followed the first-order kinetic model. L-H model confirmed the dependence of initial reaction rate on the initial humic acid concentrations and showed that the degradation efficiency decrease with the increase of initial humic acid concentrations. As a general conclusion, the results indicated that sonocatalytic degradation with nitrogen doped semiconductors could be a suitable oxidation process for removal of refractory pollutants such as humic acid from aqueous solution.

Additional file

Additional file 1: Figure S1. SEM image of pure TiO_2. **Figure S2.** EDX elemental mapping of N-doped TiO_2. (DOC 627 kb)

Competing interests

The authors declare that they have no competing interests.

Authors' contributions

HK was the main investigator, synthesized the nano-particles and drafted the manuscript. AM and SN supervised the study. RNN and MK were advisors of the study. RNN also contributed in analyzing of data. All authors read and approved the final manuscript.

Acknowledgments

This paper is a part of the results a PhD research thesis. The authors would like to thank the Center for Water Quality Research, Institute for Environmental Research, Tehran University of Medical Science for the financial support of this study (grant no. 94-33-61-20515). Authors also thank Mrs., Sheikhi and Mrs. Hoseini, the technical staffs in the chemical laboratory, for their cooperation.

Author details

[1]Department of Environmental Health Engineering, School of Public Health, Tehran University of Medical Sciences, Tehran, Iran. [2]Center for Water Quality Research, Institute for Environmental Research, Tehran University of Medical Sciences, Tehran, Iran. [3]Department of Medicinal Chemistry, Faculty of

Pharmacy and Pharmaceutical Sciences Research Center, Tehran University of Medical Sciences, Tehran, Iran. [4]Center for Solid Waste Research, Institute for Environmental Research, Tehran University of Medical Sciences, Tehran, Iran. [5]National Institute of Health Research, Tehran University of Medical Sciences, Tehran, Iran.

References

1. Rezaee R, Maleki A, Jafari A, Mazloomi S, Zandsalimi Y, Mahvi AH. Application of response surface methodology for optimization of natural organic matter degradation by UV/H_2O_2 advanced oxidation process. J Environ Health Sci Eng. 2014;12:67.
2. Mahvi A, Maleki A, Rezaee R, Safari M. Reduction of humic substances in water by application of ultrasound waves and ultraviolet irradiation. J Environ Health Sci Eng. 2010;6:233–40.
3. Sun DD, Lee PF. TiO_2 microsphere for the removal of humic acid from water: complex surface adsorption mechanisms. Sep Purif Technol. 2012;91: 30–7.
4. Bazrafshan E, Biglari H, Mahvi AH. Humic acid removal from aqueous environments by electrocoagulation process using iron electrodes. J Chem. 2012;9:2453–61.
5. Bazrafshan E, Mostafapour FK, Hosseini AR, Raksh Khorshid A, Mahvi AH. Decolorisation of reactive red 120 dye by using single-walled carbon nanotubes in aqueous solutions. J Chem. 2013;1-8.
6. Alipour V, Nasseri S, Nodehi RN, Mahvi AH, Rashidi A. Preparation and application of oyster shell supported zero valent nano scale iron for removal of natural organic matter from aqueous solutions. J Environ Health Sci Eng. 2014;12:146.
7. Ashrafi S, Kamani H, Mahvi A. The optimization study of direct red 81 and methylene blue adsorption on NaOH-modified rice husk. Desalin Water Treat. 2016;57:738-746
8. Wang G-S, Liao C-H, Wu F-J. Photodegradation of humic acids in the presence of hydrogen peroxide. Chemosphere. 2001;42:379–87.
9. Safari G, Hoseini M, Seyedsalehi M, Kamani H, Jaafari J, Mahvi A. Photocatalytic degradation of tetracycline using nanosized titanium dioxide in aqueous solution. Int J Environ Sci Technol. 2015;12:603–16.
10. Mahvi AH, Ebrahimi SJA-d, Mesdaghinia A, Gharibi H, Sowlat MH. Performance evaluation of a continuous bipolar electrocoagulation/ electrooxidation–electroflotation (ECEO–EF) reactor designed for simultaneous removal of ammonia and phosphate from wastewater effluent. J Hazard Mater. 2011;192:1267–74.
11. Pang YL, Abdullah AZ. Comparative study on the process behavior and reaction kinetics in sonocatalytic degradation of organic dyes by powder and nanotubes TiO_2. Ultrason Sonochem. 2012;19:642–51.
12. Mahvi AH, Roodbari AA, Nodehi RN, Nasseri S, Dehghani MH, Alimohammadi M. Improvement of Landfill Leachate Biodegradability with Ultrasonic Process. PLoS One. 2012;7:e27571.
13. Dobaradaran S, Nabizadeh R, Mahvi A, Mesdaghinia A, Naddafi K, Yunesian M, et al. Survey on degradation rates of trichloroethylene in aqueous solutions by ultrasound. Iranian J Environ Health Sci Eng. 2010;7:307–12.
14. Maleki A, Mahvi AH, Ebrahimi R, Zandsalimi Y. Study of photochemical and sonochemical processes efficiency for degradation of dyes in aqueous solution. Korean J Che Eng. 2010;27:1805–10.
15. Hoseini M, Safari GH, Kamani H, Jaafari J, Ghanbarain M, Mahvi AH. Sonocatalytic degradation of tetracycline antibiotic in aqueous solution by sonocatalysis. Toxicol Environ Chem. 2013;95:1680–9.
16. Daraei H, Maleki A, Mahvi AH, Zandsalimi Y, Alaei L, Gharibi F. Synthesis of ZnO nano-sono-catalyst for degradation of reactive dye focusing on energy consumption: operational parameters influence, modeling, and optimization. Desalin Water Treat. 2014;52:6745–55.
17. Mahvi A, Ghanbarian M, Nasseri S, Khairi A. Mineralization and discoloration of textile wastewater by TiO_2 nanoparticles. Desalination. 2009;239:309–16.
18. Borji SH, Nasseri S, Mahvi AH, Nabizadeh R, Javadi AH. Investigation of photocatalytic degradation of phenol by Fe (III)-doped TiO_2 and TiO_2 nanoparticles. J Environ Health Sci Eng. 2014;12:101.
19. Javid A, Nasseri S, Mesdaghinia A, Hossein Mahvi A, Alimohammadi M, Aghdam RM, et al. Performance of photocatalytic oxidation of tetracycline in aqueous solution by TiO_2 nanofibers. J Environ Health Sci Eng. 2013;11:24.
20. Pang YL, Abdullah AZ. Effect of low Fe^{3+} doping on characteristics, sonocatalytic activity and reusability of TiO_2 nanotubes catalysts for removal of Rhodamine B from water. J Hazard Mater. 2012;235:326–35.
21. Zhang S. Synergistic effects of C–Cr codoping in TiO_2 and enhanced sonocatalytic activity under ultrasonic irradiation. Ultrason Sonochem. 2012; 19:767–71.
22. Cheng X, Yu X, Xing Z, Wan J. Enhanced photocatalytic activity of nitrogen doped TiO_2 anatase nano-particle under simulated sunlight irradiation. Energy Procedia. 2012;16:598–605.
23. Pang YL, Abdullah AZ. Effect of carbon and nitrogen co-doping on characteristics and sonocatalytic activity of TiO_2 nanotubes catalyst for degradation of Rhodamine B in water. Chem Eng J. 2012;214:129–38.
24. Li X, Fan C, Sun Y. Enhancement of photocatalytic oxidation of humic acid in TiO_2 suspensions by increasing cation strength. Chemosphere. 2002;48:453–60.
25. Gurkan YY, Turkten N, Hatipoglu A, Cinar Z. Photocatalytic degradation of cefazolin over N-doped TiO_2 under UV and sunlight irradiation: prediction of the reaction paths via conceptual DFT. Chem Eng J. 2012;184:113–24.
26. Kuo Y-L, Su T-L, Kung F-C, Wu T-J. A study of parameter setting and characterization of visible-light driven nitrogen-modified commercial TiO_2 photocatalysts. J Hazard Mater. 2011;190:938–44.
27. Xie Y, Li Y, Zhao X. Low-temperature preparation and visible-light-induced catalytic activity of anatase F–N-codoped TiO_2. J Mol Catal A Chem. 2007; 277:119–26.
28. Lin X, Rong F, Fu D, Yuan C. Enhanced photocatalytic activity of fluorine doped TiO_2 by loaded with Ag for degradation of organic pollutants. Powder Technol. 2012;219:173–8.
29. Wang P, Yap PS, Lim TT. C–N–S tridoped TiO_2 for photocatalytic degradation of tetracycline under visible-light irradiation. Appl Catal A Gen. 2011;399:252–61.
30. Song L, Chen C, Zhang S, Wei Q. Sonocatalytic degradation of amaranth catalyzed by La^{3+} doped TiO_2 under ultrasonic irradiation. Ultrason Sonochem. 2011;18:1057–61.
31. Livraghi S, Elghniji K, Czoska A, Paganini M, Giamello E, Ksibi M. Nitrogen-doped and nitrogen–fluorine-codoped titanium dioxide. Nature and concentration of the photoactive species and their role in determining the photocatalytic activity under visible light. J Photochem Photobiol A Chem. 2009;205:93–7.
32. Farzadkia M, Bazrafshan E, Esrafili A, Yang J-K, Shirzad-Siboni M. Photocatalytic degradation of Metronidazole with illuminated TiO2 nanoparticles. J Environ Health Sci Eng. 2015;13:35
33. Wu Y, Xing M, Tian B, Zhang J, Chen F. Preparation of nitrogen and fluorine co-doped mesoporous TiO_2 microsphere and photodegradation of acid orange 7 under visible light. Chem Eng J. 2010;162:710–7.
34. Ananpattarachai J, Kajitvichyanukul P, Seraphin S. Visible light absorption ability and photocatalytic oxidation activity of various interstitial N-doped TiO_2 prepared from different nitrogen dopants. J Hazard Mater. 2009;168: 253–61.
35. Huang DG, Liao SJ, Liu JM, Dang Z, Petrik L. Preparation of visible-light responsive N–F-codoped TiO_2 photocatalyst by a sol–gel-solvothermal method. J Photochem Photobiol A Chem. 2006;184:282–8.
36. Kaur S, Singh V. Visible light induced sonophotocatalytic degradation of Reactive Red dye 198 using dye sensitized TiO 2. Ultrason Sonochem. 2007;14:531–7.
37. Saha S, Wang J, Pal A. Nano silver impregnation on commercial TiO_2 and a comparative photocatalytic account to degrade malachite green. Sep Purif Technol. 2012;89:147–59.
38. Jaafari J, Mesdaghinia A, Nabizadeh R, Hoseini M, Mahvi AH. Influence of upflow velocity on performance and biofilm characteristics of Anaerobic Fluidized Bed Reactor (AFBR) in treating high-strength wastewater. J Environ Health Sci Eng. 2014;12:139.
39. Salavati H, Tangestaninejad S, Moghadam M, Mirkhani V, Mohammadpoor-Baltork I. Sonocatalytic oxidation of olefins catalyzed by heteropolyanion–montmorillonite nanocomposite. Ultrason Sonoche. 2010;17:145–52.

Sonoelectrochemical mineralization of perfluorooctanoic acid using *Ti/PbO₂* anode assessed by response surface methodology

Gholamreza Bonyadinejad[1,2], Mohsen Khosravi[3], Afshin Ebrahimi[1*], Roya Nateghi[1], Seyed Mahmood Taghavi-Shahri[4] and Hamed Mohammadi[1,2]

Abstract

Background: Perfluorocarboxylic acids (PFCAs) are emerging pollutant and classified as fully fluorinated hydrocarbons containing a carboxylic group. PFCAs show intensively resistance against chemical and biological degradation due to their strong C–F bond. The Sonoelectrochemical mineralization of the synthetic aqueous solution of the perfluorooctanoic acid (PFOA) on Ti/PbO_2 anode was investigated using the response surface methodology based on a central composite design with three variables: current density, pH, and supporting electrolyte concentration.

Methods: The defluorination ratio of PFOA was determined as an indicator of PFOA mineralization. Fluoride ion concentration was measured with an ion chromatograph unit. The Ti/PbO2 electrode was prepared using the electrochemical deposition method. The ultrasonic frequency was 20 kHz.

Results: The optimum conditions for PFOA mineralization in synthetic solution were electrolyte concentration, pH, and current density of 94 mM, 2, and 83.64 mA/cm^2, respectively. The results indicated that the most effective factor for PFOA mineralization was current density. Furthermore, the PFOA defluorination efficiency significantly enhanced with increasing current density. Under optimum conditions, the maximum mineralization of PFOA was 95.48 % after 90 min of sonoelectrolysis.

Conclusions: Sonoelectrolysis was found to be a more effective technique for mineralization of an environmentally persistent compound.

Keywords: Ultrasonics, Lead dioxide, Perfluorooctanoic acid, Central composite design

Introduction

The Perfluorocarboxylic acids are classified as fully fluorinated hydrocarbons materials that contain a carboxylic group. Due to their high surface active nature, high thermal and chemical stabilities, PFCAs are extensively using in fire retardants, industrial surfactants and in fluoropolymers manufacturing processes [1]. This extensive utilization of PFCAs results in releasing of these compounds into the environment, which was estimated at 3200–7300 tons for the period of 1950–2004 by direct and indirect emissions [2]. PFCAs show intensive resistance against chemical and biological degradation. Conventional treatment methods are ineffectual for the degradation of PFCAs because it is intrinsically recalcitrant to biological and conventional chemical treatment [3]. The marvelous stability of perfluorinated compounds (PFCs) is ascribed to their strong C–F bond which makes them very persistent to most natural conditions [1, 4]. Perfluorooctanoic acid (PFOA) is one of the PFCAs' family, which is categorized as a likely potential carcinogen by The US EPA's Science Advisory Board in 2006. Recently, PFOA and its precursors have been universally detected in wildlife, water and human body [5]. For instance, the river Elbe in Germany was polluted by PFOA due to effluents discharged from wastewater treatment plant [6]. Since, PFOA has

* Correspondence: a_ebrahimi@hlth.mui.ac.ir
[1]Environment Research Center and Department of Environmental Health Engineering, School of Health, Isfahan University of Medical Sciences (IUMS), Isfahan 81676-36954, Iran
Full list of author information is available at the end of the article

toxicity to organisms and humans and persistence in the environment, a lot attention has been paid to treatment of PFOA from water which uses effective methods under moderate conditions. Lately, some chemical technologies for PFOA degradation have been reported, such as photocatalytic oxidation, direct photolysis, photochemical reduction, photochemical oxidation, thermally-induced reduction and sonochemical pyrolysis. Yet, most of the techniques could not effectively decompose PFOA [7]. The goal of destructive methods is cleaving the C–F bonds to form F- ions [2]. The electrochemical oxidation is a technology that has presented its capacity to degrade refractory organic pollutants such as emerging contaminants contained in the secondary effluents of wastewater treatment plants and also removal of various pollutants from aqueous environments [8, 9]. Electrochemical techniques generally carried out oxidatively, have the advantage of contaminant elimination without the addition of chemicals. Although electrochemical methods are easy to use and have high removal efficiency, energy consumption is the main disadvantage of these methods. Nevertheless, this disadvantage has been suppressed by the development of new anode materials [10, 11]. PbO_2 is a low-cost electrode material that can be quickly and easily prepared, and is being used by many researchers for electro-oxidation of refractory pollutants in aquatic solutions [12–15]. During the process of oxidation of polluted waters, $OH^•$ specimen is generated on the PbO_2 electrode surface, and a mechanism for the electrochemical processes occurring in the gel-crystal structure of the PbO_2 layer of the electrode has been proposed [16]. The $Pb^*O(OH)_2$ active centers placed in the hydrated PbO_2 layer on the surface of the crystalline PbO_2 anode provide electrons to the crystal district, becoming positively charged ($Pb^*O(OH)^+$ $(OH)^•$). This electric charge is neutralized according to the following reaction:

$$Pb * O(OH)^{+}(OH)^{\bullet} + H_2O \rightarrow PbO(OH)_2...(OH)^{\bullet} + H^{+} \quad (1)$$

in which hydroxyl radicals are produced in the active centers. The OH• can depart from the active centers and react with the pollutant in the aqueous solution. Thus, the PbO_2 anode is expected to perform quite well in organic pollutant mineralization. However, the main problem of PbO_2 anode is the release of poisonous ions, Pb^{2+} [17]. On the other hand, sonochemical treatment applies pyrolytic cleavages for organic pollutants degradation and is an emerging and impressive method which can be used efficiently to eliminate PFOA particularly [18]. Ultrasonic (US) treatment acts as cavitation, which not only generates plasma in water, and degrading molecules by pyrolysis, but also produces free radicals and other reactive types that can improve the amount of collisions between free radicals and contaminants. Recently, using the combination of ultrasonic method and other techniques for the treatment of organic wastewater has been largely studied. (e.g. phenol and pharmaceutical compounds) [19–21]. Optimization of the operating conditions of an experimental system and recognition of the way in which the experimental parameters affect the final output of the system are realized by using modeling techniques [22]. In addition, it is possible to determine the relationships and interactions between the variables through these techniques. In this regard, statistical methodologies, such as the response surface methodology (RSM), are suitable for studying and modeling a particular system [23]. The aim of the present study was to investigate the electrochemical mineralization of PFOA as an emerging contaminant using Ti/PbO_2 anode coupled with ultrasonic irradiation (sonoelectrochemical) and determine the optimum conditions by RSM. The effects of current density (CD), pH of the solution, and supporting electrolyte (EL) concentration were evaluated in terms of PFOA defluorination.

Materials and methods

Chemicals

Analytical-grade PFOA was purchased from Sigma Aldrich co., and used without further purification. $Pb(NO_3)_2$ (Sigma Aldrich co.), Triton X-100 (Merck co.), and $CuSO_4.5H_2O$ (Merck co.) were used for electrode preparation. Other chemicals were purchased from Merck co. The initial pH of the solutions was adjusted using sodium hydroxide and sulfuric acid. Sodium sulfate was used as the supporting electrolyte. All the solutions were prepared using deionized water.

Preparation of Ti/PbO_2 electrode

The Ti substrate with 2 mm thickness was cut into a strip (4.8 cm × 4 cm, 99.7 % Aldrich) and pre-treated according to the following procedure: the substrate was polished on 320-grit paper strips [24] to eliminate the superficial layer of TiO_2 (an electric semiconductor) and increase surface roughness (for efficient adherence of PbO_2). Then, the substrate was degreased in an ultrasonic bath of acetone for 10 min and then in distilled water for 10 min. Afterwards, the substrate was etched for 1 h in a boiling solution of oxalic acid (10 %) and rinsed with ultrapure water [25]. Finally, the cleaned Ti substrate was transferred to an electrochemical deposition cell, which contained 12 % (w/v) $Pb(NO_3)_2$ solution comprising 5 % (w/v) $CuSO_4.5H_2O$ and 3 % (w/v) surfactant (Triton X-100). The role of the surfactant was to minimize the surface tension of the solution for better wetting of the substrate and also to increase the adhesion of PbO_2 to the Ti substrate. The electrodeposition of PbO_2 was performed at a constant anodic current of 20 mA/cm^2 for 60 min at 80 °C with continuous stirring [26]. The X-ray diffraction (XRD) tests were performed using a Bruker, D8 Advance, Germany. The samples were scanned under Co Kα radiation (wavelength: 1.7890 Å) at 40 kV and 40 mA. Scanning electron microscope (SEM; Philips XI30, Netherlands) was employed to observe the surface morphology of the electrodes, which presented a typical pyramid shape similar to that reported in the literature [27].

Sonoelectrochemical mineralization of PFOA

Synthetic wastewater was prepared by dissolving PFOA in distilled water at a concentration of 50 mg/L. The sonoelectrochemical mineralization of PFOA was performed in a temperature-controlled water batch reactor (0.45 L) equipped with an ultrasonic probe (Bandelin SONOPULS,UW 3200, TT 13, Germany), a 41.12 cm^2 Ti/PbO_2 as the anode and a 80.32 cm^2 stainless steel plate as the cathode in conjunction with an adjustable power supply unit (HANI, Iran). The gap between the anode and cathode was 1 cm. The temperature of the reaction solution was kept constant at 25 ± 1 °C. The duration of all the electrolysis experiments was 90 min. The reactor was placed on a magnetic stirrer to mix its content during the experiment in order to maximize mass transport (Fig. 1).

Analytical procedure

Defluorination ratio of PFOA was determined based on the release of fluoride into the solutions. The defluorination ratio, is the indicator of PFOA mineralization [28]. Fluoride ions concentration were measured with an Ion chromatograph system (IC762, Metrohm, USA) equipped with an automatic sample injector, a degasser, a pump, a guard column (Metrosep RP guard, Metrohm), a separation column (Metrosep Anion Dual 2, Metrohm), and a conductivity detector with a suppressor device. The mobile phase was an aqueous solution containing $NaHCO_3$ (2 mM) and Na_2CO_3 (1.3 mM). The flow rate was 0.8 mL/min. The defluorination ratio (R) was calculated as Eq. 2:

$$R = \frac{C_{F^-}}{C_0 \times 15} \times 100 \qquad (2)$$

where C_{F^-} is the concentration of fluoride in mM, C_0 is the initial concentration of PFOA in mM, and the value of 15 represents the number of fluorine atoms contained in one PFOA molecule.

Experimental design

The central composite design (CCD) technique coupled with RSM has been opted for modeling and design of experimental tests [29]. The CD, EL, and pH parameters were selected as input variables. The rotatable experimental plan was performed with the three variables at five levels (−1.68, −1, 0, 1, 1.68). Table 1 shows the values and levels of the variables. Five replications were done at the center point of the design to evaluate the pure error and consequently the lack of fit. Statistica ver. 10, and R ver. 3.1.2 software were used to design and analyze the experiments. Table 2 shows the CCD matrix of the mineralization experiments.

The relationship between response Y and the three independent variables X_1, X_2, and X_3 could be approximated by quadratic polynomial Eq. 3 as follows:

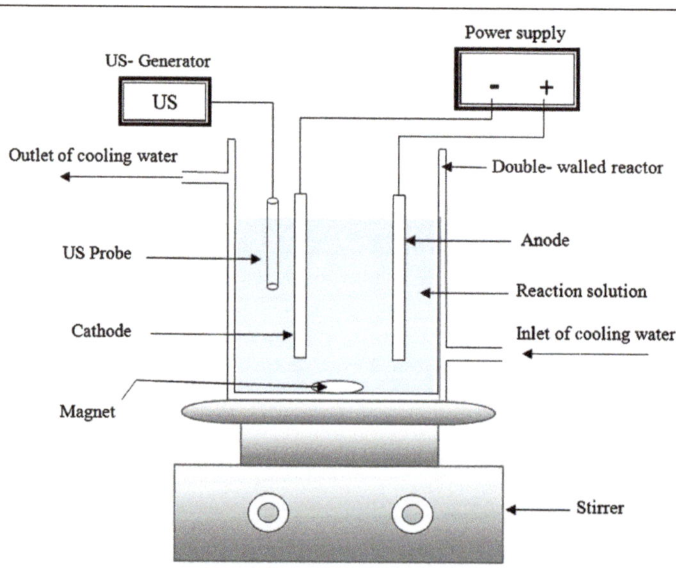

Fig. 1 Schematic of the experimental setup

Table 1 The range and codification of the independent variables (X_i) used in the experimental design

Variables	Actual values of the coded values				
	−1.68	−1	0	1	1.68
pH (X_1)	1.95	4	7	10	12.05
EL (mM) (X_2)	32.96	50	75	100	117.04
CD (mA/cm^2) (X_3)	16.36	30	50	70	83.64

$$Y = b_0 + b_1X_1 + b_2X_2 + b_3X_3 + b_{11}X_1^2 + b_{22}X_2^2 + b_{33}X_3^2 + b_{12}X_1X_2 + b_{13}X_1X_3 + b_{23}X_2X_3 \quad (3)$$

where Y is the predicted response; b_0 is a constant; b_1, b_2, and b_3 are the linear coefficients, b_{12}, b_{13}, and b_{23} are the cross-product coefficients; and b_{11}, b_{22}, and b_{33} are the quadratic coefficients. In the present study, backward variable selection was used for multiple regression modeling [30]. The assumption of final regression model was verified using the Anderson–Darling test for normality of residuals [31], Breusch–Pagan test for constant variance of residuals [32], and Durbin–Watson test for independence of residuals [33, 34]. Lack of fit test was performed to assess the fit of the final model. Validation of the final model was established using predicted R-squares (R^2), which estimates the prediction power of the model with new observations based on the leave-one-out technique [35]. The optimum values of the final model were calculated using numerical methods. In this regard, the experimental range predictors were divided into a grid and then the final model was calculated for all possible combinations of predictors in the grid.

Table 2 CCD matrix of sonoelectrochemical mineralization of PFOA

Exp. No.	pH	EL	CD	Defluorination ratio (%)	
				Observed	Predicted
1	4.00	50.00	30.00	54.16	41.90
2	4.00	50.00	70.00	87.45	81.34
3	4.00	100.00	30.00	57.47	44.16
4	4.00	100.00	70.00	89.49	83.61
5	10.00	50.00	30.00	51.89	38.57
6	10.00	50.00	70.00	85.55	78.03
7	10.00	100.00	30.00	51.46	40.84
8	10.00	100.00	70.00	87.46	80.29
9	1.95	75.00	50.00	75.71	67.15
10	12.05	75.00	50.00	69.6	61.57
11	7.00	32.96	50.00	69.12	61.36
12	7.00	117.04	50.00	74.26	65.18
13	7.00	75.00	16.36	37.73	23.20
14	7.00	75.00	83.64	91.86	89.55
15	7.00	75.00	50.00	74.32	65.36
16	7.00	75.00	50.00	75.51	65.36
17	7.00	75.00	50.00	74.66	65.36
18	7.00	75.00	50.00	73.82	65.36
19	7.00	75.00	50.00	74.32	65.36

Result and discussion

Characterization of the Ti/PbO_2 electrode

XRD pattern of the Ti/PbO_2 electrode has been represented in Fig. 2, it can be observed that PbO_2 was deposited in the form of two known polymorphs, namely, orthorhombic α-PbO_2 and tetragonal β-PbO_2, which occur naturally as scrutinyite and plattnerite, respectively. Figure 3 shows the SEM image of the surface microstructure of the Ti/PbO_2 electrode. It can be observed that the PbO_2 layer is crack free and composed of packed faceted micro crystallites. Such morphology verifies that only PbO_2 is involved in the electrochemical mineralization of the PFOA and protects the surface of the Ti substrate. Furthermore, energy-dispersive X-ray spectroscopy (EDS) analysis (data not shown) confirmed the presence of lead and oxygen atoms on the surface of the Ti/PbO_2 electrode.

CCD analysis and modeling

The interaction effect of input variables, which were statistically designed by using CCD method were studied through different combination of experimental parameters. The CCD design matrix, the predicted results and observed defluorination ratio values for the sonoelectrochemical mineralization of PFOA are shown in Table 2. Equation 4 represents the first model that was developed with all linear, quadratic, and two-way interaction of predictors:

$$Y = 6.04096 - 0.045 \times pH + 0.267 \times EL + 1.6625 \times CD - 0.00565 \times pH \times EL + 0.01108 \times pH \times CD + 0.000106 \times EL \times CD - 0.04639 \times pH^2 - 0.001247 \times EL^2 - 0.008789 \times CD^2 \quad (4)$$

The predicted R^2 of this initial model was 98.0 %. Backward elimination method was used to achieve a parsimonious model with significant predictors. In the first step, linear form of pH was removed from the initial model ($p = 0.88$). In three following steps, interactions of EL*CD ($p = 0.73$), pH*EL ($p = 0.17$), and pH*CD ($p = 0.33$) were removed from the model. Finally, linear and quadratic form of EL and CD, with quadratic form of pH, EL, and CD remain significant in the prediction model. The final model was as Eq. 5:

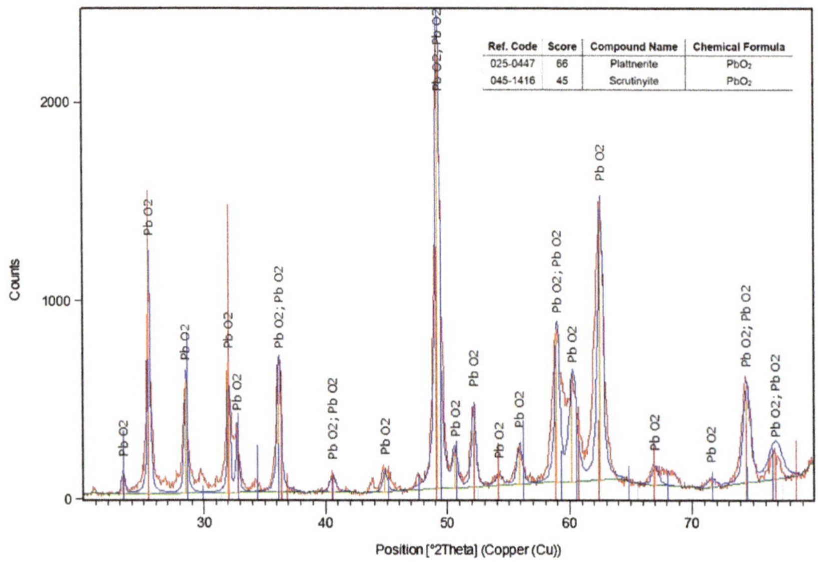

Fig. 2 XRD pattern of the prepared PbO_2 electrode

$$Y = 5.069 + 0.22311 \times EL + 1.62182 \times CD - 0.039453 \times pH^2 - 0.0011852 \times EL^2 - 0.0079442 \times CD^2 \quad (5)$$

A partial F test was performed for comparison between first model and the final model. The value of F was 0.86 with 4 and 9° of freedom, and a P-value of 0.52. Hence, the difference of these two models was not significant, although the final regression model (Eq. 5) had four predictors less than the initial model (Eq. 4). Table 3, represent the final regression model for PFOA mineralization. The P-value of Anderson-Darling, Breusch-Pagan, and Durbin-Watson tests were 0.34, 0.82 and 0.75, respectively, which confirmed the assumptions of regression model. Lack of fit test also was non-significant with P-value of 0.09 which means the final model fitted experimental data satisfactorily. (F = 4.08, degree of freedom for lack of fit = 9, degree of freedom for pure error = 4) (Fig. 4).

The final model was validated using leave-one-out technique. Predicted R-squared was 98.8 % which confirmed

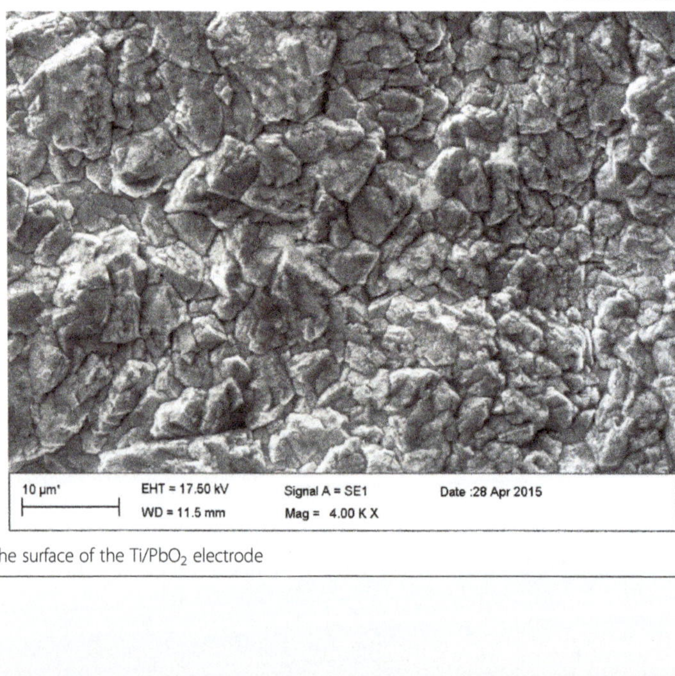

Fig. 3 SEM micrograph of the surface of the Ti/PbO_2 electrode

Table 3 Final regression model for PFOA mineralization

Variable	Coefficients	Std. Error	T statistic	P-value
(Constant)	5.07	3.36	1.51	0.155
EL	2.23×10^{-1}	7.23×10^{-2}	3.09	0.009
CD	1.62	7.57×10^{-2}	21.42	<0.001
pH^2	-3.95×10^{-2}	6.98×10^{-3}	−5.65	<0.001
EL^2	-1.18×10^{-3}	4.75×10^{-4}	−2.49	0.027
CD^2	-7.94×10^{-3}	7.42×10^{-4}	−10.71	<0.001

R-squared = 99.6 %, Adjusted R-squared = 99.4 %, Predicted R-squared = 98.8 %

external validity of the final model. Leave-one-out prediction of the final model vs. observed values of response represented in Fig. 5.

The final model was evaluated through the range of experimental predictors with a numerical method. The grid combination of predictors was ranged between 2 to 12 with 0.1 increments for pH, from 33 to 117 with 1 increments for EL and from 16 to 85 with 1 increments for CD. Model prediction was calculated for 579,600 different combination of predictors and optimum values of input parameters were pH = 2, EL = 94, CD = 83.64 which resulted in 95.48 % PFOA mineralization. Since, there are no interaction effects in the final model, effects of pH, EL, and CD on PFOA mineralization can be investigated independently. Figures 6, 7, and 8 show effects of pH, CD, and EL on PFOA mineralization, respectively, when other factors were at optimum values.

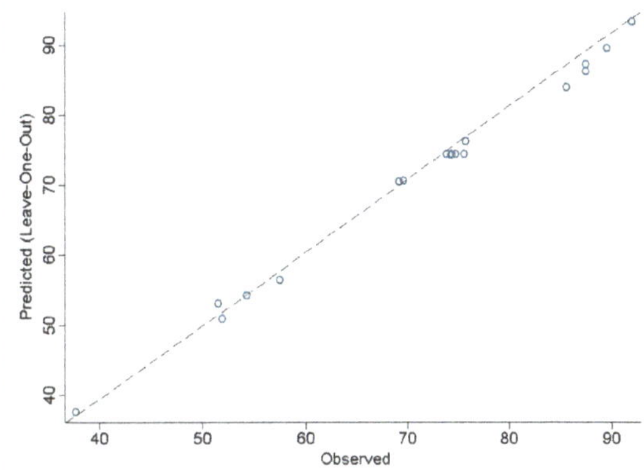

Fig. 5 Leave-one-out prediction of the final model vs. observed for PFOA mineralization

Synergistic effect

In order to evaluate the simultaneous effect of ultrasonic and electrochemical processes on PFOA mineralization, three different pretests for PFOA mineralization were done as follow: sonolysis, electrochemical and sonoelectrochemical. In all three experiments, reaction time, PFOA concentration and initial pH value were 90 min, 50 mg/L, and 7, respectively. The mineralization results were 8 %, 21 % and 73.9 % for sonolysis (frequency = 20 khz), electrochemical (CD = 50, mA/cm^2, and EL = 75, mM) and sonoelectrochemical (frequency = 20 khz, CD = 50 mA/cm^2, and EL = 75, mM) processes, respectively. This indicates

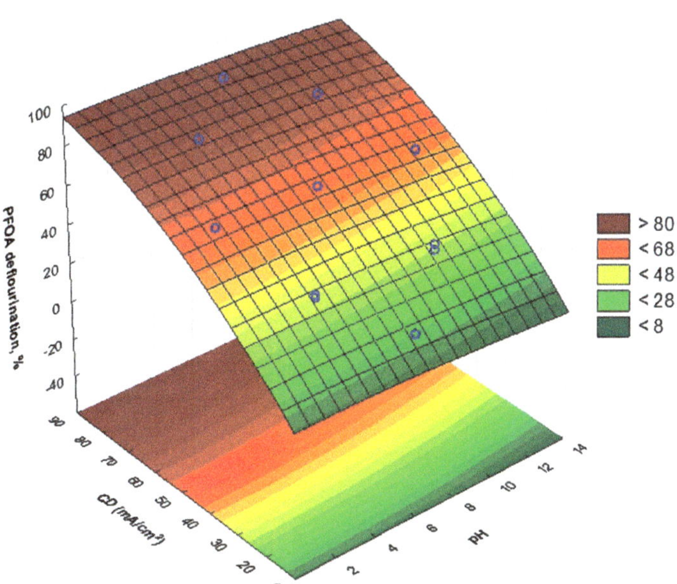

Fig. 4 3D surface plot for the PFOA deflourination as a function of CD and pH

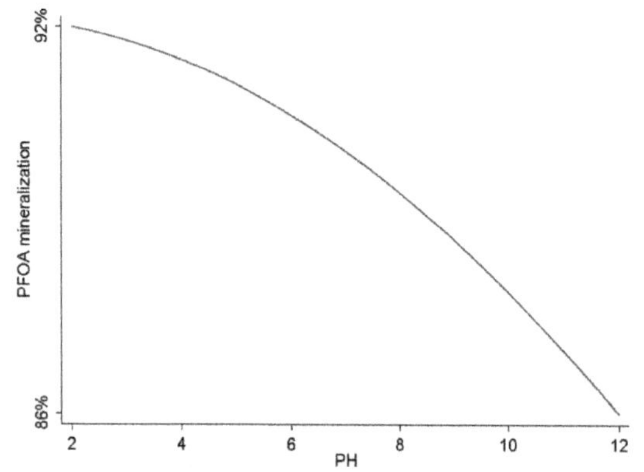

Fig. 6 Predicted PFOA mineralization using the final model vs. pH, when EL = 94 mM and CD = 83.64 mA/cm^2

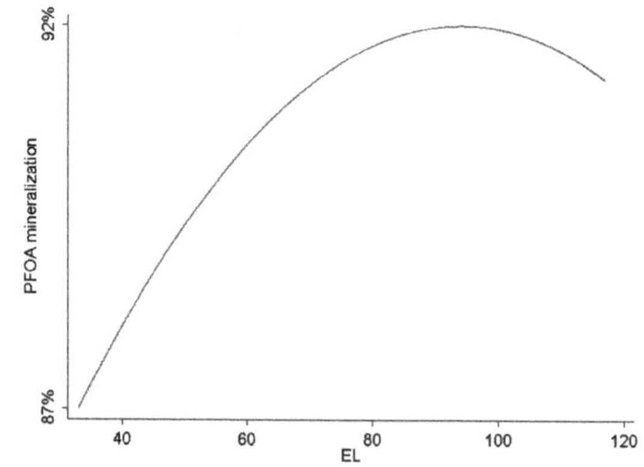

Fig. 8 Predicted PFOA mineralization using the final model vs. EL, when pH = 2 and CD = 83.64 mA/cm^2

that combination of sonolysis and electrochemical, have a remarkable synergistic effect on PFOA mineralization. Some of the basic concepts are mentioned here to clarify the sonoeletochemical, results in a higher mineralization efficiency than sonolysis and electrochemical. In the sonolysis process, the propagation of ultrasound waves through the bulk of liquid cause cavitational bubbles, which can oxidize organic substances either directly by formation of •OH by the sonolysis of water, or as a result of thermolytic reactions taking place inside [36]. The ultrasonic technique is powerful, however with respect to total input energy, there is no economical justification for using this method without the help of other techniques [37].

The synergism observed between sonolysis and electrochemical oxidation can be associated with following reasons:

1- Formation of sulfate radical by ultrasonic irradiation in the presence of sulfate ions when sodium sulfate used as supporting electrolyte [38].

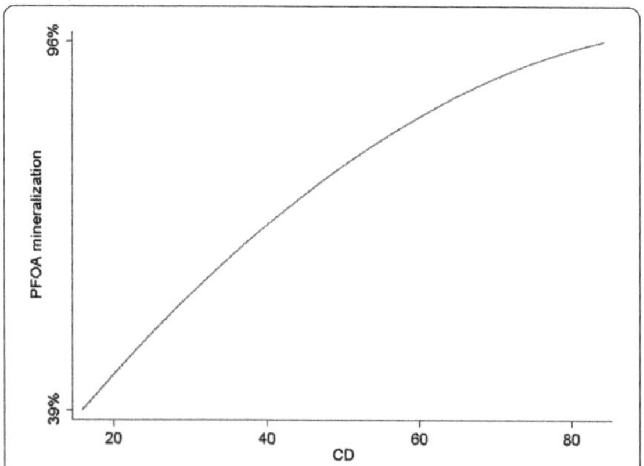

Fig. 7 Predicted PFOA mineralization using the final model vs. CD, when pH = 2 and EL = 94 mM

2- Ultrasonic waves facilitated the mass-transfer on the electrode surface, resulted in increasing diffusion of the produced hydroxyl radicals,which increased the OH radical concentration in the solution [36, 39, 40].

3- Cleaning of the electrode surface by cavitational bubbles. The mechanical effects of cavitation lead to cleaning of the electrode surface and inhibit any passive layer formation. This effect guarantee the normal electrochemical operation process with a stable electric current in the period of the treatment time [36, 40]

Effect of initial pH

Figure 4 shows the effect of initial pH of the solutions on PFOA mineralization which has been adjusted to the following values: 1.95, 4, 7, 10, and 12.05. There are many disagreements about the mechanism of influence of pH in the literature which is due to diversity of the organic structures and electrode materials, however the initial pH is one of the main factors in the oxidation process [27].

As shown in Fig. 4 and by Eq. 5, the PFOA mineralization efficiency was increased with decrease of pH. From Eq. (3), the highest and the lowest level of PFOA mineralization efficiency can be reached at an acidic (pH = 2) and alkaline (pH = 12) conditions, respectively. An increase in the PFOA mineralization with the decreasing pH from 12 to 2 can be explained as follows; first, in the alkaline conditions, sulfate radicals might react with OH resulting O^- which has a lower oxidation potential lead to PFOA mineralization rate decrease [38]. Second, the enhancement of the PFOA mineralization efficiency at pH values lower than neutral is due to the increase in oxygen over-potential that abate the anodic oxygen evolution reaction and favors the production of more potent oxidizers such as OH radicals that are appropriate for the oxidation of organic compounds [41–43]. Third, many micro bubbles formed at alkaline pH in the aqueous solution in contrast with the amounts of bubbles formed at acidic pH. This situation leads to adherence of bubbles to the sonicator's probe and therefore restrict ultrasound energy distribution through the bulk of solution [37].

However, in the strong acidic solutions, the life of anode decreases [44]. As a result, in the present study, a pH of 2 was found as the optimum pH value for maximum PFOA mineralization. Other researchers also reported similar results [45]. Equation 5 and Fig. 6 show that the difference between the minimum and maximum PFOA mineralization efficiency related to pH is 6 %, which indicates sonoelectrochemical mineralization of PFOA using PbO_2 anode is not very sensitive to the initial pH and PFOA could be mineralized under a broad range of pH. Therefore, within the scope of the present study, it can be suggested that pre-adjustment of pH with the addition of chemicals is not necessary for sonoelectrochemical mineralization of the studied compound in the full scale treatment, unless a little increase in the mineralization efficiency is logical.

Effect of CD

In the present study, the effect of CD was investigated at five levels (16.36, 30, 50, 70, and 83.64 mA/cm^2) in combination with pH and EL (Fig. 4). As shown in this figure, which is the output of the CCD, the PFOA mineralization efficiency significantly increased with the increasing CD. Respect to Eq. (5), the maximum PFOA mineralization efficiency was obtained at the CD of 83.64 mA/cm^2 in the range of investigation. Equation (5) and Fig. 7 show that the difference between the minimum and maximum PFOA mineralization efficiency related to CD is 57 %, indicating that, the most important variable for the enhancement of PFOA mineralization was CD. The main reason for the increase in efficiency with increasing current density can be attributed to increasing the number of OH specimen produced which is in well agreement with other studies [42, 44, 46]. As a result, in the present study, a CD of 83.64 mA/cm^2 was chosen as the optimum CD value for maximum PFOA mineralization.

Effect of EL

The effect of EL was investigated at five levels (32.96, 50, 75, 100, and 117.04 mM) in combination with pH and CD. It can be concluded from Fig. 8 that despite of increasing PFOA mineralization with increasing the concentration of the EL, its impact is not significant. Many researchers which have used electrochemical process for wastewater treatment, believe that the degradation rate of pollutants is not affected by electrolyte concentration [46–48], however this negligible raise in the efficiency with increasing the concentration of the electrolyte can be explained by increasing the concentration of sulphate radicals, which are generated through the irradiation of sulfate ions by ultrasonic that was mentioned before. Change the electrolyte concentration in the range of tests, leading up to a 5 % change in the PFOA mineralization efficiency.

Conclusions

In the present study, sonoelectrochemical mineralization of the PFOA was investigated and modeled by employing CCD coupled with RSM for the prediction and optimization of the PFOA mineralization in synthetic wastewater using Ti/PbO_2 as the anode and stainless steel as the cathode. The use of RSM based on CCD allowed determination of the behavior of the sonoelectrolysis on mineralization, without requiring large number of experiments, and provided sufficient information. Moreover, the CCD facilitates the process of selecting optimum conditions for defluorination. The final model was validated using the leave-one-out technique, and the predicted R^2 was 98.8, which confirmed the external validity of the model. In addition, lack of fit test was nonsignificant with a P-value of 0.09, which confirmed the fit of the final model. The results of the present study demonstrated that sonolysis and electrochemical (using Ti/PbO_2 anode) processes are not able to mineralize PFOA significantly and combination of them as sonoelectrochemical process is a suitable and an environment-friendly method for the mineralization of refractory PFOA in aqueous solution.

Abbreviations

PFOA: Perfluorooctanoic Acid; PFCA: Perfluorocarboxylic acid; AOP: Advanced Oxidation Process; EAOP: Electrochemical Advanced Oxidation Process; EO: Electrochemical Oxidation; CD: Current Density; EL: Supporting Electrolyte; RSM: Response Surface Methodology; CCD: Central Composite Design; SEM: Scanning Electron Microscope; XRD: X-Ray Diffraction.

Competing interests

The authors declare that they have no competing interests.

Authors' contributions

GB designed the study, carried out the experiments, analyzed the data and drafted the manuscript. MK participated in design of the study, electrochemical setup and contributed in the result interpretation. AE participated in the design of the study, reviewed the article and drafted the manuscript. RN helped to draft manuscript and participated in modeling process. SMTS participated in the statistical analysis and interpretation. HM helped to carry out the experiments. All authors read and approved the final manuscript.

Acknowledgments

This study is a part of a PhD approved Thesis (No. 393267) performed at Isfahan University of Medical Sciences (IUMS), Iran. The authors are thankful for the funding provided by the Department of Environmental Health Engineering and Environment Research Center, IUMS.

Author details

[1]Environment Research Center and Department of Environmental Health Engineering, School of Health, Isfahan University of Medical Sciences (IUMS), Isfahan 81676-36954, Iran. [2]Student Research Center, School of Health, IUMS, Isfahan, Iran. [3]Nanotechnology Department, University of Isfahan, Isfahan 81744-73441, Iran. [4]Research Center for Environmental Pollutants, Qom University of Medical Sciences, Qom 37136-49373, Iran.

References

1. Panchangam SC, Lin AY-C, Tsai J-H, Lin C-F. Sonication-assisted photocatalytic decomposition of perfluorooctanoic acid. Chemosphere. 2009;75(5):654–60.
2. Urtiaga A, Fernández-González C, Gómez-Lavín S, Ortiz I. Kinetics of the electrochemical mineralization of perfluorooctanoic acid on ultrananocrystalline boron doped conductive diamond electrodes. Chemosphere. 2015;129:20–6.
3. Qu Y, Zhang C-J, Chen P, Zhou Q, Zhang W-X. Effect of initial solution pH on photo-induced reductive decomposition of perfluorooctanoic acid. Chemosphere. 2014;107:218–23.
4. Liu D, Xiu Z, Liu F, Wu G, Adamson D, Newell C, et al. Perfluorooctanoic acid degradation in the presence of Fe (III) under natural sunlight. J Hazard Mater. 2013;262:456–63.
5. Zhao B, Lv M, Zhou L. Photocatalytic degradation of perfluorooctanoic acid with β-Ga 2 O 3 in anoxic aqueous solution. J Environ Sci. 2012;24(4):774–80.
6. Ahrens L, Felizeter S, Sturm R, Xie Z, Ebinghaus R. Polyfluorinated compounds in waste water treatment plant effluents and surface waters along the River Elbe. Germany Mar Pollut Bull. 2009;58(9):1326–33.
7. Song C, Chen P, Wang C, Zhu L. Photodegradation of perfluorooctanoic acid by synthesized TiO 2–MWCNT composites under 365nm UV irradiation. Chemosphere. 2012;86(8):853–9.
8. Bazrafshan E, Mahvi A, Nasseri S, Shaieghi M. Performance evaluation of electrocoagulation process for diazinon removal from aqueous environments by using iron electrodes. Iran J Environ Health Sci. 2007;4(2):127–32.
9. Bazrafshan E, Mahvi A, Nasseri S, Mesdaghinia A, Vaezi F, Nazmara S. Removal of cadmium from industrial effluents by electrocoagulation process using iron electrodes. Iran J Environ Health Sci. 2006;3(4):261–6.
10. Mondal S. Methods of dye removal from dye house effluent-an overview. Environ Eng Sci. 2008;25(3):383–96.
11. Martínez-Huitle CA, Brillas E. Decontamination of wastewaters containing synthetic organic dyes by electrochemical methods: a general review. Appl Catal B-Environmental. 2009;87(3):105–45.
12. Ju P, Fan H, Guo D, Meng X, Xu M, Ai S. Electrocatalytic degradation of bisphenol A in water on a Ti-based PbO_2–ionic liquids (ILs) electrode. Chem Eng J. 2012;179:99–106.
13. Hmani E, Elaoud SC, Samet Y, Abdelhédi R. Electrochemical degradation of waters containing O-Toluidine on PbO_2 and BDD anodes. J Hazard Mater. 2009;170(2):928–33.
14. Samet Y, Elaoud SC, Ammar S, Abdelhedi R. Electrochemical degradation of 4-chloroguaiacol for wastewater treatment using PbO_2 anodes. J Hazard Mater. 2006;138(3):614–9.
15. Abaci S, Yildiz A. Electropolymerization of thiophene and 3-methylthiophene on PbO_2 electrodes. J Electroanal Chem. 2004;569(2):161–8.
16. Panizza M, Cerisola G. Direct and mediated anodic oxidation of organic pollutants. Chem Rev. 2009;109(12):6541–69.
17. Siedlecka EM, Stolte S, Gołębiowski M, Nienstedt A, Stepnowski P, Thöming J. Advanced oxidation process for the removal of ionic liquids from water: The influence of functionalized side chains on the electrochemical degradability of imidazolium cations. Sep Purif Technol. 2012;101:26–33.
18. Moriwaki H, Takagi Y, Tanaka M, Tsuruho K, Okitsu K, Maeda Y. Sonochemical decomposition of perfluorooctane sulfonate and perfluorooctanoic acid. Environ Sci Technol. 2005;39(9):3388–92.
19. Rokhina EV, Repo E, Virkutyte J. Comparative kinetic analysis of silent and ultrasound-assisted catalytic wet peroxide oxidation of phenol. Ultrason Sonochem. 2010;17(3):541–6.
20. Merouani S, Hamdaoui O, Saoudi F, Chiha M, Pétrier C. Influence of bicarbonate and carbonate ions on sonochemical degradation of Rhodamine B in aqueous phase. J Hazard Mater. 2010;175(1):593–9.
21. Su S, Guo W, Yi C, Leng Y, Ma Z. Degradation of amoxicillin in aqueous solution using sulphate radicals under ultrasound irradiation. Ultrason Sonochem. 2012;19(3):469–74.
22. Montgomery DC. Design and analysis of experiments. 7th ed. New York: John Wiley & Sons; 2009.
23. Aquino JM, Rocha-Filho RC, Bocchi N, Biaggio SR. Electrochemical degradation of the Disperse Orange 29 dye on a β-PbO_2 anode assessed by the response surface methodology. J Environ Chem Eng. 2013;1(4):954–61.
24. Polcaro A, Palmas S, Renoldi F, Mascia M. On the performance of Ti/SnO_2 and Ti/PbO_2 anodesin electrochemical degradation of 2-chlorophenolfor wastewater treatment. J Appl Electrochem. 1999;29(2):147–51.
25. Del Río A, Benimeli M, Molina J, Bonastre J, Cases F. Electrochemical Treatment of CI Reactive Black 5 Solutions on Stabilized Doped Ti/SnO_2 Electrodes. Int J Electrochem Sci. 2012;7:13074–92.
26. Ghalwa NA, Gaber M, Khedr AM, Salem MF. Comparative Study of Commercial Oxide Electrodes Performance in Electrochemical Degradation of Reactive Orange 7 Dye in Aqueous Solutions. Int J Electrochem Sci. 2012;7:6044–58.
27. Xu L, Guo Z, Du L, He J. Decolourization and degradation of C.I. Acid Red 73 by anodic oxidation and the synergy technology of anodic oxidation coupling nanofiltration. Electrochim Acta. 2013;97:150–9.
28. Lee Y, Lo S, Kuo J, Hsieh C. Decomposition of perfluorooctanoic acid by microwaveactivated persulfate: Effects of temperature, pH, and chloride ions. Front Environ Sci Eng. 2012;6(1):17–25.
29. Myers RH, Montgomery DC, Anderson-Cook CM. Response surface methodology: process and product optimization using designed experiments. USA: John Wiley & Sons; 2009.
30. Kutner MH, Nachtsheim C, Neter J. Applied linear regression models. USA: McGraw-Hill/Irwin; 2004.
31. Anderson TW, Darling DA. A test of goodness of fit. J Am Stat Assoc. 1954;49(268):765–9.
32. Breusch TS, Pagan AR. A simple test for heteroscedasticity and random coefficient variation. Econometrica. 1979;47(5):1287-94.
33. Durbin J, Watson GS. Testing for serial correlation in least squares regression. I. Biometrika. 1950;37(3-4):409–28.
34. Durbin J, Watson GS. Testing for serial correlation in least squares regression. II. Biometrika. 1951:38(1-2):159–77.
35. Efron B, Gong G. A leisurely look at the bootstrap, the jackknife, and cross-validation. Am Stat. 1983;37(1):36–48.
36. Barros WR, Steter JR, Lanza MR, Motheo AJ. Degradation of amaranth dye in alkaline medium by ultrasonic cavitation coupled with electrochemical oxidation using a boron-doped diamond anode. Electrochim Acta. 2014;143:180–7.
37. Thi L-AP, Do H-T, Lo S-L. Enhancing decomposition rate of perfluorooctanoic acid by carbonate radical assisted sonochemical treatment. Ultrason Sonochem. 2014;21(5):1875–80.
38. Lin J-C, Lo S-L, Hu C-Y, Lee Y-C, Kuo J. Enhanced sonochemical degradation of perfluorooctanoic acid by sulfate ions. Ultrason Sonochem. 2015;22:542–7.
39. Yang B, Zuo J, Tang X, Liu F, Yu X, Tang X, et al. Effective ultrasound electrochemical degradation of methylene blue wastewater using a nanocoated electrode. Ultrason Sonochem. 2014;21(4):1310–7.
40. Ren Y-Z, Wu Z-L, Franke M, Braeutigam P, Ondruschka B, Comeskey DJ, et al. Sonoelectrochemical degradation of phenol in aqueous solutions. Ultrason Sonochem. 2013;20(2):715–21.
41. Dai Q, Shen H, Xia Y, Chen F, Wang J, Chen J. The application of a novel Ti/SnO_2–Sb_2O_3/PTFE-La-Ce-β-PbO_2 anode on the degradation of cationic gold yellow X-GL in sono-electrochemical oxidation system. Sep Purif Technol. 2013;104:9–16.
42. Niu J, Maharana D, Xu J, Chai Z, Bao Y. A high activity of Ti/SnO_2-Sb electrode in the electrochemical degradation of 2,4-dichlorophenol in aqueous solution. J Environ Sci. 2013;25(7):1424–30.
43. Neto SA, De Andrade A. Electrooxidation of glyphosate herbicide at different DSA® compositions: pH, concentration and supporting electrolyte effect. Electrochim Acta. 2009;54(7):2039–45.
44. Zhong C, Wei K, Han W, Wang L, Sun X, Li J. Electrochemical degradation of tricyclazole in aqueous solution using Ti/SnO_2–Sb/PbO_2 anode. J Electroanal Chem. 2013;705:68–74.
45. Lin H, Niu J, Ding S, Zhang L. Electrochemical degradation of perfluorooctanoic acid (PFOA) by Ti/SnO_2–Sb, Ti/SnO_2–Sb/PbO_2 and Ti/SnO_2–Sb/MnO_2 anodes. Water res. 2012;46(7):2281–9.
46. Zhou M, He J. Degradation of cationic red X-GRL by electrochemical oxidation on modified PbO_2 electrode. J Hazard Mater. 2008;153(1):357–63.
47. Niu J, Bao Y, Li Y, Chai Z. Electrochemical mineralization of pentachlorophenol (PCP) by Ti/SnO_2–Sb electrodes. Chemosphere. 2013;92(11):1571–7.
48. Shao D, Liang J, Cui X, Xu H, Yan W. Electrochemical oxidation of lignin by two typical electrodes: Ti/Sb-SnO_2 and Ti/PbO_2. Chem Eng J. 2014;244:288–95.

On the thermal effect induced in tissue samples exposed to extremely low-frequency electromagnetic field

M. Racuciu[1], S. Miclaus[2] and D. Creanga[3*]

Abstract

Background: The influence of electromagnetic exposure on mammalian tissues was approached as a public health issue aiming to reveal the putative side effect of 50 Hz industrial and domestic supply source (i) during aliments storage near such sources; (ii) in people staying couple of hours in the proximity of conducting wires.

Materials and methods: Fluorescence emission based thermal sensor was used to emphasize temperature dynamics of fresh meat samples during controlled electromagnetic exposure in Helmholtz coils adjusted to deliver 50 Hz / (4÷10) mT electromagnetic field in their inner volume. Fluoroptic temperature probe with 0.1 °C accuracy measurement and data acquisition software allowed reading temperature every second, in the tissue volume during exposure.

Results: The temperature dynamics curves of *ex-vivo* porcine tissues like liver, kidney, brain, muscle, lung, and bone, were comparatively analyzed – the choosing of the mammalian species being justified by metabolic and physiological similarities with human body. The curve slopes appear to be the same for the range of initial temperatures chosen to perform the tests (20.0 ± 0.1 °C), the temperature increase reaching around 2.0 °C for the magnetic flux density of 10 mT. Quantitative dependence was evidenced between the thermal effect and the magnetic flux density.

Conclusions: The technical interpretation is based on heating effect, on bioimpedance increasing and on water vaporization during wet sample exposure. The biomedical aspects derive from the degrading effects of food heating as well as from possible *in vivo* effects of living body exposure.

Keywords: Extremely low frequency electromagnetic field, Heating dynamics effect, Mammalian tissues

Background

During last decades there has been an increasing interest in the bioeffects of the electromagnetic fields interaction with living organisms, with focus on potential health hazard. The biological effects of industrial alternative current with 50 Hz standard frequency have been much discussed in the context of the biological response to extremely low frequency magnetic field (ELF-MF). All living organisms are continuously exposed to electromagnetic fields from industrial and domestic sources. It seems clear now that electromagnetic exposure can induce biological changes, although the precise effects are not yet well known. In recent decades, many scientific studies have confirmed that magnetic fields of extremely low frequency (ELF; frequency <300 Hz) can influence the biological systems. Data reported in the literature regarding direct effects induced by ELF-MF on cell functions are controversial and the interaction mechanisms of electromagnetic fields with biological systems are still partially understood [1].

The hypothesis that electromagnetic field may act as initiator or co-initiator of carcinogen tumours [2, 3] underlied the 1999 decision of U.S. National Institute of Environmental Health Sciences to include electromagnetic fields in the category of "possible human carcinogen". Also, in [4] it was reported the increasing of cancer rates

* Correspondence: dorina.creanga@gmail.com
[3]Biophysics and Medical Physics Laboratory, Faculty of Physics, "Alexandru Ioan Cuza" University, 11, Blvd. Carol I, Iasi 700506, Romania
Full list of author information is available at the end of the article

in children when exposed at home at 50Hz magnetic fields greater than 0.3 μT. The bioeffects of environmental constraints of electromagnetic nature are very difficult to explain since living bodies are complex dynamic systems, with many physical and chemical parameters non-homogeneously distributed and time dependent. Therefore the assumption on non-thermal mechanisms of living organism electromagnetic exposure in the lack of any relationship with general or localized temperature increase with a consequent thermal distortion of bio-molecules still remains a scientific challenge. Actually thermal effects of electromagnetic field, especially low subtle heating effects are still considered the main cause of biological damages when the focus is the major role of thermal sensitivity of enzyme catalytic activity. That is why homeostatic processes from living cells and tissues could be perturbed by couple degrees temperature rise and such small perturbation could damage the whole organism when amplify at larger scale of space and time. Injuries of cell biochemistry triggered following human electromagnetic exposure could be an actual concern for public health, biomedicine [5, 6] and food science. In contrast with radiofrequency region of the spectrum, in the ELF region, direct thermal effect in tissues was neither definitely revealed nor sufficiently analyzed up-to-now – according to International Commission on Non-Ionizing Radiation Protection [7].

Having in mind that water is the dominant element hosting all biomolecules from living matter, an unusual phenomenon of weak ELF magnetic field impact on water should underline the reported bioeffects as long as the field energy is much lower than hydrogen bonds characterizing water molecule organization; but for now, an acknowledged mechanism of influence of weak ELF magnetic field on water does not exist. Some researchers had observed the changes of water electric conductivity parameters due to action of a weak ELF magnetic field ($f \leq 50$ Hz) [8]. In [9] the authors have shown the increasing of the water evaporation rate due to the exposure to weak static magnetic fields (15 mT). Other studies led to controversial results.

In [10] the authors reported that, using a thermocouple with a precision of 0.1 °C no changes in temperature were detected for 915 MHz or 50 Hz exposures of human lymphocytes culture. Other researchers [11] reported that heat shock protein synthesis in cells exposed to 50 Hz/(0–100 μT) at 40 °C was not increased compared to cells incubated at 40 °C without magnetic field exposure. In [12] it was studied the exposure of endothelial cells culture to domestic power supply (50 Hz/700 μT) that resulted in no detectable effects on the expression of heat shock protein60 as the dominant autoantigen in endothelial cells. Thermal effect of 50 Hz/94 mT magnetic field was only reported in inert materials such as superconductors [13].

Some other reports appear to evidence opposite results such as those regarding cell apoptosis after electromagnetic exposure. So, exposure of human lymphocytes at room temperature to either 915 MHz or 50 Hz resulted in significant condensation of chromatin, as measured through the method of anomalous viscosity time dependencies [10] but no apoptosis induced by DNA morphological changes or by its fragmentation was evidenced; while electromagnetic exposure to 50 Hz/0.097 T was found able to induce and promote apoptosis of mice murine liver cells in time-effects manner as shown in [14].

In [15] it was reported the ohmic heating rate of peaches for electric pulses with frequencies varying from 50 Hz to 1 MHz, thermal damage of tissue being evaluated from electrical admittance; it was found that samples exposed to low-frequency electric field demonstrated faster electro-thermal damage rates. Apart from experimental measurements, mathematical investigations also offered interesting approaches; the theoretical analysis of tissue heating as a potential side effect of strong electric pulses, developed in [16], revealed localized tissue heating near the electrodes which is assessed mainly to the sharp radial decrease of the electric field around the needles.

The main interaction mechanism of low frequency electromagnetic fields with absorbent matter is supposed to be the Faraday induction of electric fields and associated currents, the distribution of the induced electric field depending on the conductivity of organs and tissues [7]. The maximum electric field is induced in the body when the external fields are homogeneous and directed parallel to the body axis or perpendicular to it. According to calculations on human body models [7], the maximum local peak of electric field induced by a 50 Hz magnetic field in the brain is approximately of 23–33 mV/m per mT, depending on field orientation and body model. The corresponding electric field induced in the skin is of approximately 20–60 mV/m per mT.

While most of the experiments already mentioned were focused on the *in vivo* electromagnetic exposure there is less literature regarding the effects induced in tissue samples exposed immediately after excision from animal body (*ex vivo* exposure) – which would be of interest for food processing and technology rather than for bio-electromagnetism.

Our study was designed to evidence putative effects of electromagnetic exposure in fresh food samples with practical importance for temporal food storing near 50 Hz supplied devices. The experimental work was focused on the measurement of thermal effect dynamics in *ex vivo* biological tissues (during couple of hours following withdrawn from animal body) when exposed to 50 Hz electromagnetic field (4 mT-10 mT magnetic flux density). The heating was recorded during about 3,000 s of continuous electromagnetic exposure.

Methods

Biological material

Mammalian tissue samples were consistent with specimens of porcine liver, kidney, lung, brain, muscle and bone, freshly excised, each of approximately 1 cm^3 – as resulted from direct volumetric estimation. Porcine tissue samples considered for the study had different masses, ranging within 3–4 g, because of their different densities. New tissue sample for every magnetic field exposure was used. Comparatively temperature dynamics was recorded for 1 ml deionized water as well as in the free air around the same point from the Helmholtz coil system centre.

Electromagnetic exposure

Electromagnetic exposure of tissue specimens was carried out within a Helmholtz coil system (Fig. 1) able to generate a vertical magnetic field of 50 Hz frequency and 4 mT, 6 mT, 8 mT and respectively 10 mT magnetic flux density for current intensity of 0.82 - 1.23 - 1.64 -2.05 A - as measured with a device type Simpson 260 Analog - VOM. Preliminary magnetic field measurements were carried out using a low-frequency field analyzer, NARDA EFA-300, before the mammal tissues magnetic exposure, based on the calibration curve of Helmholtz coil system presented in Fig. 2. The measurements of the magnetic field induction evidenced that within the centre of the Helmholtz coils system no significant variations of the field could be detected within 100 mm diameter area. The Helmholtz coil system consisted of two coils, each formed by 1,000 turns of 1 mm cooper wire, with a mean diameter of 260 mm and a thickness of 25 mm. The coils were mounted coaxially and placed at a mean distance of 130 mm from each other (Figs. 1, 2). The magnetic exposure was done by placing the mammal tissue sample on a glass dish (90 mm diameter), as dielectric support, in the centre of the coil system. The space chosen for the experiment was consistent with a small room arranged as thermostat with electric supply, with no windows - only double wall door; constant air temperature within the working space was displayed on thermocouples continuously during the tissue investigation project. No person was in that room during experimental recordings.

Temperature measurement

A Luxtron One fiber optic device was used to accomplish temperature recording. This one is provided with a thermal fluoroptic probe, of 1.5 mm diameter and working on the principle of fluorescence quenching in thermo-resistant phosphorescent sensor. The fiber optic probe was inserted inside the mammal tissue, in the centre of volume, through a tiny incision with the size of the probe, made at the time of its insertion. The accuracy

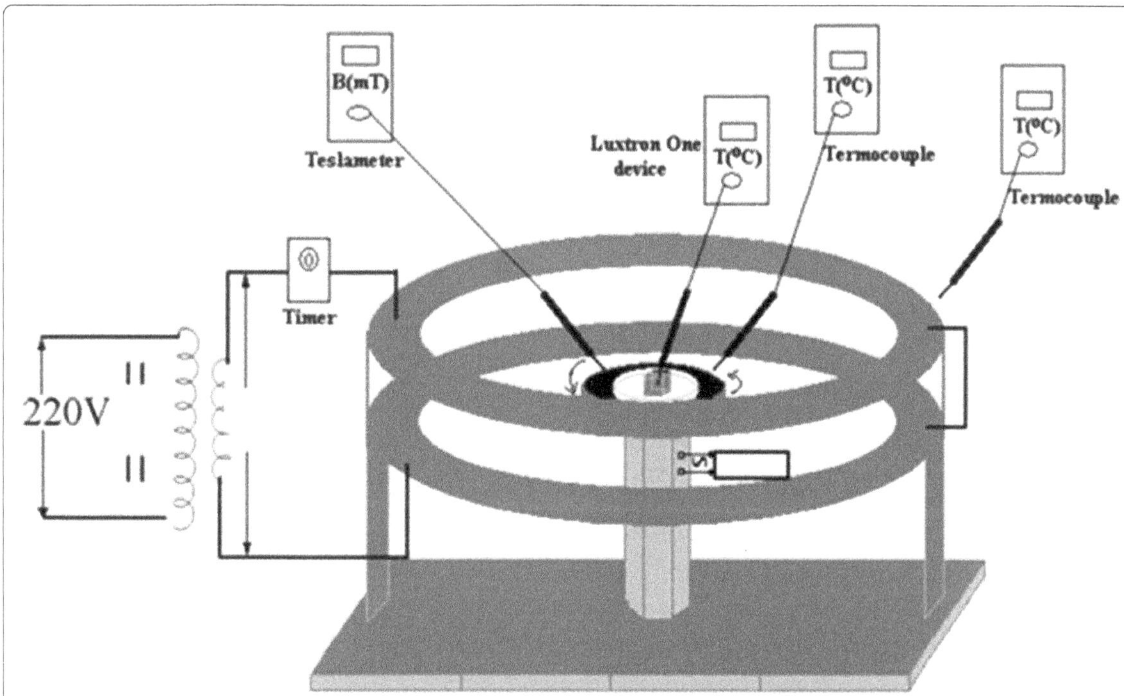

Fig. 1 Experimental set-up; coil system supplied by power transformer connected to industrial electricity grid

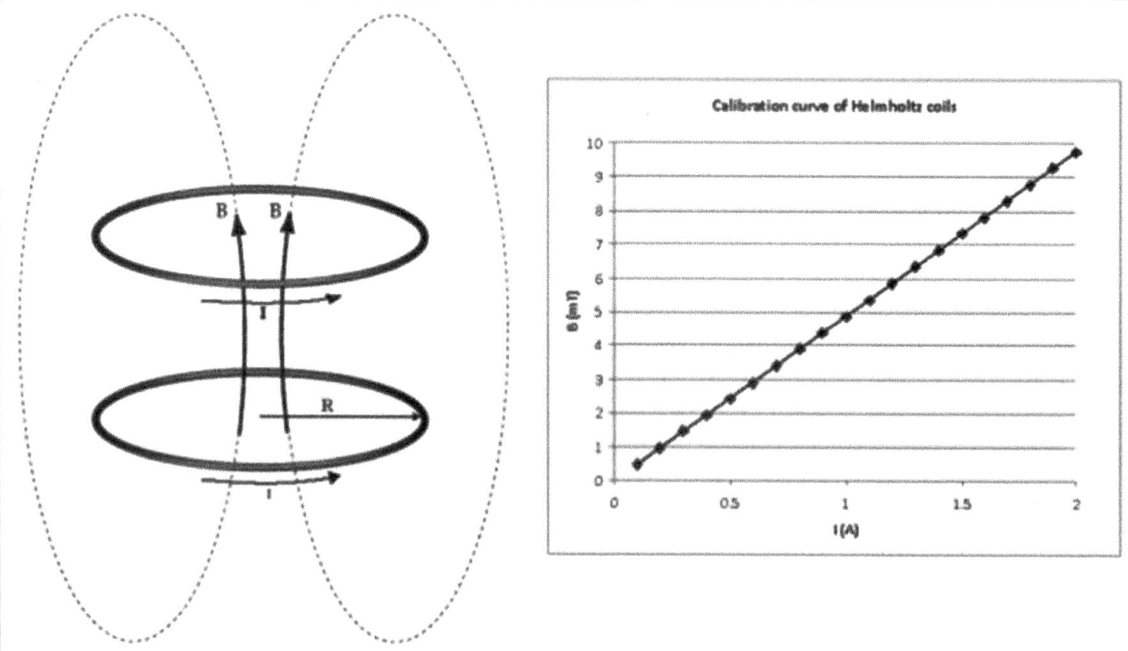

Fig. 2 Calibration curve of Helmholtz coil system using a low-frequency field analyzer, NARDA EFA-300

of temperature measurement was of ± 0.1 °C, the temperature values being recorded every second for about one hour in each tissue sample and transferred to a PC with TrueTemp3.0 program [17, 18]. Then temperature/time graphs were plotted using Origin 7.5 software.

Repeated recordings were carried out three times on similar samples extracted from each tissue – the bulk sample being kept at refrigerator (4 °C) and let to reach the environmental controlled temperature (of 20 °C) before new aliquot cutting. After thermal investigation any used tissue sample appeared less wet and structurally changed so that new incision and temperature measurement repetition seemed not reliable. Representative data series were presented and discussed below.

Results and discussions

The temperature dynamics curves were recorded for each magnetic flux density values of 4-6-8-10 mT starting from the same initial temperature (of 20.0 ± 0.1 °C). In the free air (sample missing) the temperature measured in coil system centre exhibited no variation during 3,000 s. The same occurred in tissue samples when coils system was not electrically supplied. For electromagnetically exposed liver tissue the temperature increase in Fig. 3 is presented. The curves were translated on the ordinate in order to be grouped in tissue family of curves – and this is valuable for all similar plots presented below.

For relatively low magnetic flux density, of 4 mT and 6 mT the temperature increase was of only 0.5 °C and respectively 0.7 °C but for higher magnetic flux density (8 mT, respectively 10 mT) about 1.8 °C and respectively 2.0 °C increases in liver tissue sample temperature were recorded.

Detectable variation of temperature within muscle tissue was evidenced (Fig. 4) only for 8 mT (about 0.3 °C in 3,000 s) and for 10 mT (0.5 °C), this being probably the consequence of the considerable evaporation of water (as muscle is a "wet" tissue) that partially compensated the temperature increase due to the Helmholtz coil system. As well the thermal effect of lowest magnetic flux density applied in this study could be insufficient for inducing detectable heating effect in association also with possible intrinsic peculiarities of muscle tissue. In the next graphs, for the other tissues there was also a slighter heating effect for 4 mT and 6 mT than for 8 mT and 10 mT. In lung tissue (Fig. 5) significant temperature rise was noticed for all magnetic flux densities, from 0.5 °C in the case of 4 mT to 0.7 °C in the case of 6 mT and 0.9 °C in the case of 8 mT with further increase up to 1.2 °C for 10 mT. Certain different variation trend was noticed for 6 mT graph which is probably related to the lung tissue lacunary structure – as further in the bone can be seen.

In Fig. 6 the temperature measured in kidney tissue is presented where the increase noticed for 4 mT was of about 0.6 °C, the increases for 6 mT and respectively for

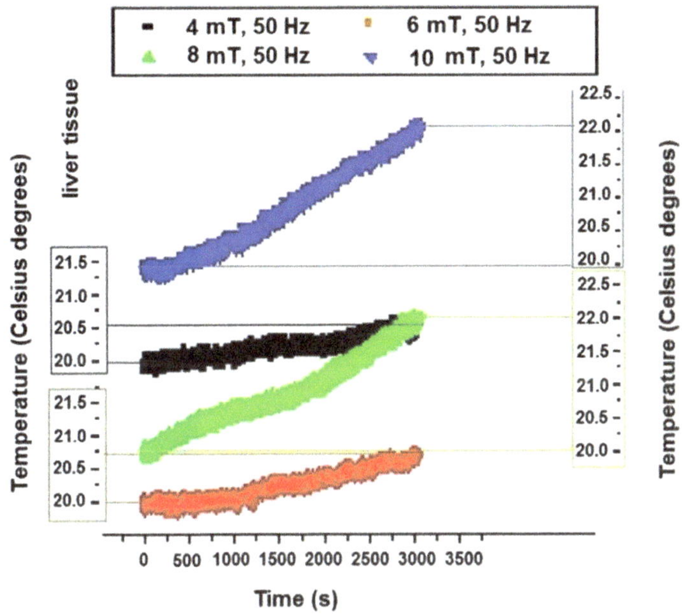

Fig. 3 Temperature variation during electromagnetic exposure of liver tissue

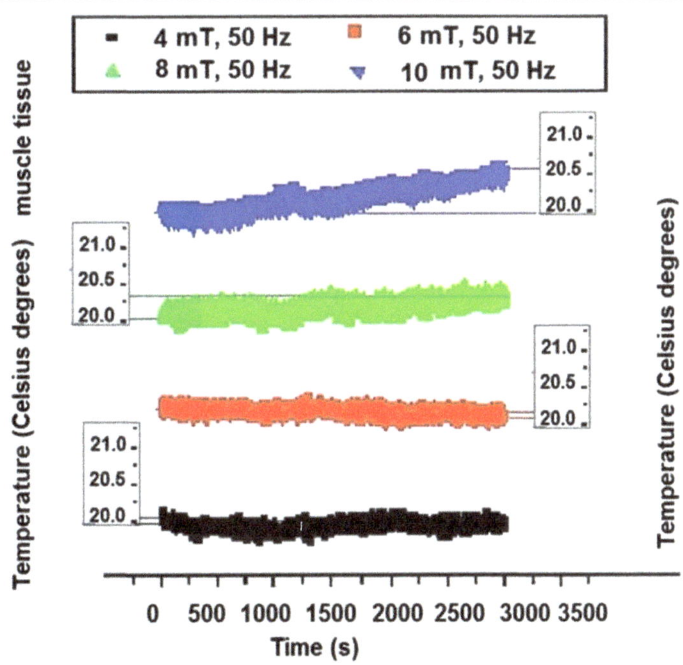

Fig. 4 Temperature dynamics in muscle tissue electromagnetically exposed

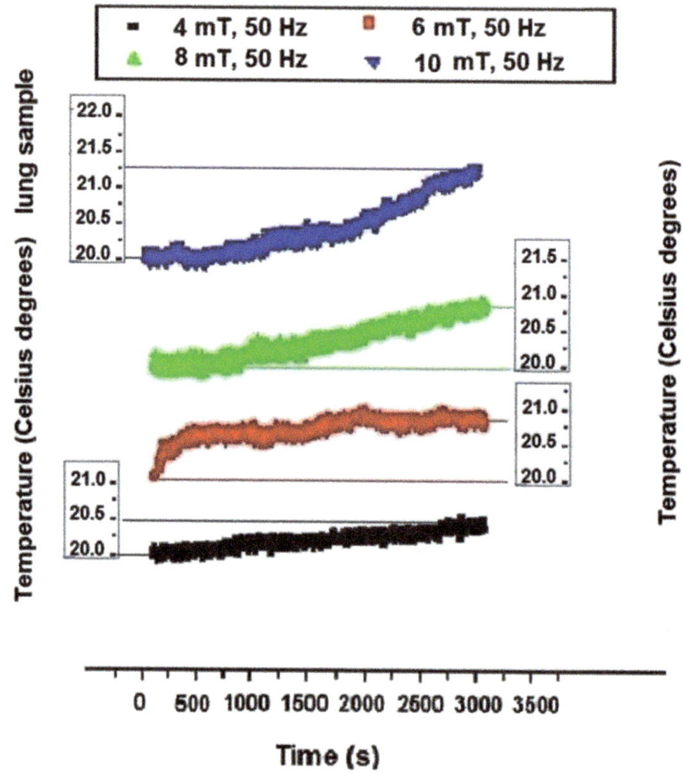

Fig. 5 Temperature dynamics in lung tissue sample electromagnetically exposed

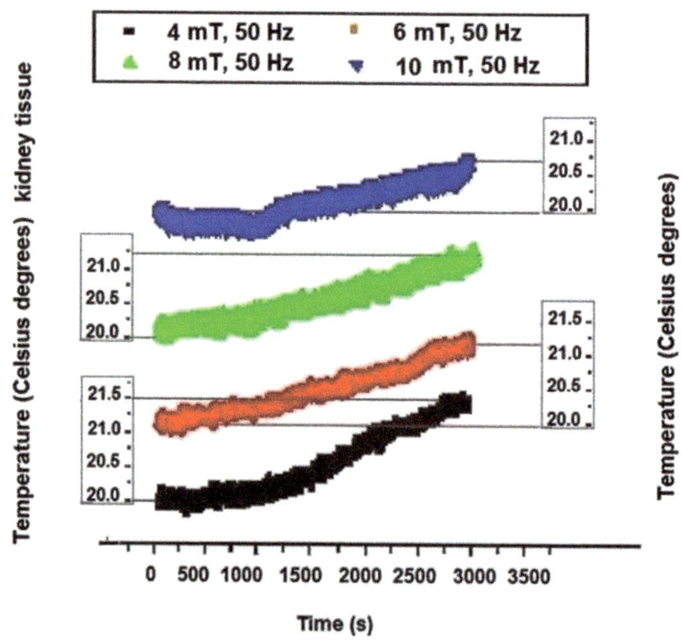

Fig. 6 Temperature recording in the electromagnetically exposed kidney sample

8 mT magnetic fields were of about 1.0 °C and 1.1 °C respectively, while finally, for 10 mT the highest 1.3 °C positive variation was recorded.

The data resulted from temperature measurement in brain tissue are given in Fig. 7.

At least 0.5 °C positive variation of temperature was recorded for 4 mT that was further progressively increased for 6 mT (at about 1.2 °C), for 8 mT (1.5 °C) as well as for 10 mT (2.2 °C). This remarkable heating amplitude could be associated with the highest fat content that seems to reduce specific heat capacity of tissues [19].

The response of bone tissue following the exposure to 50 Hz electromagnetic field in Fig. 8 is presented.

The bone tissue responded with only 0.2 °C temperature increasing for 4 mT but the positive temperature variation has increased to about 1.2 °C for 6 mT and 1.5 °C for 8 mT, reaching 1.7 °C increase for 10 mT. Specific shape of the curve first segment, corresponding to first approximately 1,000 s, could be related to the characteristic spongiest bone structure with lowest homogeneity among all tissues; early significant temperature variation during the first part of exposure was followed by saturation tendency up to about 3,000 s. The highest slope was revealed for the highest magnetic induction, i.e. 10 mT, when almost linear graph was recorded with no saturation trend. It could be presumed that 10 mT exposure induced the diminution of thermal conductivity anisotropy in the tissue so that the heat transmission from the environment resulted in the similar trend observed with the other isotropic tissue samples.

For comparison the deionized water response to the same array of magnetic flux densities is presented in Fig. 9 (the volume of 2 cm^3 was considered). From Fig. 9 it can be seen that non significant variation was recorded for 4 mT (similar with muscle tissue with highest water content) while for 6 mT the temperature increase with 0.6 °C was noticed and for 8 mT and 10 mT the increase with about 1.3 °C and 1.4 °C respectively, was reached in water.

In theory, when a sample with conductivity σ is exposed to a vertical electromagnetic field with frequency f, concentric electric current flow occurs, in a plan orthogonal to the direction of the magnetic field. If r is the radius of the sample, and B is the magnetic flux density in Tesla, then current density of the induced current J, is:

$$J = \sigma \cdot \pi \cdot f \cdot B \cdot r \qquad (1)$$

The magnetically induced electric field (E) does not depend on sample conductivity and is given by:

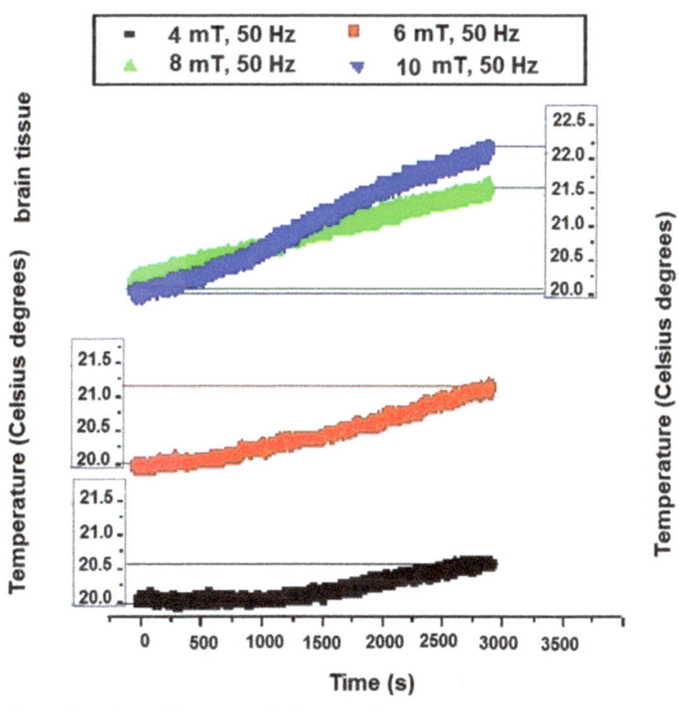

Fig. 7 The temperature recording in brain tissue electromagnetically exposed

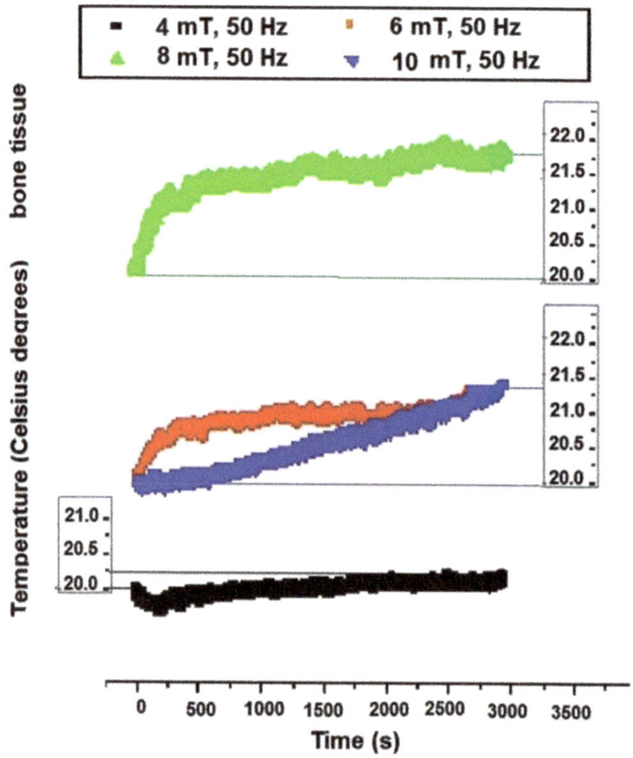

Fig. 8 The temperature recording in bone tissue electromagnetically exposed

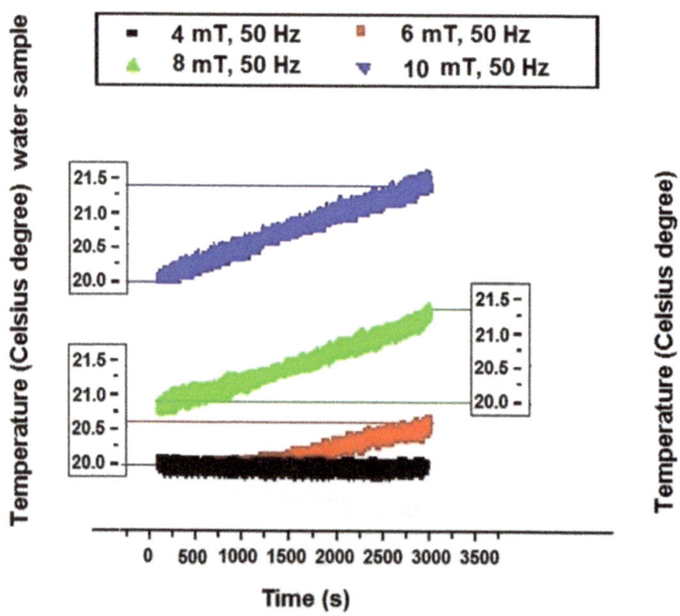

Fig. 9 Deionized water electromagnetically exposed (the volume of 2 cm^3 was considered)

$$E = \frac{J}{\sigma} = \pi \cdot f \cdot B \cdot r \tag{2}$$

The power (averaged in time) delivered to the sample unit volume is given by:

$$w = \sigma \cdot E^2 \tag{3}$$

where σ is the sample conductivity and E is the magnetically induced electric field in the sample.

Since the investigated tissue samples had around 1 cm^3 in volume, one can take an approximated value $r = 1$ cm, in order to calculate the maximum of the induced electric field corresponding to the 10 mT magnetic induction. Therefore the maximum of the magnetically induced electric field for B = 10 mT is of about 0.0157 V/m.

Assuming the same dissipation for $r < 1$ cm, the maximum power dissipation per unit volume is: $w = 2.46 \cdot \sigma \cdot 10^{-4} \frac{W}{m^3}$. Then for 1 h exposure time the energy delivered to the tissue sample of about 2 cm^3 is $\approx 0.177 \cdot \sigma \cdot 10^{-5}$ J. The conductivity (σ) for various living tissues in the extremely low frequency magnetic fields is in the range 0.02-1.5 S/m [18].

In accord with [18, 19], the electric conductivity values of the tissues at 50 Hz have been used to obtain the energy delivered by electromagnetic field to each sample (Q) divided by the mass of tissue sample (m). The dependence between these specific energy values on temperature interval corresponding to magnetic flux densities used in this experiment is presented in Fig. 10.

The regression functions fitting of the experimental graphs from Fig. 10 were calculated; thus exponential functions were found to provide best approximation (with highest corresponding correlation coefficient) of experimental curves for kidney, lung, brain and bone tissue and respectively, logarithmic function for liver and muscle tissue (Table 1).

In case of materials with definite specific heat, linear dependences would be evidenced as specific heat is constant. As we obtained obviously non-linear graphs it seems that heating provided by electromagnetic phenomena in complicated biological materials is more complex, with effects on the intimate structure of sample. In Fig. 11 temperature increasing, corresponding to magnetic induction of applied magnetic field is represented. The highest increasing of temperature appears most evident in case of brain tissue for 10 mT magnetic induction; however, as expected, this is lower in the case of water at 4 mT magnetic field exposure (with about 97 %). Bone tissue heating production profile is similar to that of brain tissue for all magnetic field inductions, but it is with about 23 % less "productive" than the first one. In Table 2 are given the logarithmic functions fitting the experimental graphs from Fig. 11 for all tissue types.

According to above results it seems that similar mechanisms undergo all tissues interaction with electromagnetic radiation, resulting in temperature increasing to the increase of electromagnetic field induction – since the same function type approximates the experimental graphs.

We assume that the heating effect could be the result of Faraday induction, i.e. of electric fields and currents which

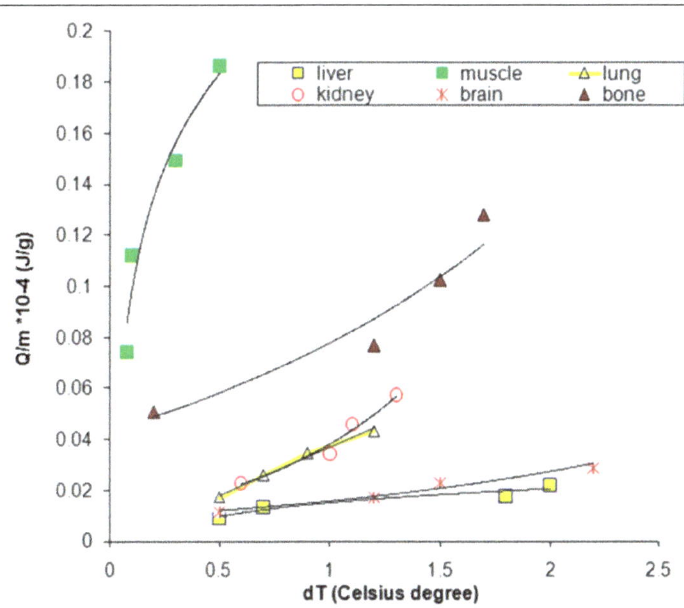

Fig. 10 Dependence of delivered electromagnetic energy divided by sample mass (*Q/m*) and temperature

Table 1 Regression functions fitting Fig. 10 graphs (*y* represents *Q/m*; *x* represents *dT*)

Mammalian tissue	Regression function	Correlation coefficient R^2
liver	$y = 0.008 \ln(x) + 0.015$	0.92
muscle	$y = 0.0535 \ln(x) + 0.2206$	0.94
kidney	$y = 0.0101e^{1.3305x}$	0.97
lung	$y = 0.0099e^{1.2934x}$	0.95
brain	$y = 0.0009e^{0.5538x}$	0.97
bone	$y = 0.0435e^{0.5798x}$	0.94

give raise to charge movement and heat production, that could be further related to electrical parameters of the tissues. When ohmic heating is considered then the current intensity through the coil system is most important, being related to Joule effect. If ohmic heating is the main cause of the temperature dynamics recorded by us then the impedance variation needs also to be considered. Electric measurements evidenced that in time, after tissue sample excision from animal body, the electrical bioimpedance of the organs increases from its *in vivo* level, multiplying twice in a few hours [20]. Related to the above hypothesis of structural changes induced during electromagnetic exposure and reflected in specific heat putative variation we believe that bioimpedance variation is also plausible. Other data were reported in literature discussing tissue changes under the influence of ELF-MF exposure; studies on porcine endothelial cells exposed to 50 Hz electromagnetic field showed the influence on heat shock protein levels and partial relocalization in the nucleus [21]; culture cell electroporation was demonstrated for 10 Hz-10 KHz ELF-MF exposures [22] – which is rather widely spread bioengineering technique, mainly for genetic purposes [23]. Electric conductivity changes in the electroporated cell culture [24–26] using ELF electromagnetic fields suggest that in the animal tissues investigated by us cellular modifications could occur with rather predictable negative consequences on the tissue stability.

Tissue structure and composition could be compromised during exposure to electromagnetic field with detectable thermal effect and with further consequences on the food quality.

Thus, it is possible to hypothesize that temperature increase effect is related both to the heat transfer toward the exposed sample and to the increase in time of excised sample bioimpedance. In the case of water no such variation of impedance could be assumed. But water evaporation during sample exposure could interfere with the other phenomena contributing to the smaller temperature rise recorded for the wet tissue of muscle.

The results presented above could be useful in understanding the electromagnetic sensitivity of human body tissues – much comparable with porcine ones – in the less studied circumstances of low frequency electromagnetic field exposure. Brain heating with about 2° in the vicinity of 50 Hz alternative current leads seems to represent a challenging issue versus the frequent utilization of domestic electrical devices like hair drier, electric shaver device etc. as well as of medical electric apparatus and installations - all supplied from the 50 Hz net (or 60 Hz in oversee countries). For human communities consuming frequently meat and derivatives, the health issues related to fresh food storing in spaces hosting electric lines or devices seems to require more attention

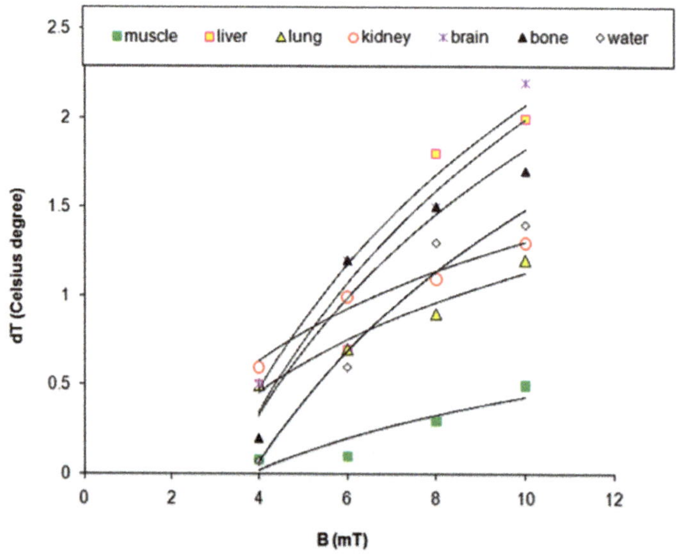

Fig. 11 Temperature increase corresponding to magnetic flux density values used in this study

Table 2 Regression functions fitting graphs from Fig. 11 (*y* represents *dT*; *x* represents *B*)

Mammalian tissue	Regression function	Correlation coefficient R^2
liver	$y = 1.80\ln(x)-2.10$	0.88
muscle	$y = 0.45\ln(x)-0.61$	0.83
kidney	$y = 0.73\ln(x)-0.38$	0.97
lung	$y = 0.73\ln(x)-0.56$	0.95
brain	$y = 1.75\ln(x)-1.95$	0.97
bone	$y = 1.6\ln(x)-1.93$	0.94
water	$y = 1.55\ln(x)-2.09$	0.96

to support indirectly prevention of subtle degradation before consuming – especially in public restaurants, student cantinas etc.

Apart from the physical considerations regarding the interaction mechanisms between biological material and electromagnetic field, life scientists and especially nutrition specialists need to pay more attention to biochemical aspects mentioned in introductory paragraph, since enzyme activity changes could trigger energetic balance perturbation with health consequences in extreme hypothetical approach.

Conclusions

Using a non-perturbing fluoroptic probe device for measuring heating of animal tissues during ELF-MF exposure, the thermal effect was revealed and analyzed. Brain, bone and liver heating dynamics resulted in over 2.0 °C temperature rise compared to the other tissues where lower heating levels were evidenced. Faraday induction and ohmic effect were supposed to be the possible causes of the recorded temperature variation with secondary changes in structural features like cell membrane integrity or tissue bioimpedance. Increased temperature was recorded for increased magnetic flux density. Further research is needed to provide deeper insight in such subtle thermal phenomena that possible affect fresh food during storage even during one hour.

Competing interests

There are no conflicts of interest neither from scientific nor from financial/non-financial point of view.

Authors' contributions

MR and DC have the main contributions in this research conception and made the experimental design including electromagnetic exposure system. MR prepared the tissue samples and carried out the measurements during sample exposure. SM contributed to experimental data acquisition and graph interpretation from engineering point of view and was involved in manuscript revising. MR and DC selected the representative data series for graphical presentation in the article body, surveyed the literature and made data interpretation. MR wrote the draft while the co-authors revised the draft critically and all authors gave the final approval of the manuscript.

Acknowledgement

The authors address many thanks to Dr. C. Goiceanu from Public Health Institute in Iasi, Romania. for technical assistance.

Author details

[1]Environmental Sciences and Physics Department, Faculty of Sciences in "Lucian Blaga" University, Dr. I. Ratiu Street, no. 5-7, Sibiu 550024, Romania. [2]Technical Sciences Department, "Nicolae Balcescu" Land Forces Academy, Revolutiei Street, no. 3-5, Sibiu 550170, Romania. [3]Biophysics and Medical Physics Laboratory, Faculty of Physics, "Alexandru Ioan Cuza" University, 11, Blvd. Carol I, Iasi 700506, Romania.

References

1. Goodman R, Blank M. Insights into electromagnetic interaction mechanisms. J Cell Physiol. 2002;192:16–22.
2. Loscher W, Mevissen M. Animal studies on the role of 50/60-hertz magnetic fields in carcinogenesis. Life Sci. 1994;54:1531–43.
3. Holmberg B. Magnetic field and cancer: animal and cellular evidence—an overview. Environ Health Perspect. 1995;103(2):63–7.
4. Tomenius L. 50-Hz electromagnetic environment and the incidence of childhood tumors in Stockholm County. Bioelectromagnetics. 1986;7(2):191–207.
5. Suhnel J, Berg H. Biological effects of electromagnetic fields (mechanisms, modeling, biological effects, therapeutic effects, international standards, exposure criteria). Bioelectrochemistry. 2003;61(1):109–10.
6. Barsam T, Monazzam MR, Haghdoost AA, Ghotbi MR, Dehghan SF. Effect of extremely low frequency electromagnetic field exposure on sleep quality in high voltage substations. Iranian J Environ Health Sci Eng. 2012;9(1):15.
7. International Commission on Non-Ionizing Radiation Protection (ICNIRP), Guidelines for Limiting Exposure to Time-Varying Electric and Magnetic Fields (1 Hz To 100 kHz) 2010 Health Physics 99(6):818–836. www.icnirp.de/PubMost.htm.
8. Ageev IM, Shishkin GG, Yeskin SM, Litvinov VN, Rybin UM. Study of weak ELF magnetic field influence on distillated water. Biomed Technol Radioel. 2007;6:59–62.
9. Holysz L, Szczes A, Chibowski E. Effects of static magnetic field on water and electrolyte solutions. J Colloid Interface Sci. 2007;316:996–1002.
10. Belyaev IY, Hillert L, Protopopova M, Tamm C, Malmgren LOG, Persson BRR, et al. 915 MHz microwaves and 50 Hz magnetic field affect chromatin conformation and 53BP1Foci in human lymphocytes from hypersensitive and healthy persons. Bioelectromagnetics. 2005;32(3):173–84.
11. Coulton LA, Harris PA, Barker AT, Pockley AG. Effect of 50 Hz electromagnetic fields on the induction of heat-shock protein gene expression in human leukocytes. Radiat Res. 2004;161:430–4.
12. Henderson BR, Pfister G, Boeck G, Kind M, Wick G. Expression levels of heat shock protein 60 in human endothelial cells *in vitro* are unaffected by exposure to 50 Hz magnetic fields. Cell Stress Chaperon. 2003;8(2):172–82.
13. Laurent P, Fagnard JF, Mathieu J, Meslin JG, Noudem JG, Cloots R, et al. Study of thermal effects in bulk RE-BCO superconductors submitted to a variable magnetic field. J Phys Conf Ser. 2006;43(1):505–8.
14. Guo H, Ge M, Li Y, Xing L, Zhang X, Yan W. Effects of 50 Hz ferromagnetic field exposure on apoptosis and proliferation of murine liver cell *in vivo*. Proceedings of the 25th Annual International Conference of IEEE (Engineering in Medicine and Biology Society). 2003;4:3263–6.
15. Shynkaryk MV, Ji T, Alvarez VB, Sastry SK. Ohmic heating of peaches in the wide range of frequencies (50 Hz to 1 MHz). J Food Sci. 2010;75(7):E493–500.
16. Lackovic I, Magjarevic R, Miklavcic D. Three-dimensional finite-element analysis of joule heating in electrochemotherapy and in vivo gene electrotransfer. IEEE Trans Dielectr Electr Insul. 2009;16(5):1338–47.
17. McGee TD. Principles and methods of temperature measurement. New Jersey: John Wiley&Sons; 1988.
18. Fraden J. Handbook of modern sensors: physics, designs, and applications. New York: Springer; 2010.
19. Faber P, Garby L. Fat content affects heat capacity: a study in mice. Acta Physiol Scand. 1995;153(2):185–7.
20. Gabriel C, Gabriel S, Corthout E. The dielectric properties of biological tissues: I. Literature survey. Phys Med Biol. 1996;41(11):2231–49.

21. Gabriel C, Peyman A, Grant EH. Electrical conductivity of tissue at frequencies below 1 MHz. Phys Med Biol. 2009;54:4863–78.
22. Spottorno J, Multigner M, Rivero G, Álvarez L, de la Venta J, Santos M. Time dependence of electrical bioimpedance on porcine liver and kidney under a 50 Hz ac current. Phys Med Biol. 2008;53(6):1701–13.
23. Bernardini C, Zannoni A, Turba ME, Bacci ML, Forni M, Mesirca P, et al. Effects of 50 Hz sinusoidal magnetic fields on Hsp27, Hsp70, Hsp90 expression in porcine aortic endothelial cells (PAEC). Bioelectromagnetics. 2007;28(3):231–7.
24. Zhan Y, Cao Z, Bao N, Li J, Wang J, Geng T, et al. Low-frequency ac electroporation shows strong frequency dependence and yields comparable transfection results to dc electroporation. J Control Release. 2012;160(3):570–6.
25. Lebovka N, Ghimi P, Vorobiev E. Does electroporation occur during the ohmic heating of food? J Food Sci. 2005;70(5):308–11.
26. Pavlin M, Kandušer M, Reberšek M, Pucihar G, Hart FX, Magjarevićcacute R, et al. Effect of cell electroporation on the conductivity of a cell suspension. Biophys J. 2005;88(6):4378–90.

PERMISSIONS

All chapters in this book were first published in JEHSE, by BioMed Central; hereby published with permission under the Creative Commons Attribution License or equivalent. Every chapter published in this book has been scrutinized by our experts. Their significance has been extensively debated. The topics covered herein carry significant findings which will fuel the growth of the discipline. They may even be implemented as practical applications or may be referred to as a beginning point for another development.

The contributors of this book come from diverse backgrounds, making this book a truly international effort. This book will bring forth new frontiers with its revolutionizing research information and detailed analysis of the nascent developments around the world.

We would like to thank all the contributing authors for lending their expertise to make the book truly unique. They have played a crucial role in the development of this book. Without their invaluable contributions this book wouldn't have been possible. They have made vital efforts to compile up to date information on the varied aspects of this subject to make this book a valuable addition to the collection of many professionals and students.

This book was conceptualized with the vision of imparting up-to-date information and advanced data in this field. To ensure the same, a matchless editorial board was set up. Every individual on the board went through rigorous rounds of assessment to prove their worth. After which they invested a large part of their time researching and compiling the most relevant data for our readers.

The editorial board has been involved in producing this book since its inception. They have spent rigorous hours researching and exploring the diverse topics which have resulted in the successful publishing of this book. They have passed on their knowledge of decades through this book. To expedite this challenging task, the publisher supported the team at every step. A small team of assistant editors was also appointed to further simplify the editing procedure and attain best results for the readers.

Apart from the editorial board, the designing team has also invested a significant amount of their time in understanding the subject and creating the most relevant covers. They scrutinized every image to scout for the most suitable representation of the subject and create an appropriate cover for the book.

The publishing team has been an ardent support to the editorial, designing and production team. Their endless efforts to recruit the best for this project, has resulted in the accomplishment of this book. They are a veteran in the field of academics and their pool of knowledge is as vast as their experience in printing. Their expertise and guidance has proved useful at every step. Their uncompromising quality standards have made this book an exceptional effort. Their encouragement from time to time has been an inspiration for everyone.

The publisher and the editorial board hope that this book will prove to be a valuable piece of knowledge for researchers, students, practitioners and scholars across the globe.

LIST OF CONTRIBUTORS

Akiko Ido, Tsuyoshi Nakanishi and Hisamitsu Nagase
Laboratory of Hygienic Chemistry and Molecular Toxicology, Gifu Pharmaceutical University, 1-25-4 Daigaku-nishi, Gifu 501-1196, Japan

Miki Niikawa
Laboratory of Hygienic Chemistry and Molecular Toxicology, Gifu Pharmaceutical University, 1-25-4 Daigaku-nishi, Gifu 501-1196, Japan
Laboratory of Organic Chemistry, Gifu Pharmaceutical University, 1-25-4 Daigaku-nishi, Gifu 501-1196, Japan

Shinji Ishihara, Yoshinari Sawama, Yasunari Monguchi and Hironao Sajiki
Laboratory of Organic Chemistry, Gifu Pharmaceutical University, 1-25-4 Daigaku-nishi, Gifu 501-1196, Japan

Mehdi Rezaeimahmoudi and Alireza Gharegozlu
Department of GIS & RS, School of Environment and Energy, Science and Research Branch, Islamic Azad University, Tehran, Iran

Abdolreza Esmaeli
Plasma and Nuclear Fusion Research School, Nuclear Science and Technology Research Institute, AEOI, Tehran, Iran

Hassan Shabanian
Department of GIS & RS, School of Geography, University of Tehran, Tehran, Iran

Ladan Rokni
Department of Tourism Planning, School of Human Geography, University of Tehran, Tehran, Iran

Hesham M Sharaf
Department of Zoology, Faculty of Science, Zagazig University, Zagazig, Egypt

Abdalla M Shehata
Chemistry Department, Faculty of Science, Suez Canal University, Al-arish, Egypt

Mohammad Mehdi Amin and Bijan Bina
Environment Research Center, Isfahan University of Medical Sciences (IUMS) and Department of Environmental Health Engineering, School of Health,IUMS, Isfahan, Iran
Student Research Center, School of Health, Isfahan University of Medical Sciences, Isfahan, Iran

Amir Rahimi
Chemical Engineering Department, College of Engineering, University of Isfahan, Isfahan, Iran

Mohsen Heidari
Environment Research Center, Isfahan University of Medical Sciences (IUMS) and Department of Environmental Health Engineering, School of Health, IUMS, Isfahan, Iran
Student Research Center, School of Health, Isfahan University of Medical Sciences, Isfahan, Iran
Department of Environmental Health Engineering, Faculty of Health, Hormozgan University of Medical Sciences, Bandar Abbas, Iran

Fazel Mohammadi Moghadam
Department of Environmental Health Engineering, School of Health, Shahrekord University of Medical Sciences, Shahrekord, Iran

Shamas Tabraiz, Saeed Ahmad and Muhammad Bilal Asif
Department of Environmental Engineering, University of Engineering & Technology, Taxila, Pakistan

Iffat Shehzadi
Department of Psychology, University of Gujrat, Pakistan

Shahla Rezaei and Hamed Tahmasbi
Department of Pharmaceutical Biotechnology, Faculty of Pharmacy and Biotechnology Research Center, Tehran University of Medical Sciences, P.O. Box 14155-6451, Tehran 1417614411, Iran

Mehdi Mogharabi and Mohammad Ali Faramarzi
Department of Pharmaceutical Biotechnology, Faculty of Pharmacy and Biotechnology Research Center, Tehran University of Medical Sciences, P.O. Box 14155-6451, Tehran 1417614411, Iran
Pharmaceutical Sciences Research Center, Tehran University of Medical Sciences, Tehran 1417614411, Iran

Alieh Ameri
Department of Medicinal Chemistry, Faculty of Pharmacy, Kerman University of Medical Sciences, Kerman, Iran

Hamid Forootanfar
Department of Pharmaceutical Biotechnology, Faculty of Pharmacy, Kerman University of Medical Sciences, Kerman, Iran

Mohammad Reza Khoshayand
Department of Drug and Food Control, Faculty of Pharmacy and Pharmaceuticals Quality Assurance Research Center, Tehran University of Medical Sciences, Tehran 1417614411, Iran

Masoumeh Ravanipour
Department of Environmental Health Engineering, School of Public Health, Tehran University of Medical Sciences, Tehran, Iran

Ali Esrafili, Mahdi Farzadkia and Roshanak Rezaei Kalantary
Department of Environmental Health Engineering, School of Public Health, Iran University of Medical Sciences, Tehran, Iran

Anoushiravan Mohseni-Bandpi
Department of Environmental Health Engineering, School of Public Health, Shahid Beheshti University of Medical Sciences, Tehran, Iran

Samireh Hashemi-Najafabadi
Department of Chemical Engineering, Biotechnology Group, Tarbiat Modares University, Tehran, Iran

Seyed Ahmad Mirbagheri, Majid Bagheri, Siamak Boudaghpour and Majid Ehteshami
Department of Civil Engineering, K.N. Toosi University of Technology, Vanak square, Tehran, Iran

Zahra Bagheri
Department and Faculty of Basic Sciences, PUK University, Kermanshah, Iran

Meenu Chhabra
Department of Biology, Indian Institute of Technology-Jodhpur, Jodhpur 342011, Rajasthan, India

Saroj Mishra and Trichur Ramaswamy Sreekrishnan
Department of Biochemical Engineering and Biotechnology, Indian Institute of Technology Delhi, Hauz-Khas, New-Delhi 110016, India

Jaber Aazami and Abbas Esmaili-Sari
Department of Environment, Faculty of Natural Resources, Tarbiat Modares University, Tehran, Iran

Asghar Abdoli
Department of Biodiversity and Ecosystem Management, Environmental Research Institute, Shahid Beheshti University, Tehran, Iran

Hormoz Sohrabi
Department of Forestry, Faculty of Natural Resources, Tarbiat Modares University, Tehran, Iran

Paul J Van den Brink
Department of Aquatic Ecology and Water Quality Management, Wageningen University, Wageningen University and Research Centre, Wageningen, The Netherlands. 5Alterra, Wageningen University and Research Centre, Wageningen, The Netherlands

Roopesh Jain and Archana Tiwari
School of Biotechnology, Rajiv Gandhi Proudyogiki Vishwavidyalaya, Airport Bypass Road, Bhopal, Madhya Pradesh, India

Abdullah Aksu
Department of Chemical Oceanography, Institute of Marine Science and Management, Istanbul University, Vefa 34134-00, Istanbul, Turkey

Kamyar Yaghmaeian and Ramin Nabizadeh
Department of Environmental Health, School of Public Health, Tehran University of Medical Sciences, Tehran, Iran
Center for Solid Waste Research, Institute for Environmental Research, Tehran University of Medical Sciences, Tehran, Iran

Nematallah Jaafarzadeh
Environmental Technologies Research Center, Ahvaz Jundishapur University of Medical Sciences, Ahvaz, Iran

Golbarg Dastforoushan
Department of Environmental Engineering, Islamic Azad University, Tehran, Iran

Jalil Jaafari
Department of Environmental Health, School of Public Health, Guilan University of Medical Sciences, Rasht, Iran

Fatemeh Shamsi Mahmoudabadi, Mohammad Firoozabadi and Anoshirvan Kazemnejad
Faculty of Medical Science, Tarbiat Modares University, Tehran, Iran

Saeideh Ziaei
Reproductive Health Department, Tarbiat Modares University, P. O. Box:14115–111, Tehran, Iran

Anupama Vijaya Nadaraja, Prajeesh Gangadharan Puthiyaveettil and Krishnakumar Bhaskaran
Environmental Technology, CSIR-National Institute for Interdisciplinary Science & Technology, Thiruvananthapuram 695019, India

Naficeh Sadeghi, Mohammad Reza Oveisi, Mannan Hajimahmoodi, Masoomeh Behzad and Sahereh Saadatmand
Department of Drug and Food Control, School of Pharmacy, Tehran university of Medical Sciences, Tehran, Iran

Behrooz Jannat and Fatemeh Sadeghi
Halal research center, Ministry of Health and Medical Education, Tehran, Iran

Abdolazim Behfar
Department of Drug and Food Control, Faculty of Pharmacy, Jondishapour Ahvaz University of Medical Sciences, Ahvaz, Iran

Roshanak Rezaei Kalantary, Ali Esrafili and Fatemeh Rashid Ashmagh
Department of Environmental Health Engineering, School of Public Health, Iran University of Medical Sciences, Tehran, Iran

Anoushiravan Mohseni-Bandpi
Department of Environmental Health Engineering, School of Public Health, Shahid Beheshti University of Medical Sciences, Tehran, Iran

Simin Nasseri
Department of Environmental Health Engineering, School of Public Health and Center for Water Quality Research, Institute forEnvironmental Research, Tehran University of Medical Sciences, Tehran, Iran

Sahand Jorfi
Environmental Technology Research Center, Ahvaz Jondishapour University of MedicalSciences, Ahvaz, Iran
School of Public Health, Ahvaz Jondishapour University of MedicalSciences, Ahvaz, Iran

Mahsa Ja'fari
Petroleum University of technology, Abadan faculty of petroleum engineering, Abadan, Iran

Madhuri M Sahasrabudhe
Department of Microbiology, Maulana Azad College, Aurangabad, MS, India

Rijuta G Saratale
Department of Biotechnology, Shivaji University, Kolhapur, MS, India

Ganesh D Saratale
Department of Environmental Biotechnology, Shivaji University, Kolhapur, MS, India
Department of Biochemistry, Shivaji University, Kolhapur, MS, India

Girish R Pathade
H.V. Desai College, Pune, MS, India

L. D. Naidu and S. Saravanan
Department of Chemical Engineering, National Institute of Technology, Tiruchirapalli 15, India

Mukesh Goel
Center for Environmental Engineering, PRIST University, Thanjavur, India

S. Periasamy
Department of Textile Technology, PSG College of Technology, Coimbatore, India

Pieter Stroeve
Department of Chemical Engineering, University of California Davis, Davis, CA 95616, USA

Abdolhossein Parizanganeh and Abbasali Zamani
Environmental Science Research Laboratory, Department of Environmental Sciences, Faculty of Science, University of Zanjan, Zanjan 45371-38791, Iran

Vahid Bijnavand and Behzad Taghilou
Department of Environmental Science, Faculty of Science, University of Zanjan, Zanjan 45371-38791, Iran

R Hiranmai Yadav
School of Natural Resources Management and Environmental Sciences, College of Agriculture and Environmental Sciences, Haramaya University, P.O. Box #337, Dire Dawa, Ethiopia

Kamyar Yaghmaeian, Reza Khosravi Mashizi, Amir Hossein Mahvi, Mahmood Alimohammadi and Shahrokh Nazmara
Department of Environmental Health Engineering, School of Public Health, Tehran University of Medical Sciences, Tehran, Iran

Simin Nasseri and Simin Nasseri
Department of Environmental Health Engineering, School of Public Health, Tehran University of Medical Sciences, Tehran, Iran
Center for Water Quality Research (CWQR), Institute for Environmental Research (IER), Tehran University of Medical Sciences, Tehran, Iran

Allahbakhsh Javid and Hamed Gharibi
School of Public Health, Shahroud University of Medical Sciences, Shahroud, Iran

Alireza Mesdaghinia
Department of Environmental Health Engineering, School of public Health, Tehran University of Medical Sciences, Tehran, Iran

Amir Hossein Mahvi
Department of Environmental Health Engineering, School of public Health, Tehran University of Medical Sciences, Tehran, Iran
Center for Solid Waste Research, Institute for Environmental Research, Tehran University of Medical Sciences, Tehran, Iran

Mahmood Alimohammadi
Department of Environmental Health Engineering, School of public Health, Tehran University of Medical Sciences, Tehran, Iran

Hossein Kamani and Ramin Nabizadeh Nodehi
Department of Environmental Health Engineering, School of Public Health, Tehran University of Medical Sciences, Tehran, Iran

Mehdi Khoobi
Department of Medicinal Chemistry, Faculty of Pharmacy and Pharmaceutical Sciences Research Center, Tehran University of Medical Sciences, Tehran, Iran

Amir Hossein Mahvi
Department of Environmental Health Engineering, School of Public Health, Tehran University of Medical Sciences, Tehran, Iran
Center for Solid Waste Research, Institute for Environmental Research, Tehran University of Medical Sciences, Tehran, Iran
National Institute of Health Research, Tehran University of Medical Sciences, Tehran, Iran

Gholamreza Bonyadinejad and Hamed Mohammadi
Environment Research Center and Department of Environmental Health Engineering, School of Health, Isfahan University of Medical Sciences (IUMS), Isfahan 81676-36954, Iran
Student Research Center, School of Health, IUMS, Isfahan, Iran

Mohsen Khosravi
Nanotechnology Department, University of Isfahan, Isfahan 81744-73441, Iran

Afshin Ebrahimi and Roya Nateghi
Environment Research Center and Department of Environmental Health
Engineering, School of Health, Isfahan University of Medical Sciences (IUMS), Isfahan 81676-36954, Iran

Seyed Mahmood Taghavi-Shahri
Research Center for Environmental Pollutants, Qom University of Medical Sciences, Qom 37136-49373, Iran

M. Racuciu
Environmental Sciences and Physics Department, Faculty of Sciences in "Lucian Blaga" University, Dr. I. Ratiu Street, no. 5-7, Sibiu 550024, Romania

S. Miclaus
Technical Sciences Department, "Nicolae Balcescu" Land Forces Academy, Revolutiei Street, no. 3-5, Sibiu 550170, Romania

D. Creanga
Biophysics and Medical Physics Laboratory, Faculty of Physics, "Alexandru Ioan Cuza" University, 11, Blvd. Carol I, Iasi 700506, Romania

Index